Proof Theory of Modal Logic

APPLIED LOGIC SERIES

VOLUME 2

SCOPE OF THE SERIES

Logic is applied in an increasingly wide variety of disciplines, from the traditional subjects of philosophy and mathematics to the more recent disciplines of cognitive science, computer science, artificial intelligence, and linguistics, leading to new vigor in this ancient subject. Kluwer, through its Applied Logic Series, seeks to provide a home for outstanding books and research monographs in applied logic, and in doing so demonstrates the underlying unity and applicability of logic.

Proof Theory of Modal Logic

edited by

HEINRICH WANSING

University of Leipzig,
Institute of Logic and Philosophy of Science

KLUWER ACADEMIC PUBLISHERS
DORDRECHT / BOSTON / LONDON

Library of Congress Cataloging-in-Publication Data

Proof theory of modal logic / edited by Heinrich Wansing.
p. cm. -- (Applied logic series ; v. 2)
Proceedings of a workshop held at the University of Hamburg, Nov. 19-20, 1993.
Includes index.
ISBN 0-7923-4120-1
1. Modal logic--Congresses. 2. Proof theory--Congresses. I. Wansing, H. (Heinrich) II. Series.
QA9.46.P76 1996
511.3--dc20 96-9019

ISBN 0-7923-4120-1

Published by Kluwer Academic Publishers,
P.O. Box 17, 3300 AA Dordrecht, The Netherlands.

Kluwer Academic Publishers incorporates
the publishing programmes of
D. Reidel, Martinus Nijhoff, Dr W. Junk and MTP Press.

Sold and distributed in the U.S.A. and Canada
by Kluwer Academic Publishers,
101 Philip Drive, Norwell, MA 02061, U.S.A.

In all other countries, sold and distributed
by Kluwer Academic Publishers Group,
P.O. Box 322, 3300 AH Dordrecht, The Netherlands.

Logo design by L. Rivlin

Printed on acid-free paper

Printed in the Netherlands

EDITORIAL PREFACE

The editors of the Applied Logic Series are pleased to present the second volume in the series, this one on Proof Theory and Modal Logic. In recent years the topic of multi-modal logics has had significant applications in a variety of disciplines, including Artificial Intelligence, theoretical computer science, philosophy, linguistics, Peano arithmetic, and generalized quantifiers. In this volume, leading researchers examine the proof theory of this rich and active field.

The Editors

CONTENTS

Preface ix

I Standard Proof Systems

JÖRG HUDELMAIER
A Contraction-free Sequent Calculus for *S*4 3

GRIGORI MINTS, VLADIMIR OREVKOV AND TANEL TAMMET
Transfer of Sequent Calculus Strategies to Resolution for S4 17

HAROLD SCHELLINX
A Linear Approach to Modal Proof Theory 33

TOMASZ SKURA
Refutations and Proofs in S4 45

II Extended Formalisms

EWA ORLOWSKA
Relational Proof Systems for Modal Logics 55

NUEL BELNAP
The Display Problem 79

MARCUS KRACHT
Power and Weakness of the Modal Display Calculus 93

HEINRICH WANSING
A Proof-theoretic Proof of Functional Completeness for Many Modal and Tense Logics 123

RAJEEV GORÉ
On the Completeness of Classical Modal Display Logic 137

CLAUDIO CERRATO
Modal Sequents 141

KOSTA DOŠEN AND ZORAN PETRIĆ
Modal Functional Completeness 167

SIMONE MARTINI AND ANDREA MASINI
A Computational Interpretation of Modal Proofs 213

SZABOLCS MIKULÁS
Gabbay-style Calculi 243

III Translation-based Proof Systems

H. J. OHLBACH, R. SCHMIDT AND U. HUSTADT
Translating Graded Modalities into Predicate Logics 253

OLIVIER GASQUET AND ANDREAS HERZIG
From Classical to Normal Modal Logics 293

Index 313

PREFACE

This book deals with formal, mechanizable reasoning in modal logics, that is logics of necessity, possibility, belief, time, computations etc. It is therefore of vigorous interest for various interrelated disciplines like Philosophy, Artificial Intelligence, Computer Science, Logic, Cognitive Science and Linguistics. *Proof Theory of Modal Logic* is the first book focusing entirely on matters of *modal* proof theory and for the first time brings together various directions in the development of generalized sequent-style proof systems.

Though early this century modal logic started as a purely syntactic enterprise, modal proof theory did not undergo such a tremendous development as was experienced by modal model theory after the emergence of possible worlds semantics around 1960. In the 1980s several distinguished modal logicians explicitly noted this discrepancy and since that time, various generalizations of the hitherto standard proof-theoretic formalisms have been developed and investigated. Suggestions for generalizing the ordinary notion of sequent have for instance led to higher-level, higher-arity, and higher-dimensional Gentzen-type proof systems for modal logics. Moreover, modal correspondence theory has been applied in order to use theorem provers for classical predicate logic. General proof-theoretic frameworks have also been devised, like Nuel Belnap's Display Logic and Dov Gabbay's theory of Labelled Deductive Systems. This lively development is still ongoing.

Proof Theory of Modal Logic is the proceedings volume of a workshop on the proof theory of modal logic, held at the University of Hamburg on November 19–20, 1993. The aim of this workshop was to bring together various, up to then, more or less unrelated directions of research in modal proof theory and thereby to stimulate significant progress in the field. The book consists of 15 research papers, divided into three parts. In the first part, papers are collected which give a profound description of powerful proof-theoretic methods as applied to the normal modal logic $S4$. Part II is devoted to generalizations of the standard proof-theoretic formats. This part contains papers either dealing with or related to (i) relational proof systems, (ii) Display Logic, or, more generally, approaches incorporating modal operators as structural elements, (iii) higher-level and higher-dimensional proof systems, and (iv) axiomatic presentations using the 'irreflexivity-rule'. The third part presents new and important results on semantics-based proof systems for modal logics.

The workshop culminating in the present volume was generously sponsored by the Stiftung Volkswagenwerk (Hannover). I would like to express my gratitude to the Volkswagen Foundation for making the workshop possible and to Christopher Habel and the graduate programme in Cognitive Science (Graduiertenkolleg Kognitionswissenschaft) at the University of Hamburg for their additional support. I am grateful to Annette Tschernig for her assistance during the workshop. Special thanks go to (i) Dov Gabbay for supporting the publication of this book, (ii) Jane Spurr for preparing the final manuscript, and (iii) the colleagues who acted as referees in the process of preparing this manuscript, namely:

Patrick Blackburn	Makoto Kanazawa	Luiz C. Pereira
Valentin Goranko	Marcus Kracht	Krister Segerberg
Rajeev Goré	Joachim Lambek	Valentin Shehtman
Wiebe van der Hoek	Andrea Masini	Max Urchs
Jan Jaspars	Grigori Mints	Richard Zach

Heinrich Wansing
Leipzig, September 1995

PART I

STANDARD PROOF SYSTEMS

JÖRG HUDELMAIER

A CONTRACTION-FREE SEQUENT CALCULUS FOR $S4$

Theorem proving in the modal logic S4 is notoriously difficult, because in conventional sequent style calculi for this logic lengths of deductions are not bounded in terms of the length of their endsequent. This means that the usual depth first search strategy for backwards construction of deductions of given sequents may give rise to infinite search paths and is not guaranteed to terminate. Thus using such a search strategy prevents us not only from obtaining a decision procedure for the logic in question, but even from arriving at a complete proof procedure. There are two well known approaches for overcoming this problem: both approaches rely on the fact that all formulas which occur in a deduction are subformulas of the endsequent and that out of these formulas one may only form finitely many "essentially different" sequents. Thus although there is no bound on the length of *all* deductions of a given sequent, we know that for any given sequent there is a number such that if the sequent is deducible at all, then it has a deduction of length smaller than this number. Hence by only considering deductions of appropriately bounded length one may obtain a decision procedure. But due to the fact that using such a procedure one is forced to construct many redundant inferences—one will, for instance, have to consecutively apply the same inference many times—this approach is considered rather inefficient. Instead the usual technique for deciding provability of formulas in S4 is based on loop checking: If "essentially the same" sequent occurs twice on a branch of a constructed deduction, then there is a shorter deduction with the same endsequent which does not show this redundancy, and one may backtrack. Although more efficient in terms of run time than the previous approach this loop checking method requires quite involved implementation techniques. Now in the context of intuitionistic propositional logic recently a third approach has been found, which is based on so called contraction free sequent calculi, i.e. calculi for which there is a certain measure such that for all rules of the calculi all premisses have smaller measure then the conclusion (cf. [1, 3]). Thus in a contraction free calculus *all* deductions of a given sequent are bounded in length by some function of the length of their endsequent and a decision procedure is obtained by simple depth first backwards application of the rules. In this paper we show that there is a contraction free sequent calculus for S4, too. This calculus is more complicated than the corresponding calculi for intuitionistic propositional logic, but still it gives rise to a simpler decision procedure for S4 than conventional methods.

H. Wansing (ed.), Proof Theory of Modal Logic, 3–15.

1 INTRODUCTION

We consider a language of sequents, i.e. pairs of multisets of formulas built up from propositional variables, the Boolean connectives $\neg$ and $\vee$ and the modal connective $\Box$.

We start from a calculus LM_0 for the modal logic S4 consisting of axioms of the form $M, a \Rightarrow a, N$, where a is a propositional variable, the well known Boolean rules

$$\mathrm{E}\neg\frac{M \Rightarrow N, v}{M, \neg v \Rightarrow N} \qquad \mathrm{I}\neg\frac{M, v \Rightarrow N}{M \rightarrow N, \neg v}$$

$$\mathrm{E}\vee\frac{M, u \Rightarrow N \quad M, v \Rightarrow N}{M, u \vee v \Rightarrow N} \qquad \mathrm{I}\vee\frac{M \Rightarrow N, u, v}{M \Rightarrow N, u \vee v}$$

and the two modal rules

$$\mathrm{E}\Box\frac{M, \Box v, v \Rightarrow N}{M, \Box v \Rightarrow N} \qquad \mathrm{I}\Box\frac{M^0 \Rightarrow v}{M \Rightarrow N, \Box v}$$

where M^0 results from M by omitting all formulas not of the form $\Box v$.

Obviously the so called weakening rule is an admissible rule of this calculus: If a sequent $M \Rightarrow N$ is deducible in LM_0 by a deduction of length n, then both the sequent $M \Rightarrow N, v$ and the sequent $M, v \Rightarrow N$ are deducible by deductions of length $\leq n$. Moreover it is immediately clear that the Boolean rules of LM_0 and the rule E$\Box$ are invertible: If a conclusion of one of these rules is deducible in LM_0 by a deduction of length n, then all its premisses are deducible by deductions of length $\leq n$. Therefore the following holds:

LEMMA 1 *a) Every* LM_0*-deduction of a sequent* $M, v, v \Rightarrow N$ *may be transformed into a deduction of the sequent* $M, v \Rightarrow N$ *of smaller or equal length.*
b) Every LM_0*-deduction of a sequent* $M \Rightarrow v, v, N$ *may be transformed into a deduction of the sequent* $M \Rightarrow v, N$ *of smaller or equal length.*

Proof. a) is true of axioms, and if $M, v, v \Rightarrow N$ is the conclusion of an inference I different from I$\Box$ with principal formula different from v, then the induction hypothesis applies to the premisses and by an application of I to the transformed premisses the sequent $M, v \Rightarrow N$ may be obtained. If $M, v, v \Rightarrow N$ is the conclusion of an I$\Box$-inference, then either both occurrences of v are contained in M^0, in which case the induction hypothesis applies to it and $M, v \Rightarrow N$ may be obtained as before, or neither occurrence of v is in M^0 and one of them may be introduced by weakening. If $M, v, v \Rightarrow N$ is the conclusion of a Boolean rule B with principal formula v, then the premisses may be transformed according to the inversion principle for the Boolean rules and the resulting sequents may be transformed by the

induction hypothesis; finally to these transformed sequents the rule B may be applied again, thereby obtaining a deduction of $M, v \Rightarrow N$. If $M, v, v \Rightarrow N$ is the conclusion of an E$\Box$-inference with principal formula v, then v is of the form $\Box u$ and the premiss reads $M, \Box u, \Box u, u \Rightarrow N$; the rule E$\Box$ applied to this gives the required deduction of $M, v \Rightarrow N$.
b) is also true of axioms, and if $M \Rightarrow v, v, N$ is the conclusion of an inference I different from I$\Box$, then the deduction of $M \Rightarrow v, N$ is constructed as above. But if $M \Rightarrow v, v, N$ is the conclusion of an I$\Box$-inference, then from its premiss we directly obtain $M \Rightarrow v, N$ by a different application of I$\Box$ with the same principal formula, viz. by introducing only one occurrence of v into the conclusion. ■

This shows that the calculus LM$_0$ is equivalent to the more frequently encountered S4-calculi for sequents made up of pairs of *sets* of formulas (cf.[2]), and in particular the so called cut rule is also a derived rule of LM$_0$: if two sequents $M \Rightarrow c$ and $M, c \Rightarrow v$ are deducible by LM$_0$, then so is the sequent $M \Rightarrow v$.

2 REDUCTION TO CLAUSAL FORM

In order to determine deducibility of sequents by our calculus LM$_0$ we may restrict the language to so called *clausal sequents*: there is a simple procedure which associates to every sequent s of the full language a clausal sequent $C(s)$, such that s is derivable by LM$_0$ if and only if $C(s)$ is derivable by LM$_0$.

DEFINITION 2 (Cf. [4]) *a) A modal literal is either a propositional variable, a negated propositional variable, a formula of the form $\Box a$, where a is a propositional variable or a formula of the form $\neg\Box a$, where a is a propositional variable.*
b) A modal clause is a formula of the form $l_0 \vee (l_1 \vee \ldots) \ldots)$ or $\Box(l_0 \vee l_1 \vee \ldots) \ldots))$, where the l_i are modal literals.
c) A clausal sequent is a sequent of the form $c_1, \ldots, c_m \Rightarrow a_1, \ldots, a_n$, where the c_i are modal clauses and the a_i are propositional variables.

For simplifying notation we let the expression $[v_0, \ldots, v_n](n \geq 0)$ denote the formula $v_0 \vee v_1 \vee (\ldots \vee v_n) \ldots)$. Thus a formula $[v_0, \ldots, v_n]$ or $\Box[v_0, \ldots, v_n]$ is a modal clause iff all the v_i are modal literals.

Then the following holds:

LEMMA 3 *The sequents in the first column of the following table are deducible by* LM$_0$ *if and only if the corresponding sequents of the second column are deducible:*

$M \Rightarrow v, N$	$M, \neg v \Rightarrow N$
$M, [A, \neg\neg v, B] \Rightarrow N$	$M, [A, v, B] \Rightarrow N$
$M, [A, \neg(u \vee v), B] \Rightarrow N$	$M, [A, \neg p, B], [p, \neg u], [p, \neg v] \Rightarrow N$
$M, [A, u \vee v, B] \Rightarrow N$	$M, [A, u, v, B] \Rightarrow N$
$M, [A, \neg\Box\neg v, B] \Rightarrow N$	$M, [A, \neg\Box p, B], \Box[p, v] \Rightarrow N$
$M, [A, \neg\Box(u \vee v), B] \Rightarrow N$	$M, [A, \neg\Box p, B], \Box[p, \neg u], \Box[p, \neg v] \Rightarrow N$
$M, [A, \neg\Box\Box v, B] \Rightarrow N$	$M, [A, \neg\Box v, B] \Rightarrow N$
$M, [A, \Box\neg v, B] \Rightarrow N$	$M, [A, \Box p, B], [\neg p, \neg v] \Rightarrow N$
$M, [A, \Box(u \vee v), B] \Rightarrow N$	$M, [A, \Box p, B], \Box[\neg p, u, v] \Rightarrow N$
$M, [A, \Box\Box v, B] \Rightarrow N$	$M, [A, \Box v, B] \Rightarrow N$
$M, \Box[A, , \neg\neg v, B] \Rightarrow N$	$M, \Box[A, v, B] \Rightarrow N$
$M, \Box[A, \neg(u \vee v), B] \Rightarrow N$	$M, \Box[A, \neg p, B], \Box[p, \neg u], \Box[p, \neg v] \Rightarrow N$
$M, \Box[A, u \vee v, B] \Rightarrow N$	$M, \Box[A, u, v, B] \Rightarrow N$
$M, \Box[A, \neg\Box\neg v, B] \Rightarrow N$	$M, \Box[A, \neg\Box p, B], \Box[p, v] \Rightarrow N$
$M, \Box[A, \neg\Box(u \vee v), B] \Rightarrow N$	$M, \Box[A, \neg\Box p, B], \Box[p, \neg u], \Box[p, \neg v] \Rightarrow N$
$M, \Box[A, \neg\Box\Box v, B] \Rightarrow N$	$M, \Box[A, \neg\Box v, B] \Rightarrow N$
$M, \Box[A, \Box\neg v, B] \Rightarrow N$	$M, \Box[A, \Box p, B], \Box[\neg p, \neg v] \Rightarrow N$
$M, \Box[A, \Box(u \vee v), B] \Rightarrow N$	$M, \Box[A, \Box p, B], \Box[\neg p, u, v] \Rightarrow N$
$M, \Box[A, \Box\Box v, B] \Rightarrow N$	$M, \Box[A, \Box v, B] \Rightarrow N$

(Here the formulas p on the right hand sides are propositional variables which do not occur on the corresponding left hand sides.)

Proof. This lemma is well known: for instance from a deduction of $M, \Box[A, \neg\Box(u \vee v), B] \Rightarrow N$ we obtain the required deduction of $M, \Box[A, \neg\Box p, B], \Box[p, \neg u]$ $\Box[p, \neg v] \Rightarrow N$ by a cut with the LM_0-deducible sequent $\Box[A, \neg\Box p, B], \Box[p, \neg u], \Box[p, \neg v] \Rightarrow \Box[A, \neg\Box(u \vee v), B]$, whereas given a deduction of $M, \Box[A, \neg\Box p, B], \Box[p, \neg u], \Box[p, \neg v] \Rightarrow N$ we obtain the required deduction of $M, \Box[A, \neg\Box(u \vee v), B] \Rightarrow N$ by changing all occurrences of p to $u \vee v$ and cutting out the two LM_0-decucible formulas $\Box[u \vee v, \neg u]$ and $\Box[v \vee v, \neg v]$ from the resulting endsequent. ∎

From this lemma we easily obtain a procedure for reducing a given sequent of arbitrary form to a clausal sequent. This procedure consists of two steps:
a) Applying the property expressed by the first row of the above table to the formulas v_i on the right hand side of a given sequent $a_1, m \ldots, a_m, \Box d_1, \ldots, \Box d_n \Rightarrow v_1, \ldots, v_q, p_1, \ldots, p_r$ (where the a_i are not of the form $\Box b$, the v_i are not propositional variables and the p_i are propositional variables), we arrive at a new sequent which has only propositional variables on its right hand side and which has at most twice as many connectives as the given sequent.
b) Applying the remaining rows to the resulting sequent $[a_1], \ldots, [a_m], \Box[d_1], \ldots, \Box[d_n], [\neg v_1], \ldots, [\neg v_q] \Rightarrow p_1, \ldots, p_r$ it is easily seen, that at each step the number of connectives inside one of the square brackets of this sequent, which do not belong to modal literals, decreases. This number is bounded by the number of

all connectives of the sequent which is itself bounded by twice the number of connectives of the original sequent. Since at each step the number of connectives of a sequent is at most increased by 1, this shows the required

LEMMA 4 *There is a procedure which converts any sequent s into a clausal sequent $C(s)$ such that the number of connectives of $C(s)$ is at most four times the number of connectives of s and s is deducible by LM_0 iff $C(s)$ is deducible.*

Now unfortunately in an LM_0-deduction of a clausal sequent there will usually occur sequents which aren't clausal, viz. sequents occurring as premisses of applications of E¬ and having a formula $\Box a$ on their right hand side. But the following lemma shows that we may combine such applications of E¬ with an immediately preceding application of I□ into a single new rule, thus preserving clausal forms for all sequents of a deduction:

LEMMA 5 *There is a transformation sending every* LM_0*-deduction of a given sequent into a deduction of the same sequent such that in the new deduction every premiss of an application of* E¬ *with principle formula, $\neg\Box v$ is the conclusion of an application of* I□, *with principle formula $\Box v$.*

Proof. We consider a maximal application I of E¬ with principal formula $\neg\Box v$ and premiss s for which $\Box v$ is not the principal formula of the inference leading to s and we use recursion on the maximal number of successive sequents preceding s in which $\Box v$ occurs on the right hand side to turn it into an E¬-inference having the required property: If this number is 1, then either s is an axiom or $\Box v$ must have disappeared because of some application of I□ with principal formula different from $\Box v$ and in both cases the E¬-inference I may be dropped. If this number is greater than 1 and the inference leading to s is either a Boolean inference with principal formula not of the form $\neg\Box w$ or an E□-inference, then this inference may be shifted down past the E¬-inference, thereby lowering the recursion parameter. But if the inference leading to s is another E¬-inference J with principal formula $\neg\Box w$, then its premiss is the conclusion of an I□-inference with principal formula $\Box w$ and again the inference I may be dropped. Using this technique we will eventually turn all E¬-inferences of a given deduction into inferences of the form required by this lemma. ■

Using this lemma we may now replace any pair of inferences consisting of an application of I□ with principal formula $\Box v$ and an application of E¬ with principle formula $\neg\Box v$ immediately following it by a new inference leading directly from the premiss of the I□-inference to the conclusion of the E¬-inference. This gives us the new calculus LM_1 which has the axioms and the rules E□ of LM_0 and the following divided E¬-rule:

$$\text{E}\neg s\frac{M \Rightarrow a, N}{M, \neg a \Rightarrow N}\ (a \text{ a propositional variable}) \qquad \text{E}\neg d\frac{M^0 \Rightarrow a}{M, \neg\Box a \Rightarrow N}$$

Now in an LM_1-deduction of a clausal sequent there occur only clausal sequents, and the rules just mentioned are the only rules of LM_0 which are applicable to clausal sequents. Therefore since LM_1 shall only be used for clausal sequents, it does not need any further rules; in particular it does not need any I-rules for the following lemma to hold:

LEMMA 6 *The calculi* LM_0 *and* LM_1 *are equivalent as regards deducibility of clausal sequents.*

Therefore the rules $E\neg s$ and $E\vee$ are invertible rules of this calculus, too, and we may safely extend the $E\vee$-rule to disjunctions of arbitrary length, showing the following:

LEMMA 7 *The following calculus* LM_2 *is equivalent to* LM_1*:* LM_2 *has the axioms and the rule* $E\neg d$ *of* LM_1 *and it has the rules* $E\vee$ *and* $E\Box$ *in the form*

$$E\vee\,\frac{\begin{array}{c} M, a_1 \Rightarrow N \;\ldots\; M, a_p \Rightarrow N \\ M \Rightarrow b_1, N \;\ldots\; M \Rightarrow b_q, N \\ M, \Box c_1 \Rightarrow N \;\ldots\; M, \Box c_r \Rightarrow N \\ M, \neg\Box d_1 \Rightarrow N \;\ldots\; M, \neg\Box d_s \Rightarrow N \end{array}}{M, v \Rightarrow N}$$

$$E\Box\,\frac{\begin{array}{c} M, \Box v, a_1 \Rightarrow N \;\ldots\; M, \Box v, a_p \Rightarrow N \\ M \Box v \Rightarrow b_1, N \;\ldots\; M \Box v \Rightarrow b_q, N \\ M, \Box v, \Box c_1 \Rightarrow N \;\ldots\; M, \Box v, \Box c_r \Rightarrow N \\ M, \Box v, \neg\Box d_1 \Rightarrow N \;\ldots\; M, \Box v, \neg\Box d_s \Rightarrow N \end{array}}{M, \Box v \Rightarrow N}$$

where v *is the formula* $[a_1, \ldots, a_p, \neg b_1, \ldots, \neg b_q, \Box c_1, \ldots, \Box c_r, \neg\Box d_1, \ldots, \neg\Box d_s]$*, the* a_i, b_j, c_k *and* d_l *are propositional variables,* $p + q + r + s > 0$ *and* v *is meant to represent any permutation of its subformulas.*

3 THE CALCULUS

Here we note, that the rule $E\neg s$ of LM_1 is a special case of the rule $E\vee$ and that due to the invertibility of the $E\neg s$-rule the formulas b_i in the rules $E\vee$ and $E\Box$ may be put to the right hand side of the premisses in the second row, whereas the formulas $\neg\Box d_i$ have to remain on the left hand side, because the rule $E\neg d$ is not directly invertible. Still we show in the next lemma that any LM_2-deduction may be transformed in such a way that these formulas can immediately be put to the right hand sind, too. For this purpose we call the premisses in the first and second rows of the $E\vee$- and $E\Box$-rules α-premisses, those in the third row β-premisses and those in the fourth row γ-premisses.

LEMMA 8 *There is a transformation sending any* LM_2*-deduction of a given sequent into another deduction of the same sequent, such that in the new deduction every* γ*-premiss is the conclusion of an application of* E$\neg$*d.*

Proof. In order to construct such a deduction from a given arbitrary deduction we consider a maximal γ-premiss $s = M, \Box v, \neg\Box d \Rightarrow N$ of an inference I of our deduction, which isn't the conclusion of an application of E$\neg d$ (here $v = [a_1, \ldots, a_p, \neg b_1, \ldots, \neg b_q, \Box c_1, \ldots, \Box c_r, \neg\Box d_1, \ldots, \neg\Box d_s, \neg\Box d]$), and we use recursion on the maximal number of successive sequents preceding s in which $\neg\Box d$ occurs: if it is 0, then either s is an axiom, or it is the conlcusion of an applciation of E$\neg d$ with principal formula different from $\neg\Box d$. In both cases the inference whose premiss s is may be dropped. If this number is greater than 0, and the inference J leading to s is an application of E$\neg s$, then we may shift J down past I in the usual way and lower the recursion parameter. But if J is by an application of a multipremiss rule, e.g. E$\Box$, then we proceed as follows:

Let w be $[e_1, \ldots, e_t, \neg f_1, \ldots, \neg f_u, \Box g_1, \ldots, \Box g_v, \neg\Box h_1, \ldots, \neg\Box h_w]$, let M be $L, \Box w$ and let $L, \Box w, \Box v \Rightarrow N$ be the conclusion of I, let $L, \Box w, \Box v, a_i \Rightarrow N$ and $L, \Box w, \Box v \Rightarrow b_j, N$ be its α-premisses, $L, \Box w, \Box v, \Box c_k \Rightarrow N$ its β-premisses and $L, \Box w, \Box v, \neg\Box d_l \Rightarrow N$ and s its γ-premisses: then the premisses of J are of the form $L, \Box w, \Box v, \neg\Box d, e_{i'} \Rightarrow N$ resp. $L, \Box w, \Box v, \neg\Box d \Rightarrow f_{j'}, N$ resp. $L, \Box w, \Box v, \neg\Box d, \Box g_{k'} \Rightarrow N$ resp. $L, \Box w, \Box v, \neg\Box d, \neg\Box h_{l'} \Rightarrow N$. Now since I was a maximal inference not obeying the lemma, the sequents $L, \Box w, \Box v, \neg\Box d, \neg\Box h_{l'} \Rightarrow N$ are immediately preceded by sequents $L^0, \Box w, \Box v \Rightarrow b_{l'}$. Thus we may shift the J-inference below the I-inference:

From the I-premisses different from s we obtain by weakening sequents $L, \Box w, \Box v, a_i, e_{i'} \Rightarrow N$, sequents $L, \Box w, \Box v, e_{i'} \Rightarrow, b_j N$, sequents $L, \Box w, \Box v, \Box c_k, e_{i'} \Rightarrow N$ and sequents $L, \Box w, \Box v, \neg\Box d_l, e_{i'} \Rightarrow N$ and we combine these with the J-premiss $L, \Box w, \Box v, \neg\Box d, e_{i'} \Rightarrow N$ to form the premisses of an E$\Box$-inference leading to $L, \Box w, \Box v, e_{i'} \Rightarrow N$. Similar E$\Box$-inferences provide us with the sequents $L, \Box w, \Box v \Rightarrow f_{j'}, N$ and $L, \Box w, \Box v, \Box g_{k'} \Rightarrow N$, and combining these with the sequents $L, \Box w, \Box v, \neg\Box h_{l'} \Rightarrow N$ which result from $L^0, \Box w, \Box v \Rightarrow b_{l'}$ we obtain all th premisses of an E$\Box$-inference leading to $L, \Box w, \Box v \Rightarrow N$, i.e. the conclusion of I. Now in the new deduction the recursion parameter has decreased and at the same time—although the total number of E$\Box$-inferences has increased—the sum of the recursion parameters of inferences on the single branches of our deduction has not increased. Therefore we may successively eliminate all the E$\Box$-inferences in our deduction which violate the condition expressed in this lemma.

Now as before we may combine any E$\vee$- resp. E$\Box$-inference with the E$\neg$d inferences leading to its γ-premisses into a single new inference and we obtain the calculus LM_3 consisting of the usual axioms and the two rules

$$\text{E}\vee\frac{\begin{array}{ccc} M, a_1 \Rightarrow N & \ldots & M, a_p \Rightarrow N \\ M \Rightarrow b_1, N & \ldots & M \Rightarrow b_q, N \\ M, \Box c_1 \Rightarrow N & \ldots & M, \Box c_r \Rightarrow N \\ M^0 \Rightarrow d_1 & \ldots & M^0 \Rightarrow d_s \end{array}}{M, v \Rightarrow N}$$

$$\text{E}\Box\frac{\begin{array}{ccc} M, \Box v, a_1 \Rightarrow N & \ldots & M, \Box v, a_p \Rightarrow N \\ M \Box v \Rightarrow b_1, N & \ldots & M \Box v \Rightarrow b_q, N \\ M, \Box v, \Box c_1 \Rightarrow N & \ldots & M, \Box v, \Box c_r \Rightarrow N \\ M^0, \Box v \Rightarrow d_1 & \ldots & M^0, \Box v \Rightarrow d_s \end{array}}{M, \Box v \Rightarrow N}$$

which are to be read as before, except that in case $p + q + s = 0$, the number r has to be > 1. Now the E∨-rule also comprises the E¬d-rule of LM_2 thus we have shown:

LEMMA 9 *The calculi* LM_2 *and* LM_3 *are equivalent.*

Now the rule E∨ isn't invertible any more, but it still holds that if its conclusion is deducible, then so are its α- and β-premisses. This shows:

LEMMA 10 *Every* LM_3*-deduction of a sequent* $M, v, v \Rightarrow N$ *may be transformed into a deduction of the sequent* $M, v \Rightarrow N$ *of smaller or equal length.*

Proof. This is true of axioms, and if $M, v, v \Rightarrow N$ is the conclusion of an inference I with principal formula different from v, then the premisses of this inference contain either both occurrences of v or none and to the former ones the induction hypothesis applies. Thus by an application of I to these transformed premisses and to those which contain no occurrence of v the sequent $M, v \Rightarrow N$ may be obtained. If $M, v, v \Rightarrow N$ is the conclusion of an E∨-inference with principal formula v, then the above mentioned restricted inversion principle for E∨ may be applied to the α- and β–premisses and then these premisses may be transformed according to the induction hypothesis, whereas the γ-premisses do not contain v. Thus by applying E∨ to the transformed α- and β-premisses and to the original γ-premisses we arrive at the required deduction of $M, v \Rightarrow N$. Finally if our sequent is the conclusion of an E□-inference with principal formula v, then the induction hypothesis directly applies to all premisses and the required sequent may be derived by a similar application of E□ to the transformed premisses. ■

Now while for the rule E∨ all premisses have smaller length than the conclusion, this is not the case for E□. Therefore we need the following

LEMMA 11 *a) Every* LM$_3$*-deduction of a sequent* $M, \Box[A, u] \Rightarrow N$ *or a sequent* $M, \Box[A, \Box u] \Rightarrow N$ *may be transformed into a deduction of the sequent* $M, \Box u \Rightarrow N$ *of smaller or equal length.*
b) A sequent $M, \Box[a, \Box u], \Box u \Rightarrow N$ *is deducible by* LM$_3$ *iff the sequent* $M, \Box u \Rightarrow N$ *is deducible.*

Proof. a) This is true for axioms, and it is trivially preserved under inferences with principal formulas different from $\Box[A, u]$ resp. $\Box[A, \Box u]$. For an E$\Box$-inference with principal formula $\Box[a, u]$ resp. $\Box[A, \Box u]$ one premiss contains u resp. $\Box u$ and to this premiss the induction hypothesis applies and yields a sequent with two occurrences of u resp. $\Box u$. Thus by an application of the preceding lemma we obtain deductions of the required sequents.
b) If $M, \Box[A, \Box u], \Box u \Rightarrow N$ is deducible, then (a) and Lemma 10 show that $M, \Box u \Rightarrow N$ is deducible too. If this latter sequent is deducible, then weakening shows that $M, \Box[A, \Box u], \Box u \Rightarrow N$ is deducible. ■

From this follows:

LEMMA 12 *The calculus* LM$_4$ *consisting of the usual axioms, the rule* E$\vee$ *and the two* E$\Box$*-rules*

$$\text{E}\Box\text{s}\ \frac{\begin{array}{c} M, \Box v, a_1 \Rightarrow N \ \ldots\ M, \Box v, a_p \Rightarrow N \\ M, \Box v \Rightarrow b_1, N \ \ldots\ M, \Box v \Rightarrow b_q, N \\ M, \Box c_1 \Rightarrow N \ \ldots\ M, \Box c_r \Rightarrow N \end{array}}{M, v \Rightarrow N}$$

$$\text{E}\Box\text{d}\ \frac{\begin{array}{c} M, \Box\bar{v}, a_1 \Rightarrow N \ \ldots\ M, \Box\bar{v}, a_p \Rightarrow N \\ M\Box\bar{v} \Rightarrow b_1, N \ \ldots\ M\Box\bar{v} \Rightarrow b_q, N \\ M, \Box v, \Box c_1 \Rightarrow N \ \ldots\ M, \Box v, \Box c_r \Rightarrow N \\ M^0, \Box v \Rightarrow d_1 \ \ldots\ M^0, \Box v \Rightarrow d_s \end{array}}{M, \Box v \Rightarrow N}$$

where v *is the formula* $[a_1, \ldots, a_p, \neg b_1, \ldots, \neg b_1, \Box c_1, \ldots, \Box c_r, \neg\Box d_1, \ldots, \neg\Box d_s]$, $s = 0$ *for* E$\Box s$, $s > 0$ *for* E$\Box$d *and* $\bar{v}$ *is the formula* $[a_1, \ldots, a_p, \neg b_1, \ldots, \neg b_q, \Box c_1, \ldots, \Box c_r]$*, deduces all the sequents which* LM$_3$ *deduces. (Henceforth we shall call formulas* v *with* $s = 0$ *shallow formulas and those with* $s > 0$ *deep formulas.)*

Proof. All we have to show, is that the E$\Box$-rule of LM$_3$ is admissible for this new calculus: so given all the premisses of an application of E$\Box$, Lemma 11 allows us to transform all its β-premisses into a form suitable for premisses of our new E$\Box$-rules. Moreover if the principal formula v of the given application of E$\Box$ is deep, then the same lemma allows us to transform this formula into $\bar{v}$ in all α-premisses. Thus we may deduce the conclusion of any application of E$\Box$ from its premisses by applying either the rule E$\Box$s or E$\Box$d. ■

This shows one half of the

LEMMA 13 *The calculi* LM$_3$ *and* LM$_4$ *are equivalent.*

Proof. To prove the other direction we have to show that both E$\Box$-rules of LM$_4$ are admissible for LM$_3$: for E$\Box$s this follows directly from Lemma 11. For E$\Box$d we rely on the equivalence of LM$_3$ and LM$_0$: suppose we are given all the premisses of an application of E$\Box$d, then from the γ-premisses $M^0, \Box v \Rightarrow d_l$, we may deduce in LM$_0$ the sequents $M, \Box v, \neg\Box d_l \Rightarrow N$, and from the β-premises we may as before deduce the sequents $M, \Box v, \Box c_k \Rightarrow N$. Thus by suitable applications of the rules E$\neg$ and E$\vee$ from all premisses we may deduce the sequent

$$M, \ldots, \vee\neg(\neg\Box\bar{v} \vee \neg a_i) \vee \ldots \vee \neg(\neg\Box\bar{v} \vee \neg b_j) \vee \ldots$$
$$\vee\neg(\neg\Box v \vee \neg\Box c_k) \vee \ldots\ldots \vee \neg(\neg\Box v \vee \neg\neg\Box d_l) \vee \ldots \Rightarrow N.$$

But we may also deduce

$$M, \Box v \Rightarrow \ldots \vee \neg(\neg\Box\bar{v} \vee \neg a_i) \vee \ldots \vee \neg)\neg\Box\bar{v} \vee \neg b_j) \vee \ldots$$
$$\ldots \vee \neg(\neg\Box v \vee \neg\Box c_k) \vee \ldots \vee \neg(\neg\Box v \vee \neg\neg\Box d_l) \vee \ldots, N.$$

Thus by an application of the cut rule we may deduce the required conclusion $M, \Box v \Rightarrow N$ of E$\Box$d. ■

Now for LM$_4$ the β-premisses of both E$\Box$-rules and the α-premisses of E$\Box$d are shorter than the conclusion, whereas for the γ-premisses of both rules and the α-premisses of E$\Box$s this does not hold. We deal with the latter premisses first.

LEMMA 14 *There is a transformation which converts every* LM$_4$*-deduction of a given sequent into another* LM$_4$*-deduction of the same sequent in which no α-premiss of an application of* E$\Box$s *is the conclusion of* E$\Box$d.

Proof. Given an arbitrary LM$_4$-deduction w.l.o.g. we consider a maximal α-premiss $s = M, \Box v, a \Rightarrow N$ of an application of E$\Box$s, which is the conclusion of an application of E$\Box$d (here $v = [a, a_1, \ldots, a_p \neg b_1, \ldots, b_q, \Box c_1, \ldots, \Box c_r]$), and we use recursion on the maximal number of successive applications of E$\Box$d preceding s: this number cannot be 0, and thus we let $w = [e_1, \ldots, e_t, \neg f_1, \ldots, \neg f_u, \Box g_1, \ldots, \Box g_v, \neg\Box h_1, \ldots, \neg\Box h_w]$ be the principal formula of the application J of E$\Box$d leading to s, we let M be $L, \Box w$ and we let $L, \Box w, \Box v \Rightarrow N$ be the conclusion of I, the sequents s and $L, \Box w, \Box v, a_i \Rightarrow N$ and $L, \Box w, \Box v \Rightarrow b_j, N$ its α-premisses, $L, \Box w, \Box v, \Box c_k \Rightarrow N$ its β-premisses: then the premisses of J are of the form $L, \Box w, \Box v, a, e_{i'} \Rightarrow N$ resp. $L, \Box w, \Box v, a \Rightarrow f_{j'}, N$ resp. $L, \Box w, \Box v, a, \Box g_{k'} \Rightarrow N$ resp. $L^0, \Box w, \Box v, \Rightarrow h_{l'}$ and the inference J is shifted down past I as follows:

We use the given sequent $L, \Box w, \Box v, a, e_{i'} \Rightarrow N$ and the sequents $L, \Box w, \Box v, a_i, e_{i'} \Rightarrow N$ and $L, \Box w, \Box v, e_{i'} \Rightarrow b_j, N$ and $L, \Box w, \Box v, \Box c_k, e_{i'} \Rightarrow N$

(where these latter sequents are obtained from the corresponding I-premisses by weakening) as premisses of an application of E□s leading to $L, \Box w, \Box v, e_{i'} \Rightarrow N$. Similarly we obtain deductions of the sequents $L, \Box w, \Box v \Rightarrow f_{j'}, N$ resp. $L, \Box w, \Box v, \Box g_{k'} \Rightarrow N$ and from these and $L^0, \Box w, \Box v, \Rightarrow h_{l'}$ using an E□d-inference with principal formula $\Box w$ we arrive at a deduction of $L, \Box w, \Box v \Rightarrow N$ where the recursion parameter has decreased by 1. Now the number of E□s-inferences may have increased in this new deduction, but the maximal sum of recursion parameters on any branch of our deduction cannot have increased. Hence we may in this way eliminate all the α-premisses of E□s which are conclusions of applications of E□d. ■

This lemma is applied for proving completeness of the calculus LM$_5$ which in addition to the modal operator □ uses a new operator ○ to be substituted for □ in certain situations: LM$_5$ has besides the usual axioms the rules E∨, E□s and E□d in the following form:

$$
\text{E}\vee\ \frac{\begin{array}{c} M^2, a_1 \Rightarrow N \ \ldots\ M^2, a_p \Rightarrow N \\ M^2, \Rightarrow b_1, N \ \ldots\ M^2, \Rightarrow b_q, N \\ M^2, \Box c_1 \Rightarrow N \ \ldots\ M^2, \Box c_r \Rightarrow N \\ M^3 \Rightarrow d_1 \ \ldots\ M^3 \Rightarrow d_s \end{array}}{M, v \Rightarrow N}
$$

$$
\text{E}\Box\text{s}\ \frac{\begin{array}{c} M^1, \bigcirc v, a_1 \Rightarrow N \ \ldots\ M^2, \bigcirc v, a_p \Rightarrow N \\ M^1, \bigcirc v \Rightarrow b_1, N \ \ldots\ M^1, \bigcirc v \Rightarrow b_q, N \\ M^2, \Box c_1 \Rightarrow N \ \ldots\ M^2, \Box c_r \Rightarrow N \end{array}}{M, v \Rightarrow N}
$$

$$
\text{E}\Box\text{d}\ \frac{\begin{array}{c} M^2, \Box\bar{v}, a_1 \Rightarrow N \ \ldots\ M^2, \Box\bar{v}, a_p \Rightarrow N \\ M^2, \Box\bar{v} \Rightarrow b_1, N \ \ldots\ M^2, \Box\bar{v} \Rightarrow b_q, N \\ M^2, \Box c_1 \Rightarrow N \ \ldots\ M^2, \Box c_r \Rightarrow N \\ M^3, \Box v \Rightarrow d_1 \ \ldots\ M^3, \Box v \Rightarrow d_s \end{array}}{M, \Box v \Rightarrow N}
$$

where M^1 results from M by replacing any formula $\Box v$ with deep v by $\bigcirc v$, M^2 results from M by replacing any formula $\bigcirc v$ by $\Box v$, and M^3 results from M by omitting all non modalized formulas, i.e. all formulas not of the form $\Box v$ or $\bigcirc v$. There is no rule for introducing the operator ○, thus any application of one of the rules of LM$_5$ becomes a valid application of the corresponding LM$_4$-rule if we replace any ○ in all premisses and in the conclusion by a □. This shows that if $M, \bigcirc v \Rightarrow N$ is deducible by LM$_5$, then $M, \Box v \Rightarrow N$ is deducible by LM$_5$ and by LM$_4$; and in general: if $M \Rightarrow N$ is deducible by LM$_5$, then $M^2 \Rightarrow N$ is deducible by LM$_5$ and by LM_4—one half of the equivlanence of LM_4 and LM_5. For the proof of the other direction we call a formula $\Box v$ distant in a deduction d of a sequent $m, \Box v \Rightarrow N$ iff below any conclusion of an inference with principal formula $\Box v$ there is a conclusion of an E$\vee$-inference or a β-premiss. Then we show

LEMMA 15 *Any* LM_4*-deduction d of a sequent* $M \Rightarrow N$ *may be transofrmed into an* LM_5*-deduction of the sequent* $M^4 \Rightarrow N$*, where* M^4 *results from* M *by replacing any number of formulas* $\Box v$ *distant in d by* $\bigcirc v$*.*

Proof. Suppose we are given an LM_4-deduction of a sequent s. We may assume that it has the property expressed by Lemma 14. If the final inference of this deduction is an application of E$\vee$, then by the induction hypothesis the deductions of all premisses may be transformed into LM_5-deductions of the same sequents and these are the premisses of an E$\vee$-inference of LM_5-inference leading to the required sequent. If the final inference is an application of E$\Box$d, then by the induction hypothesis the α- and β-premisses are deducible by LM_5 and these are the corresponding premisses of an E$\Box$d-inference of LM_5 leading to the required sequent. But the γ-premisses necessary for this inference are obtained from the given γ-premisses by replacing sufficiently many distant formulas $\Box v$ by $\bigcirc v$, because all the distant formulas of s are distant in all γ-premisses, too. Therefore the required sequent is deducible by LM_5. If the final inference I is an application of E$\Box$s with principle formula $\Box v$, where $v = [a_1, \ldots, a_p, \neg b_1, \ldots, \neg b_q, \Box c_1, \ldots, \Box c_r]$ and the formula $\Box v$ is not distant in an α-premiss preceding P and such that one of these α-premisses is the conclusion of another inference with principle formula $\Box v$. In this case we may drop the inference I. Otherwise the formula $\Box v$ is distant in all α-premisses of I, and also all the distant formulas of our sequent $M, \Box v \Rightarrow N$ in its deduction d are distant formulas in all α-premisses too. Now again by the property of Lemma 14 in the α-premisses at least all formulas $\Box w$ with w deep are distant. Thus by the induction hypothesis all sequents $M^4, \Box v, a_i \Rightarrow N$ resp. $M^4, \Box v \Rightarrow b_j, N$ are deducible, and they are the α-premisses of an LM_5-application of E$\Box$s leading to $M^4, \Box v \Rightarrow N$. Moreover as before the transformed deductions of the β-premisses again yield the β–premisses of a corresponding LM_5-inference leading to the required conclusion $M, \Box v \Rightarrow N$. Thus this latter sequent is deducible by LM_5. ■

The calculus LM_5 shows one single obstacle to contraction freeness, viz. the presence of the principal formula of an E$\Box$d-inference in the γ-premisses. This obstacle is removed by considering the calculus LM having the usual axioms, the rules E$\vee$ and E$\Box$s of LM_5 and the rule E$\Box$d in the form

$$\mathrm{E\Box d}\ \frac{\begin{array}{c} M^2, \Box\bar{v}, a_1 \Rightarrow N \ \ldots\ M^2, \Box\bar{v}, a_p \Rightarrow N \\ M^2, \Box\bar{v} \Rightarrow b_1, N \ \ldots\ M^2, \Box\bar{v} \Rightarrow b_q, N \\ M^2, \Box c_1 \Rightarrow N \ \ldots\ M^2, \Box c_r \Rightarrow N \\ M^3, \bigcirc v \Rightarrow d_1 \ \ldots\ M^3, \bigcirc v \Rightarrow d_s \end{array}}{M, \Box v \Rightarrow N}$$

In order to proof completeness of this calculus we call a formula $\Box v$ distant in a deduction d of a sequent $M, \Box v \Rightarrow N$ iff v is deep and below any conclusion of an inference with principal formula $\Box v$ there is a conclusion of an E$\vee$-inference or an α–premiss or a β-premiss. Then we reproof Lemma 15 as:

LEMMA 16 *Any* LM$_5$*-deduction d of a sequent* $M \Rightarrow N$ *may be transformed into an* LM*-deduction of the sequent* $M^4 \Rightarrow N$, *where* M^4 *results from M by replacing any number of formulas* $\Box v$ *distant in d by* $\bigcirc v$.

Proof. If the last rule applied in a given LM$_5$-deduction of our sequent $s = M \Rightarrow N$ is E$\vee$, then all premisses are deducible by LM by the induction hypothesis, and moreover they are the LM-premisses of an inference leading to $M^4 \Rightarrow N$. Therefore this sequent is deducible by LM, too. If the last rule applied is E$\Box$s, then the induction hypothesis gives us LM-deductions of all β-premisses, and these are the required LM-premisses for an inference leading to $M^4 \Rightarrow N$. But in the α-premisses all deep formulas are marked by a $\bigcirc$. Thus we may also use the α-premisses of our LM$_5$-inference as α-premisses of an LM-inference leading to $M^4 \Rightarrow N$. Finally if the last inference is by an application of E$\Box$d with principal formula $\Box v$ and there is a γ-premiss P of this inference in which $\Box v$ is not distant, then there is a chain of γ-premisses preceding P which contains another occurrence of P. In this case the last inference may be dropped. Otherwise the formula $\Box v$ is distant in the deductions of all γ-premisses, and furthermore all formulas distant in our given deduction of s are distant in all γ-premisses, too. Therefore by the induction hypothesis we obtain all the γ-premisses necessary for an LM-application of E$\Box$d leading to $M^4 \Rightarrow N$. But the α- and β-premisses for such an inference are obtained as before, hence in all cases we arrive at an LM-deduction of the required sequent $M^4 \Rightarrow N$. ■

The calculus LM now has the desired property that there is a measure μ, such that in every one of its rules the measure of the conclusion is greater than the measures of all premisses: namely we may take $\mu(s)$ for a sequent s to be (the total number of connectives of s times (the number of $\Box$'s plus the number of $\bigcirc$'s)) minus the number of $\bigcirc$'s, and moreover there holds the

THEOREM 17 *A sequent is valid in S4 if and only if it is deducible by LM.*

University of Tübingen, Germany.

REFERENCES

[1] R. Dyckhoff. Contraction free sequent calculi for intuitionistic logic. *Journal of Symbolic Logic*, **57**, 795–807, 1992.

[2] M. Fitting. *Proof Methods for Modal and Intuitionistic Logics*. Reidel, Dordrecht, 1983.

[3] J. Hudelmaier. Bounds for cut elimination in intuitionistic propositional logic. *Archive for Mathematical Logic*, **31**, 331–353, 1992.

[4] G. Mints. Gentzen-type systems and resolution rules. Part I. In: P. Martin-Löf, G. Mints (eds.), *COLOG-88*, Lecture Notes in Computer Sci. 417, pp. 198–231, Springer-Verlag, Berlin, 1988.

GRIGORI MINTS, VLADIMIR OREVKOV AND TANEL TAMMET

TRANSFER OF SEQUENT CALCULUS STRATEGIES TO RESOLUTION FOR S4

1 INTRODUCTION

This paper presents a general scheme [10, 11] of transforming a cutfree Gentzen-type system into a resolution type system, preserving the structure of derivations. This is a direct extension of the method introduced by Maslov [7] for classical predicate logic. In this way completeness of strategies is first established for the Gentzen-type system, and then transferred to resolution. The method is illustrated here for the propositional S4 in a way very similar to [13]. An adaptation of the Maslov's method to S4 and some of the more elaborate strategies mentioned below are presented in [17]. Let us recapitulate some material from [10, 11, 12].

The main idea of Maslov's method can be summarized as follows:

A resolution derivation of the goal clause g from a list Γ of input clauses can be obtained as the result of deleting Γ from the Gentzen-type cutfree derivation of the sequent $\Gamma \Rightarrow g$.

Recall [10] what we mean by a resolution method for a formal system **C**. Such a method is determined by specifying:

1. A class of formulas called clauses.
2. A method of reduction of any formula F of the system **C** to a finite list Γ_F of clauses.
3. An inference rule R called the resolution rule for deriving clauses.
4. The derivation process by forward chaining so that all derivable objects are consequences of initial clauses, and garbage removal from the search space is possible.

The resolution method is sound and complete iff for any formula F, the derivability of F in **C** is equivalent to derivability of the goal clause g from Γ_F using the rule R. Reduction of an arbitrary formula F to the form $\Gamma_F \Rightarrow g$ where Γ_F is a finite set of clauses and g is a goal (variable or $\emptyset$) is based on the familiar depth-reducing transformation by introduction of new variables.

H. Wansing (ed.), Proof Theory of Modal Logic, 17–31.

2 GENTZEN-TYPE FORMULATIONS OF S4

Consider a Gentzen-type formulation of the predicate calculus S4 from [11] suitable for pruning superfluous formulas (cf. below). Its derivable objects are *sequents* $\Gamma \Rightarrow \Delta$, where Γ, Δ are finite (possibly empty) lists of formulas of the language considered. The order of formulas in Γ, Δ will always be disregarded, hence Γ, Δ etc. are treated as multisets (number of occurrences of formulas is important).

Gentzen-type system GS4

Axioms: $A \Rightarrow A$.
Inference rules:

$$\frac{A_1, \Gamma_1 \Rightarrow \Delta_1; \ldots ; A_n, \Gamma_n \Rightarrow \Delta_n}{A_1 \vee \cdots \vee A_n, \Gamma \Rightarrow \Delta} \quad (\vee\Rightarrow)$$

where $\Gamma_1 \cup \ldots \cup \Gamma_n = \Gamma$, $\Delta_1 \cup \ldots \cup \Delta_n = \Delta$.

$$\frac{\Gamma \Rightarrow \Delta, A_{i_1}, \ldots, A_{i_m}}{\Gamma \Rightarrow \Delta, A_1 \vee \cdots \vee A_n} (\Rightarrow\vee) \qquad \frac{\Gamma \Rightarrow \Delta}{\Gamma', \Gamma \Rightarrow \Delta, \Delta'} (weakening)$$

where $i_1 < \ldots < i_m \leq n$.

$$(\Rightarrow \neg) \frac{A, \Gamma \Rightarrow \Delta}{\Gamma \Rightarrow \Delta, \neg A} \qquad \frac{\Gamma \Rightarrow \Delta, A}{\neg A, \Gamma \Rightarrow \Delta} (\neg \Rightarrow)$$

Modal rules:

$$(\Box \Rightarrow) \frac{A, (\Box A)^0, \Gamma \Rightarrow \Delta}{\Box A, \Gamma \Rightarrow \Delta} \qquad (\Rightarrow \Box) \frac{\Box\Gamma \Rightarrow A}{\Box\Gamma \Rightarrow \Box A}$$

where superscript 0 means possible absence of the formula. So in fact we have two versions of the rule $(\Box \Rightarrow)$:

$$\frac{A, \Gamma \Rightarrow \Delta}{\Box A, \Gamma \Rightarrow \Delta} \qquad \frac{A, \Box A, \Gamma \Rightarrow \Delta}{\Box A, \Gamma \Rightarrow \Delta}$$

Let us fix terminology concerning Gentzen-type systems. In the sequent $\Gamma \Rightarrow \Delta$ the *left-hand side* Γ and *right-hand side* Δ are sometimes called *antecedent* and *succedent*. In each inference rule the sequent written under the line is the *conclusion*, and the sequents over the line are *premises*. The formula shown explicitly in the conclusion, for example $A_1 \vee \cdots \vee A_n$ in $(\vee \Rightarrow)$ or Γ', Δ' in (weakening), is the *main formula*, the formulas shown explicitly in the premises, for example $A_1, \ldots, A_n$ in the rule $(\vee \Rightarrow)$, are *side* formulas, and the remaining formulas, for example $\Gamma_1, \Delta_1, \ldots, \Gamma_n, \Delta_n, \Gamma, \Delta$ in $(\vee\Rightarrow)$ or Γ, Δ in (*weakening*), are *parametric* formulas.

Pruned derivation is one containing weakenings only immediately preceding the last sequent of the derivation (cf. [5]). A derivation is *p-inverted* if propositional connectives are immediately analyzed. More precisely, if a sequent S is not an axiom and contains a formula of the form $A \vee B, \neg A$, then one of these formulas is the main formula of the inference L introducing S. Moreover, if some of the side formulas of the inference L is again of this form, one of them is the main formula of the inference immediately preceding L.

Recall the definition of the sign of a subformula occurrence in a formula: *positive subformulas* occur within the scope of even number (for example zero) of occurrences of negations and the premises of implication. Non-positive occurrences are *negative*. The proof of part (a) of the following statement can be found in [3], the proof of (b) is standard, (c) and (d) are well-known.

THEOREM 1

(a) *Formula F is derivable in S4 iff the sequent $\Rightarrow F$ is derivable in GS4.*

(b) *Any provable sequent has a p-inverted pruned derivation.*

(c) *The derivation of a sequent S uses only rules for connectives occurring in S, or more precisely, the succedent and antecedent rules corresponding to positive or negative occurrences of connectives.*

(d) *A list Γ of formulas is inconsistent in S4 iff the sequent $\Gamma \Rightarrow$ (with empty right-hand side) is derivable in GS4.*

2.1 Bottom-up Proof Search

Cut-free systems like GS4 suggest proof search from goal to subgoals. Initially one has goal formula (or sequent). Each subsequent step replaces some of the current goals by subgoals from which it can be derived until all leaves are axioms. Subgoals are written on the top of the goal, so the whole procedure is working *bottom up*.

EXAMPLE 2 Here are stages of proof search in GS4 where weakening is attempted only for testing axioms.

$$\begin{array}{ccc} \mathit{Goal} & \mathit{Step1} & \mathit{Step2} \\ \Rightarrow \neg\Box a \vee \Box(a \vee b) & \dfrac{\Rightarrow \neg\Box a, \Box(a \vee b)}{\Rightarrow \neg\Box a \vee \Box(a \vee b)} & \cfrac{\Box a \Rightarrow \Box(a = veeb)}{\cfrac{\Rightarrow \neg\Box a, \Box(a \vee b)}{\Rightarrow \neg\Box a \vee \Box(a \vee b)}} \end{array}$$

Step3

$$\cfrac{\cfrac{\cfrac{\Box a \Rightarrow a \vee b}{\Box a \Rightarrow \Box(a \vee b)}}{\Rightarrow \neg\Box a, \Box(a \vee b)}}{\Rightarrow \neg\Box a \vee \Box(a \vee b)}$$

Step4

$$\cfrac{\cfrac{\cfrac{\cfrac{\Box a \Rightarrow a}{\Box a \Rightarrow a \vee b}}{\Box a \Rightarrow \Box(a \vee b)}}{\Rightarrow \neg\Box a, \Box(a \vee b)}}{\Rightarrow \neg\Box a \vee \Box(a \vee b)}$$

Step5

$$\cfrac{\cfrac{\cfrac{\cfrac{\cfrac{a \Rightarrow A}{\Box a \Rightarrow a}}{\Box a To a \vee b}}{\Box a \Rightarrow \Box(a \vee b)}}{\Rightarrow \neg\Box a, \Box(a \vee b)}}{\Rightarrow \neg\Box a \vee \Box(a \vee b)}$$

3 INVERSE METHOD

3.1 System $S4_F$

Instead of the bottom-up proof search described in the previous section, we consider here, following Maslov [6], *top-down* proof search by direct chaining.

We begin with axioms, derive everything possible by one inference (application of an inference rule), then derive everything possible by two inferences etc. until the goal formula appears. Thus the proof search proceeds in the direction opposed to bottom-up search described in the previous section, which was used in most advanced programs during the first period of automated deduction. That is why Maslov used the term *inverse method.* Both the language and rules of the present formulation are determined by the goal formula F. Introduce new distinct propositional variables called *labels* l_A for all subformulas A of F which are not *literals* (propositional variables and their negations). For literals L put $l_L = L$. *Derivable objects* are sequents $\Gamma \Rightarrow \Delta$ where Γ, Δ are lists of labels.
Axioms and inference rules of the system $S4_F$ are obtained from the axioms and inference rules of $GS4$ by replacing all formulas by their labels.

EXAMPLE 3 Let $F = \neg\Box a \vee \Box(a \vee b)$.

Here is a derivation of $\Rightarrow l_F$ in the system $S4_F$.

$$\cfrac{\cfrac{\cfrac{\cfrac{\cfrac{a \Rightarrow a}{l_{\Box a} \Rightarrow a}}{l_{\Box a} \Rightarrow l_{a \vee b}}}{l_{\Box a} \Rightarrow l_{\Box(a \vee b)}}}{\Rightarrow l_{\neg\Box a}, l_{\Box(a \vee b)}}}{\Rightarrow l_{\neg\Box a \vee \Box(a \vee b)}}$$

DEFINITION 4 *Notation* $d :\ A_1, \ldots, A_n \Rightarrow B$ *means that* d *is a derivation of the sequent* $A_1, \ldots, A_n \Rightarrow B$.

If $A_1, \ldots, A_n, B$ *are subformulas of* F, *and* $d :\ A_1, \ldots, A_n \Rightarrow B$ *then* l_d *denotes the result of replacing all formulas (sequent members) in* d *by their labels.*

EXAMPLE 5 Let d be the derivation in GS4 of the formula F from the Example 2 , i.e. the result of the last step 5. Then l_d is a derivation from the Example 3.

THEOREM 6 *If* $d: A_1, \ldots, A_n \Rightarrow B_1, \ldots, B_m$ *is a derivation in* $GS4$, *then it is isomorphic to* $l_d: l_{A_1}, \ldots, l_{A_n} \Rightarrow l_{B_1}, \ldots, l_{B_m}$ *as a labelled tree.*

Proof. Proof is obvious. ■

COROLLARY 7 $GS4 \vdash F$ iff $S4_F \vdash l_F$

Proof. Take $d :\Rightarrow F$ and $l_d :\Rightarrow l_F$. ■

Our derivable objects are clauses up to replacing $l_1, \ldots, l_k \Rightarrow l_{k+1}, \ldots, l_n$ by $\neg l_1 \vee \ldots \vee \neg l_k \vee l_{k+1} \ldots \vee l_n$. It will be seen that this view is fruitful.

3.2 *Strategies of Inverse Method*

First of all, the language can be restricted. The following signed subformula property is obvious.

LEMMA 8 *If* $d :\Rightarrow F$ *is a derivation in GS4, and sequent*

$$A_1, \ldots, A_n \Rightarrow B_1, \ldots B_m$$

occurs in d, then $A_1, \ldots, A_n$ *are negative subformulas of* F, *and* $B_1, \ldots B_m$ *are positive subformulas of* F.

From now on the language of the system $S4_F$ is restricted to sequents of the form

(1) $l_{A_1}, \ldots, l_{A_n} \Rightarrow l_{B_1}, \ldots, l_{B_m}$

where $A_1, \ldots, A_n$ are negative subformulas of F, and $B_1, \ldots B_m$ are positive subformulas of F .

The previous Lemma shows that this restriction is still *complete*: if F is derivable in $S4$, then l_F is derivable in $S4_F$.

EXAMPLE 9 Complete proof search for the formula $F = \Box a \vee \Box \neg a$ in $S4_F$.

In view of the restriction (1), only the axiom

$$a \Rightarrow a$$

is possible. No antecedent inference rule is applicable, since there are no composite (non-atomic) negative subformulas of F. The rule $\Rightarrow \Box$ is not applicable to a sequent containing more than one formula not beginning with $\Box$. Hence only one succedent rule ($\Rightarrow \neg$) is applicable to the axiom , resulting in

$$\Rightarrow a, \neg a$$

The only rule possible now under restriction (1), is $\Rightarrow \Box$, but it is not applicable since there are two succedent formulas. This concludes the search: F is not derivable.

Inversion Strategy and Subsumption

Recall that the propositional rules of the system GS4 (Section 2) are *invertible* if the rules for $\vee$ are taken in their non-pruned versions :

$$\frac{\Gamma \Rightarrow \Delta, A, B}{\Gamma \Rightarrow \Delta, A \vee B} \qquad \frac{A, \Gamma \Rightarrow \Delta \qquad B, \Gamma \Rightarrow \Delta}{A \vee B, \Gamma \Rightarrow \Delta}$$

The conclusion of a propositional rule is derivable iff all premises are derivable. This motivates the following definition.

DEFINITION 10 *The label* l_A *is* p-invertible *if* A *has one of the forms* $B \vee C, \neg D$.
p-inversion strategy *for* $S4_F$ *is the following restriction: if a sequent is not an axiom and contains invertible label, then it is obtained by a propositional rule.*

Note a consequence of the p-inversion strategy for direct chaining: a non-p-invertible rule cannot be applied if the result contains a p-invertible label. For example, sequent $l_{\Box A}, l_{C \vee D} \Rightarrow l_{E \vee G}$ can be obtained (according to inversion strategy) by the rules $\vee \Rightarrow, \Rightarrow \vee$, but not by the rule $\Box \Rightarrow$. Hence this sequent cannot be generated from the sequent $l_A, l_{C \vee D} \Rightarrow l_{E \vee G}$.

LEMMA 11 *p-inversion strategy is complete.*

Completeness of the p-inversion strategy for GS4 is stated in Theorem 1(b), and is preserved by our transformation from GS4 to $S4_F$. $\Box$

Note that the rule $\Rightarrow \Box$

$$\frac{\Box\Gamma \Rightarrow A}{\Box\Gamma \Rightarrow \Box A}$$

is also invertible. This motivates the following definition.

DEFINITION 12 *The label* l_A *is* invertible *in a sequent* S *if* A *has one of the forms* $B \vee C, \neg D$ *or* A *is* $\Box B$ *and* S *is* $l_{\Box A_1} \vee \ldots \vee l_{\Box A_n} \vee l_{\Box B}$.
Inversion strategy *for* $S4_F$ *is the following restriction: if a sequent is not an axiom and contains invertible label, then it is obtained by a rule introducing such a label*

LEMMA 13 *Inversion strategy is complete.*

Completeness of the corresponding inversion strategy for GS4 is proved in a standard way: cf. Section 8B.1 in [3]. It is preserved by our transformation from GS4 to $S4_F$. $\Box$

DEFINITION 14 *Sequent* $\Gamma \Rightarrow \Delta$ subsumes *the sequent* $\Gamma' \Rightarrow \Delta'$ *if* $\Gamma \subseteq \Gamma', \Delta \subseteq \Delta'$.

Note. It is also possible to extend this definition using implication $\Box A \Rightarrow A$, but this requires some care.

Subsumption strategy allows one to delete from the search space any clause which is subsumed by the remaining clauses. It is complete for the same reason as the analogous strategies for resolution, but its compatibility with other strategies should be verified in each particular case.

4 RESOLUTION

We reproduce some material from [8, 9, 10] and connect it with Section 3.

4.1 Modal clauses

Recall from the section 3.1 that the label l_A stands for a subformula A of the goal formula F. This can be expressed by postulating equivalence

$$\Box(l_A \leftrightarrow A)$$

or two strict implications

$$(2) \quad \Box(l_A \to A) \qquad \Box(A \to l_A)$$

Most often only one of the implications (2) will be preserved: $A \to l_A$ for positive subformulas A, and $l_A \to A$ for negative A. Some of the implications (2) can be simplified, for example $a \vee b \to c$ can be replaced by two implications $a \to c$, $b \to c$. Finally, implications $a \to D$ are replaced by clauses $\neg a \vee D$.

We define *literals* as atoms and their negations and denote them by $l, l_1, \ldots$. *Modal literals* are by definition expressions of the form l, $\Box l$, $\Diamond l$ where $\Diamond$ is treated as $\neg\Box\neg$. Modal literals are denoted by $L, M, N, L_1, M_1, N_1, \ldots$. Complements are defined by $\neg\Box l = \Diamond\neg l$, $\neg\Diamond l = \Box\neg l$ in a natural way.

Propositional clauses are disjunctions of literals. *Modal clauses* (or simply clauses) are disjunctions of modal literals. *Initial* modal clauses are expressions of the form $\Box C$ where C is a modal clause. As usual, clauses are treated as sets of literals, i.e. up to the order and number of occurrences of literals. Hence there is an implicit contraction rule. To simplify notation we require that C in an initial modal clause contain at least two terms. Reduction of arbitrary formula to a clause form is given by the standard depth-reducing transformation.

THEOREM 15 *For any formula F one can construct (by introduction of new variables) the list X_F of initial clauses and a propositional variable g such that*

$$\vdash_{S4} F \qquad iff \qquad \vdash_{S4} \& X_F \to g$$

Proof. This is Theorem 3.1 of [10]. ■

4.2 *Resolution Calculus* $RS4$

Derivable objects: modal clauses and initial modal clauses, the latter only as initial objects of derivations. We are interested in derivability relations $X \vdash C$ where X is a set of initial clauses, C is a modal clause.

Axioms: $L \vee \neg L$ for modal literals L which occur both positively and negatively in $X \vdash C$.

Inference rules:

$$(R)\ \frac{D \vee L \qquad E \vee \neg L}{D \vee E} \qquad \frac{l \vee D}{\Diamond l \vee D}\ (\Diamond) \qquad \frac{l \vee \Diamond D}{\Box l \vee \Diamond D}\ (\Box) \qquad \frac{\Box D}{D}\ (\Box^-)$$

Here D, E denote modal clauses. $\Diamond D$ stands for the result of prefixing $\Diamond$ to all literals in a propositional clause D, while $\Box D$ is the result of prefixing $\Box$ to the whole of D.

Notes.

1. The rule $(\Box^-)$ can be restricted to apply only for bootstrapping initial clauses. Cf. Section 4.4.

2. All rules are obviously valid for derivability from $\Box$-formulas in S4.

3. Important derived inference is the clash rule [2] where the whole clause is resolved:

$$\frac{L_1 \vee \ldots \vee L_n \qquad D_1 \vee \neg L_1 \qquad \ldots \qquad D_n \vee \neg L_n}{D_1 \vee \ldots \vee D_n}$$

4. In our present formulation (but not in the formulation of the section 4.4 below) it is possible to avoid postulating tautologies $L \vee \neg L$ as the axioms.

5. Subformula property for $RS4$: a literal occurs in a derivation of $X \vdash C$ only if it occurs in X, C up to $\neg, \Box$. The latter can happen when rules $(\Diamond), (\Box)$ are applied.

The notation for derivability in RS4 is $X \vdash_R C$

EXAMPLE 16 Here is a derivation of $\Box(\neg a \vee \Diamond b),\ \Box(a \vee c) \vdash_R \Diamond b \vee \Box c$.

$$\cfrac{\cfrac{\cfrac{\Box(\neg a \vee \Diamond b)}{\neg a \vee \Diamond b} \qquad \cfrac{\Box(a \vee c)}{a \vee c}}{\Diamond b \vee c}}{\Diamond b \vee \Box c}$$

4.3 Maslov Transformation

The translation GR into $RS4$-derivations is defined here for pruned $GS4$ -derivation $d :\Rightarrow F$ where F is a formula. The first step transforms d into a $S4_F$-derivation $l_d :\Rightarrow l_F$ (cf. Definition 4 and Theorem 6 above). Note that any sequent in l_d has the form

(3) $l_{A_1}, \ldots, l_{A_n} \Rightarrow l_{B_1}, \ldots, l_{B_m}$

where B_i are positive subformulas of F, and A_j are negative subformulas.

We obtain GR(d) by replacing sequents (3) by

$$\neg l_{A_1} \vee \ldots \vee \neg l_{A_n} \vee l_{B_1} \vee \ldots \vee l_{B_m}$$

and adding necessary input clauses to construct correct resolution inferences. Some contractions intervene since clauses are handled as sets, but they are easily dealt with. Details are in [10, 11]. Cf. also next section.

The transformation RG from a derivation in RS4, $d : X_F \vdash g$, into the derivation in GS4 can also be easily defined if d uses only $GR(d)$-rules from the Section 4.4 below.

THEOREM 17 (Soundness and Completeness Theorem) *Let F be a modal formula, and X_F, g be as in Theorem 15.*

(a) If d $:\Rightarrow F$ is a derivation in GS4, then GR(d) : $X'_F \vdash g$ (or $X'_F \vdash \emptyset$) is a derivation in RS4 , where X'_F is a sublist of X_F.

(b) If d : $X_F \vdash g$ (or $X_F \vdash \emptyset$) in RS4 then RG(d) : $X_F \Rightarrow g$ is the derivation in GS4.

(c) $\vdash_{S4} F$ iff $X_F \vdash_{RS4} g$.

Proof. Cf. Theorem 2.4 in [11]. ∎

EXAMPLE 18 Let $F = \neg\Box a \vee \Box(a \vee b)$, and let $d :\Rightarrow F$ be the derivation from the Example 3. Then l_d is presented in Example 5, and $GR(d)$ is the following derivation

$$\dfrac{\dfrac{\Box(l_F \vee \neg l_{\Box(a\vee b)})}{l_F \vee \neg l_{\Box(a\vee b)}} \qquad \dfrac{\dfrac{\Box(l_F \vee \neg l_{\neg\Box a})}{l_F \vee \neg l_{\neg\Box a}} \qquad \dfrac{\dfrac{\Box(l_{\Box a} \vee l_{\neg\Box a})}{l_{\Box a} \vee l_{\neg\Box a}} \quad d' : \neg l_{\Box a} \vee l_{\Box(a\vee b)}}{l_{\neg\Box a} \vee l_{\Box(a\vee b)}}}{l_F \vee l_{\Box(a\vee b)}}}{l_F}$$

where $d' : \neg l_{\Box a} \vee l_{\Box(a \vee b)}$ is

$$\dfrac{\dfrac{\Box(\neg l_{\Box a} \vee \Box l_a)}{\neg l_{\Box a} \vee \Box l_a} \qquad \dfrac{\dfrac{\Box(l_{\Box(a\vee b)} \vee \neg\Box l_{a\vee b})}{l_{\Box(a\vee b)} \vee \neg\Box l_{a\vee b}} \qquad \dfrac{\dfrac{\dfrac{\Box(l_{a\vee b} \vee \neg l_a)}{l_{a\vee b} \vee \neg l_a} \qquad \dfrac{\neg l_a \vee l_a}{\neg\Box l_a \vee l_a}}{\neg\Box l_a \vee l_{a\vee b}}}{\neg\Box l_a \vee \Box l_{a\vee b}}}{\neg\Box l_a \vee l_{\Box(a\vee b)}}}{\neg l_{\Box a} \vee l_{\Box(a\vee b)}}$$

4.4 *Strategies of Resolution. Transfer from Gentzen-type Systems*

We see that the structure of $GR(d)$ is the same as the structure of d. This allows us to carry over strategies. Note that some strategies known in the classical case are obviously incomplete. One example is hyperresolution. A rich source of complete strategies is the transfer from Gentzen-type systems via the transformation GR and its analogs. It allows us in particular to prove the completeness of a powerful analog of hyperresolution due to V. Orevkov [15]. Let us first describe in more detail the structure of the derivation $GR(d)$ from the previous theorem. *Derived objects* are clauses

(4) $\quad \neg l_{A_1^-} \vee \ldots \neg l_{A_n^-} \vee l_{B_1^+} \vee \ldots l_{B_m^+}$

where A_i^- are negative subformulas of the goal formula F, and B_j^+ are positive subformulas of F. In the following list we denote resolution-type rules by the same symbols as the rules of GS4 from which they are derived.
Note. Only the following rules are actually used in $GR(d)$.

$GR(d)$**-rules** :

$$(\Rightarrow \vee)\ \dfrac{\Box(l_{A_1 \vee A_2} \vee \neg l_{A_i}) \quad D \vee l_{A_i}}{D \vee l_{A_1 \vee A_2}}$$

$$(\vee \Rightarrow)\ \dfrac{\Box(\neg l_{A_1 \vee A_2} \vee l_{A_1} \vee l_{A_2}) \quad D \vee \neg l_{A_1} \quad D' \vee \neg l_{A_2}}{\neg l_{A_1 \vee A_2} \vee D \vee D'}$$

$$(\Rightarrow \neg)\ \dfrac{\Box(l_{\neg A} \vee l_A) \quad D \vee \neg l_A}{D \vee l_{\neg A}} \qquad (\neg \Rightarrow)\ \dfrac{\Box(\neg l_{\neg A} \vee \neg l_A) \quad D \vee l_A}{D \vee \neg l_{\neg A}}$$

$$(\neg\Box)\ \dfrac{\neg l_A \vee D}{\neg\Box l_A \vee D} \qquad (\Box)\ \dfrac{l_A \vee \neg\Box l_1 \vee \ldots \neg\Box l_n}{\Box l_A \vee \neg\Box l_1 \vee \ldots \neg\Box l_n}$$

$$(\Box \Rightarrow)\ \dfrac{\Box(\neg l_{\Box A} \vee \Box l_A) \quad \neg\Box l_A \vee D}{\neg l_{\Box A} \vee D} \qquad (\Rightarrow \Box)\ \dfrac{\Box(l_{\Box A} \vee \neg\Box l_A) \quad \Box l_A \vee D}{l_{\Box A} \vee D}$$

Note that only some instances of the rule (R) are needed, and rule $(\Box^-)$ is subsumed into other rules. Note also that all rules except the rules $(\Box)$, $(\neg\Box)$ have initial clause as one of the premises. In any derivation of the form $GR(d)$ the rule $\neg\Box$ is always combined with the rule ($\Box \Rightarrow$), and the rule $\Box$ is always combined with the rule ($\Rightarrow \Box$). One can replace these four rules by the two rules:

$$(\Box \Rightarrow)\ \frac{\Box(\neg l_{\Box A} \vee \Box l_A) \quad \neg l_A \vee D}{\neg l_{\Box A} \vee D} \qquad (\Rightarrow \Box)\ \frac{\Box(l_{\Box A} \vee \neg \Box l_A) \quad l_A \vee D_\Box}{l_{\Box A} \vee D_\Box}$$

where $D_\Box$ is $\neg l_{\Box A_1} \vee \ldots \neg l_{\Box A_n}$. We transfer to $GR(d)$-rules notation of *main* and *side* formulas from the sequent calculi. *Main* literal of each rule except $(\neg\Box), (\Box)$ is the literal l_A explicitly shown in the conclusion (maybe negated). For example, main literal of both rules $\Rightarrow \vee, \vee \Rightarrow$ is $L_{A_1 \vee A_2}$. Main literal of both rules $(\neg\Box), (\Box)$ is $\Box l_A$. *Side literal* of a rule is any literal l_A shown explicitly in the non-initial premise of the rule. More precisely, side literal of the rule $\Rightarrow \vee$ is l_{A_i}, side literals of the rule $\vee \Rightarrow$ are l_{A_1}, l_{A_2}, and side literal of the next four rules is l_A, and side literal of the two remaining rule is $\Box l_A$. *Main literal* of an axiom $L \vee \neg L$ is L. *Main or side formula* of an axiom or rule with main or side literal l_A is A.

Restriction to clauses of the form (4) and to the rules used in $GR(d)$ already constitutes a strategy. S. Maslov called similar restriction for classical logic *inverse method strategy* of the resolution method. The reason for this is clear from the section 3.

Atomic axioms

Axioms can be restricted to $l_A \vee \neg l_A$ for atomic formulas A, since similar restriction for the axioms $A \Rightarrow A$ is complete for $GS4$.

Inversion strategy for RS4 is the restriction: if a clause contains at least one literal $\pm l_A$ with $A = \neg B, C \vee D$ (non-atomic B), or is of the form $l_{\Box A} \vee D_\Box$ then this clause is obtained by one of the rules $\Rightarrow \neg \Rightarrow$, $\Rightarrow \vee \Rightarrow$, $\Rightarrow \Box$.

LEMMA 19 *Inversion strategy is complete for RS4*

Proof. As in Section 3.2, take a pruned inverted derivation in $GS4$, and apply Maslov Transformation GR. ■

Unique occurrence strategy This is a strategy combining several inferences into one block and allowing to delete intermediate inferences in the block from the search space. A general scheme of the strategy as well as its form for S4 was introduced by A. Voronkov in [17]. An analogous strategy for linear logic [4] was independently introduced in [16] and was one of the decisive factors of the success of Tammet's prover for linear logic.

A positive literal l_A is *dischargeable*, if A has exactly one positive occurrence in the goal formula F, $A \neq F$ and A does not occur as an argument of $\Box$.

A negative literal $\neg l_A$ is *dischargeable*, if A has exactly one negative occurrence in the goal formula F, and either A occurs as an argument of $\Box$ or A occurs as an argument of $\neg$ and $A \neq \Box A'$.

Unique Occurrence Strategy. If a clause C contains a dischargeable literal l_A or $\neg l_A$ then immediately apply a rule with l_A or $\neg l_A$ as a side literal to C and delete C itself from the search space.

THEOREM 20 *Unique occurrence strategy is complete.*

Proof. The proof can be found in [17]. ■

Focusing strategy [15]. This strategy is known for linear logic [1], cf. also [16]. It was first discovered for bottom-up proof search in classical and intuitionistic logic in 1974 by V. Orevkov [15].

DEFINITION 21 *Let d be a derivation of $X_F \vdash l_F$ according to $GR(d)$-rules and $\mathcal{I}$ be an inference (application of an inference rule) in d, S be one of the premises of $\mathcal{I}$, and l_B be the side literal of $\mathcal{I}$ in S. Inference $\mathcal{I}$ is* $(+)$ -focused *on S, if S is an axiom or the following conditions are satisfied.*

(a) if $B = \neg C, C \vee D$, or ($B = \Box C$ and l_B is not negated in S), then S is the conclusion of the inference with the main literal l_B.

(b)if $B = \Box C$, l_B is negated in S and $\mathcal{I}$ is $(\Box \Rightarrow)$, then S is the conclusion of the inference with the main literal l_B.

(c) if B is atomic, and l_B occurs in S unnegated, then S is an axiom.

Derivation d is $(+)$ -focused, if it is $(+)$ -focused on all premises of all its inferences except possibly $(\Rightarrow \Box)$.

$(-)$ *-focusing* is defined by changing sign in the clause (c) of the previous definition:

(c') if B is atomic, and $\neg l_B$ occurs in S , then S is an axiom.

+-focusing strategy: generate only derivations +-focused on all subformulas.

–focusing strategy is defined similarly.

Note. Compare clause (a) above with the inversion strategy.

THEOREM 22 *The $(+)$ -focusing is complete, as well as the $(-)$ -focusing.*

Proof. We establish completeness of similar strategy for a modification of $GS4$ and apply Maslov transformation. ■

Gentzen-type system $GS4*$ is obtained from $GS4$ by restricting formulas in the axioms to be atomic, and replacing the rule $\Rightarrow \vee$ by two rules:

$$\frac{\Gamma \Rightarrow \Delta, A_i, A_1 \vee \ldots \vee A_n}{\Gamma \Rightarrow \Delta, A_1 \vee \ldots \vee A_n} \quad \Rightarrow \vee \quad \frac{\Gamma \Rightarrow \Delta, A_i}{\Gamma \Rightarrow \Delta, A_1 \vee \ldots \vee A_n}$$

Definition of the $(+)$ -focusing for $GS4*$-derivations is very similar to one given above for $RS4$:

DEFINITION 23 *Let d be a derivation of a formula F in $GS4*$, $\mathcal{I}$ be an inference in d, S be one of the premises of $\mathcal{I}$, and l_B be the side literal of $\mathcal{I}$ in S. Inference $\mathcal{I}$ is* $(+)$ -focused *on S, if S is an axiom or the following conditions are satisfied.*

(a) if $B = \neg C, C \vee D$*, or (*B *is a succedent formula of the form* $\Box C$*), then* S *is the conclusion of the inference with the main formula* B*.*

(b)if L *is* $(\Box \Rightarrow)$ *and* $B = \Box C$*, then* S *is the conclusion of the inference with the main literal* B*.*

(c) if B *is atomic succedent formula, then* S *is an axiom.*

Derivation d *is* $(+)$ *-focused, if it is* $(+)$ *-focused on all premises of all its inferences except possibly* $(\Rightarrow \Box)$*.*

Completeness for GR-rules is reduced to the following proposition.

THEOREM 24 *Every derivation* d *in* $GS4*$ *can be transformed into a* $(+)$ *-focused derivation with the same endsequent.*

Proof. We can assume d to be pruned and all axioms atomic. Uppermost non- $(+)$ -focused inferences are normalized by permuting them maximally upward into the place where they are alrėady $(+)$ -focused. More precisely, we use induction on the size $n(d)$ of non-focused part of given pruned derivation d, that is on the number n of inferences which are below any premise of a non-focused inference. Pick up one of the uppermost non- $(+)$ -focused inferences. Assume (to save notation) that it is $(\Rightarrow \vee)$-inference with main formula $A \vee B$ and side formula A. Since the derivation d is pruned, the side formula A is traceable up the derivation to main formulas of some inferences or axioms:

$$\begin{array}{ccc} \dfrac{\Delta' \Rightarrow \Pi', A' \quad \Delta'' \Rightarrow \Pi'', A''}{\Delta \Rightarrow \Pi, A} & & A \Rightarrow A \\ | & & / \\ \dfrac{\Gamma \Rightarrow \Sigma, A}{\Gamma \Rightarrow \Sigma, A \vee B}\ L & & \end{array}$$

where the part ending in $\Gamma \Rightarrow \Sigma, A$ is $(+)$ -focused. Permuting inference L upward to the places shown explicitly results in the following derivation :

$$\begin{array}{ccc} \dfrac{\dfrac{\Delta' \Rightarrow \Pi', A' \quad \Delta'' \Rightarrow \Pi'', A''}{\Delta \Rightarrow \Pi, A}}{\Delta \Rightarrow \Pi, A \vee B}\ L & & \dfrac{A \Rightarrow A}{A \Rightarrow A \vee B}\ L \\ | & & / \\ & \Gamma \Rightarrow \Sigma, A \vee B & \end{array}$$

where the part ending in $\Gamma \Rightarrow \Sigma, A \vee B$ is $(+)$ -focused, and the induction parameter is decreased at least by 1.

In the case of two-premise inference, permutation should be repeated. ■

To prove the completeness for resolution rule it remains to apply Maslov transformation. □

EXAMPLE 25 The following inference satisfies the clause (c) in the definition of +-focusing.

$$\frac{\Box(a \vee l_B \vee \neg l_{a\vee B}) \quad \neg a \vee a \quad \neg l_B \vee D}{a \vee \neg l_{a\vee B} \vee D}$$

Recall from the section 3.1 that $a = l_a$ for atomic a.

4.5 Problems

1. Find reasonable extension of Maslov transformation to propositional logics containing induction: dynamic logic, temporal logic of linear time etc. The difficulty : cut-free Gentzen-type derivations in these logics are graphs with cycles, not trees. Hence any syntactically admissible clause can be initial.

2. Extend the present approach to Ohlbach translations of modal logics [14].

ACKNOWLEDGMENTS

This paper was written during visit of Vladimir Orevkov to CSLI (Stanford University) supported by a grant from Cooperation in Applied Science and Technology (CAST) Program of the National Research Council of USA.

Grigori Mints, *Stanford University, USA*
Vladimir Orevkov, *Steklov Institute of Mathematics, Russia*
Tanel Tammet, *Chalmers University, Sweden*

REFERENCES

[1] J. Andreoli, R. Paresci. Linear Objects: Logical processes with built-in inheritance. *New Generation Computing*, Special issue, Selected Papers of ICLP'90.

[2] C. Chang, R. Lee. *Symbolic Logic and Mechanical Theorem Proving*. Academic Press, New York, 1973.

[3] H. Curry. *Foundations of Mathematical Logic*. McGraw Hill, New York, 1963.

[4] J.-Y. Girard. Linear logic, *Theoretical Computer Science*, **50**, 1–102, 1987.

[5] S. C. Kleene. Two Papers on the Predicate Calculus, *Memoirs of the American Mathematical Society*, 10, 1952.

[6] S. Maslov. Inverse method of establishing deducibility for logical calculi. *Proc. Steklov Inst. of Math.*, 98, 1968.

[7] S. Maslov. The connection between tactics of inverse method and the resolution method. In A. O.Slisenko (ed.), *Zapiski Nauchnykh Seminarov LOMI*. 16, 1969. English translation in: J.Siekmann, G. Wrightson (eds.), *Automated Reasoning Part II*, pp. 264–270, Springer-Verlag, Berlin, 1983.

[8] G. Mints. Resolution Calculi for Non-classical Logics (Russian), *9-th Soviet Symposium in Cybernetics*, pp. 34–36, 1981.

[9] G. Mints. Resolution Calculi for Non-classical Logics (Russian). *Semiotika i Informatika*, **25**, 120–135, 1985.

[10] G. Mints. Gentzen-type Systems and Resolution Rule. Part I. Propositional Logic. In: P. Martin-Löf, G. Mints (eds.), *COLOG-88*, Lecture Notes in Computer Sci. 417, pp. 198–231, Springer-Verlag, 1990.

[11] G. Mints. Gentzen-type Systems and Resolution Rule. Part II. *Logic Colloquium'90*, Lecture Notes in Logic, 2, pp. 163–190, Springer-Verlag, 1994.

[12] G. Mints. Resolution Calculus for the First Order Linear Logic, *Journal of Logic, Language and Information*, **2**, 59–83, 1993.

[13] G.Mints. Resolution strategies for the intuitionistic logic. In: B. Mayoh, J. Penjam, E. Tyugu (eds.), *Constraint Programming*, NATO ASI Series, pp. 289–311, Springer Verlag, 1994.

[14] H. J.Ohlbach. A resolution calculus for modal logics. In: E. Lusk, R. Overbeek (eds.), *Proc. of CADE '88*, Lecture Notes in Computer Sci. 310, pp. 500–516, Springer-Verlag, Berlin, 1988.

[15] V. Orevkov. A specialization of the form of derivations in Gentzen calculi, and some of its applications. *J. Soviet Mathematics*, **6**, 1976.

[16] T. Tammet. Proof Search Strategies in Linear Logic. *Journal of Automated Reasoning*, **12**, 273–304, 1994.

[17] A. Voronkov. Theorem proving in non-standard logics based on the inverse method. In: *Automated Deduction-11*. Lecture Notes in Computer Science, 607, pp. 648–662, Springer-Verlag, Berlin, 1992.

HAROLD SCHELLINX

A LINEAR APPROACH TO MODAL PROOF THEORY

We show how to extend the proof theoretical analysis of sequent derivations in classical logic by means of linear logic, as exposed in [2], to sequent derivations in **S4**, by constructing embeddings of **S4** into classical linear logic that are correct with respect to *proofs* (i.e. define *linear decorations* of **S4**-derivations). Among the immediate corollaries are cut elimination for **S4**, and reduction strategies that are strongly normalizing.

"dedicated to dj"

1 AIMS

Linear logic[1] bans the structural rules of weakening and contraction from the formulation of classical logic as a sequent calculus, and re-introduces them in modalized form: structural manipulation to the *left* of the entailment-sign is allowed only for formulas prefixed by "!", to the *right* only for those prefixed by "?". ('Linear' logicians refer to !, ? as the *exponentials*.) The resulting calculus is a proof theoretical jewel, which combines the deep symmetries of classical, with the computational properties (strong normalization, confluence) of intuitionistic logic, without loss of expressive power.

If one forgets the linear 'type' of the connectives appearing in a derivation in linear logic (i.e. one replaces $\{\&, \otimes\}$ by $\wedge$, $\{\oplus, \wp\}$ by $\vee$, and $\multimap$ by $\rightarrow$), and erases all exponentials, then (modulo possible repetitions of sequents) the result (π's *skeleton*, $\mathsf{sk}(\pi)$) is a correct derivation in sequent calculus for *classical* logic. Conversely each derivation in intuitionistic or classical sequent calculus occurs as the skeleton of a derivation in linear logic: there exist *modal translations* $(\cdot)^\circ$ of formulas into linear formulas, obtained by replacing each connective by one of its linear analogues and prefixing each subformula by a modality (i.e. a (possibly empty) string

[1] [3]. A good introduction is [8].

H. Wansing (ed.), Proof Theory of Modal Logic, 33–43.

of exponentials), whose inductive application to a sequent derivation π in classical logic results in a correct derivation π°, satisfying $\mathsf{sk}(\pi^\circ) = \pi$, in linear logic.[2]

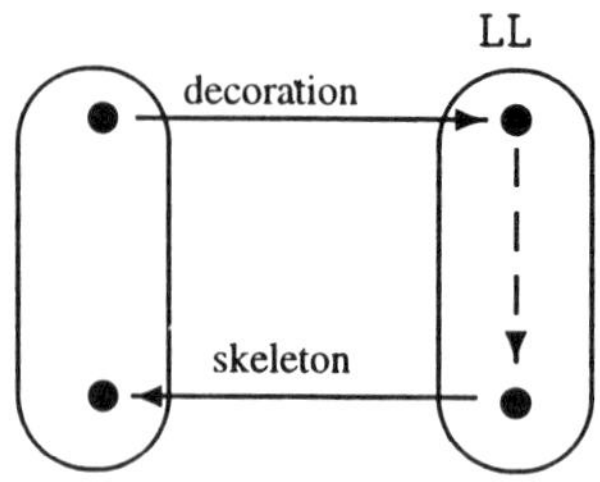

Thus we are able to study properties of derivations in intuitionistic or classical sequent calculus via their linear decorations. For example, cut elimination for classical and intuitionistic (second order) logic becomes an immediate corollary to cut elimination for linear (second order) logic, as reductions of the linear decoration of a proof π become, via the method of 'reflection' or 'pull-back', reductions of the original: one decorates the original, reduces the decoration in linear logic, and takes the skeleton of the result (cf. the diagram above). This procedure defines reductions of classical derivations that are strongly normalizing and confluent. (PROOF: The reductions of the decoration have these properties.)

Our method moreover enables us to identify possible obstacles to a decent computational interpretation of cut elimination in classical sequent calculus, and to formulate solutions as suggested by linear logic. For all details, and much more, we refer the reader to [2].

The purpose of this note is to point out that this 'linear analysis' of derivations and their normalization is not limited to sequent calculus for classical logic, but applies, virtually unchanged, to the (standard) sequent calculus formulation of the modal logic **S4** as well.

Though classical linear logic is essentially different from **S4**, there are also profound resemblances between the systems. And it probably is largely due to linear logic's rather unconventional notational apparatus, that many still appear puzzled when confronted with the fact that, in some sense, linear logic might be deemed a *fragment* of **S4**. Consider the introduction rules for the exponentials (the 'dereliction' and the 'contextual' rules) in table 1. When reading $\Box$ for '!' and $\Diamond$ for '?' one finds precisely the modal introduction rules one needs to add to classical sequent calculus in order to obtain a sequent calculus formulation of **S4**: the only formal difference between the implicational fragment of **S4** and the calculus of table 1 is

[2] In fact, as shown in [5], there are essentially *two* such translations. We refer to them as t and q.

Identity axiom and cut rule:

$$\text{Ax}\ \ A \Rightarrow A \qquad \text{cut}\ \frac{\Gamma \Rightarrow \Delta, A \qquad A, \Gamma' \Rightarrow \Delta'}{\Gamma, \Gamma' \Rightarrow \Delta, \Delta'}$$

Logical rules:

$$\text{L}\multimap \frac{\Gamma \Rightarrow \Delta, A \qquad B, \Gamma' \Rightarrow \Delta'}{\Gamma, \Gamma', A \multimap B \Rightarrow \Delta, \Delta'} \qquad \text{R}\multimap \frac{\Gamma, A \Rightarrow B, \Delta}{\Gamma \Rightarrow A \multimap B, \Delta}$$

Exponential structural rules:

$$\text{W!}\frac{\Gamma \Rightarrow \Delta}{\Gamma, !A \Rightarrow \Delta} \qquad \text{W?}\frac{\Gamma \Rightarrow \Delta}{\Gamma \Rightarrow ?A, \Delta} \qquad \text{C!}\frac{\Gamma, !A, !A \Rightarrow \Delta}{\Gamma, !A \Rightarrow \Delta} \qquad \text{C?}\frac{\Gamma \Rightarrow ?A, ?A, \Delta}{\Gamma \Rightarrow ?A, \Delta}$$

Exponential contextual rules:

$$\text{L?}\frac{!\Gamma, A \Rightarrow ?\Delta}{!\Gamma, ?A \Rightarrow ?\Delta} \qquad \text{R!}\frac{!\Gamma \Rightarrow A, ?\Delta}{!\Gamma \Rightarrow !A, ?\Delta}$$

Exponential dereliction rules:

$$\text{R?}\frac{\Gamma \Rightarrow A, \Delta}{\Gamma \Rightarrow ?A, \Delta} \qquad \text{L!}\frac{\Gamma, A \Rightarrow \Delta}{\Gamma, !A \Rightarrow \Delta}$$

Table 1. The implicational fragment of classical linear logic

that in the latter the use of structural rules is restricted, at the left to formulas prefixed by '!' ($\Box$), at the right to formulas prefixed by '?' ($\Diamond$). So taking the skeleton of a derivation π in linear logic, as described above, in fact is a *compound* operation: forgetting the linear 'typing' of the logical connectives takes us to **S4**; erasing the exponentials then leads from **S4** to classical logic.

Hence every derivation in linear logic has, let's call it a *modal skeleton*, which is a sequent calculus derivation in **S4**.

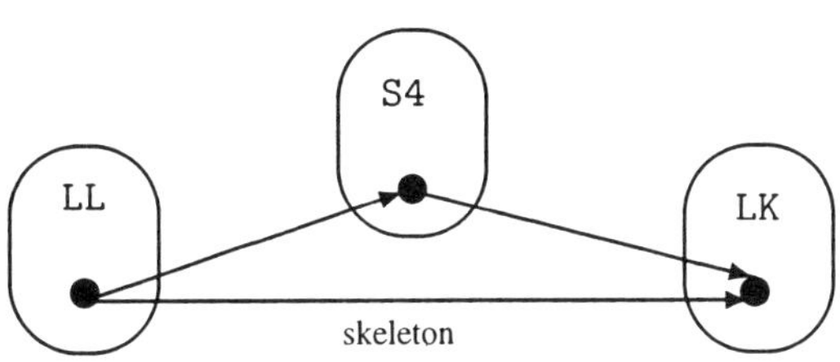

But how about the converse? Does every **S4**-derivation occur as the modal skeleton of a proof in linear logic?[3] If we take 'modal skeleton' in the above, strict sense, this clearly can *not* be the case. A derivation like that of $p, \Diamond p \Rightarrow p$ consisting in an (atomic) instance of the identity-axiom followed by a weakening on $\Diamond p$ could never be realized in linear logic: $p, ?p \Rightarrow p$ simply is not derivable. However, in a slightly more devious sense, in which 'modal skeleton' will come to mean "forget some exponentials, but keep others", the answer to the question turns out to be affirmative.

2 MEANS

Of course! If we want to interpret **S4**-proofs in linear logic, we have to be able to distinguish the **S4**-modalities (which, as usual, are about '*what might and what must be the case*') from the linear exponentials, which keep track of the management of the structural rules within the proof under consideration, and, dually, inform us on its future behaviour under cut elimination. (By the way, as we are talking proof theory, cut-free (normal, direct) proofs will not interest us very much; only 'how we get there'[4] by means of operating on 'real' proofs: yes, yes, cuts indeed, lots of cuts!)

How to distinguish between modalities serving a different purpose, but obeying the same (introduction) laws? Well, recall that these (contrary to those for the logical connectives) do *not* determine the modalities uniquely (modulo provable equivalence). So, in linear logic (as in **S4**) we can add a second, a third, fourth, etc., pair of modalities to the language, plus, for each, the usual left/right introduction rules to the calculus. These new pairs of modalities will be *independent* of the original one (and of each other). This often is considered a vice, but for our purposes it is a virtue. (It will however turn out to be convenient for the fulfilment of our task to nevertheless stipulate some kind of relation between these different exponentials.)

We introduced 'linear logics' using (ordered) sets of distinct (indexed or 'coloured') exponentials in [1]. Let us recall their definition: given some preorder $\mathsf{R} = (R, \preceq)$, the language of the multicolour linear logic R-**LL** consists in the non-modal part of the language of usual linear logic, extended with pairs of indexed exponentials $\underset{a}{!}, \underset{a}{?}$, one for each element of R. To the usual non-modal rules of linear logic we add

1. for each pair of indexed exponentials the usual dereliction rules $\mathrm{R}\underset{a}{?}$ and $\mathrm{L}\underset{a}{!}$;
2. structural rules to the left for formulas prefixed by $\underset{a}{!}$, structural rules to the right for those prefixed by $\underset{a}{?}$, eventually restricted to indices from subsets $\mathcal{W}$

[3] It is important to observe that this is more than merely asking for an embedding of **S4** into linear logic, correct with respect to provability. That in a such sense **S4** may be embedded in linear logic is shown in [4], by means of a simple extension of the Grishin-Ono translation of classical into linear logic (which can be looked up e.g. in [8]).

[4] Us 'linear folks' call this: *dynamics*.

(allowing weakening) and $\mathcal{C}$ (allowing contraction) of R, that are *upwardly closed* with respect to $\preceq$ (i.e. if $a \in \mathcal{W}(\mathcal{C})$ and $a \preceq b$ than also $b \in \mathcal{W}(\mathcal{C})$);

3. rules for the introduction of indexed exclamation marks on the right, namely

$$\mathrm{R}\underset{z}{!}\ \frac{\underset{x_1}{!}G_1,\ldots,\underset{x_n}{!}G_n \Rightarrow A, \underset{y_1}{?}D_1,\ldots,\underset{y_m}{?}D_m}{\underset{x_1}{!}G_1,\ldots,\underset{x_n}{!}G_n \Rightarrow \underset{z}{!}A, \underset{y_1}{?}D_1,\ldots,\underset{y_m}{?}D_m},$$

provided that $z \preceq x_i, z \preceq y_j$ for all x_i, y_j, and, under the same condition, a rule for the introduction of indexed question marks on the left:

$$\mathrm{L}\underset{z}{?}\ \frac{\underset{x_1}{!}G_1,\ldots,\underset{x_n}{!}G_n, A \Rightarrow \underset{y_1}{?}D_1,\ldots,\underset{y_m}{?}D_m}{\underset{x_1}{!}G_1,\ldots,\underset{x_n}{!}G_n, \underset{z}{?}A \Rightarrow \underset{y_1}{?}D_1,\ldots,\underset{y_m}{?}D_m}.$$

It is easy to verify that all **LL**-properties (cut elimination, strong normalization, etcetera) are inherited by R-**LL**. (PROOF: Note that if we forget about the indices, a derivation in R-**LL** is just a derivation in **LL**; we therefore eliminate cuts as if we were in **LL**, which is correct from an R-**LL** point of view due to transitivity of $\preceq$ and the closure conditions on $\mathcal{W}$ and $\mathcal{C}$.)

As a matter of fact, for each preorder R and choice of $\mathcal{W}$ and $\mathcal{C}$, the set of derivations in the calculus R-**LL** corresponds to a certain (not necessarily proper) subset of the collection of all **LL**-derivations, a subset satisfying certain restrictions on the use of exponential rules and which is closed under cut.

In order to interpret **S4** we are going to use a particularly simple preorder, namely the usual $\leq$-relation on the set $\{0, 1\}$. So we have two pairs of exponentials: $\underset{0}{!}, \underset{0}{?}$ and $\underset{1}{!}, \underset{1}{?}$ (the former can be introduced in a context containing the latter, but the converse is forbidden). Moreover (but this is not essential) it turns out that we can take $\mathcal{W} = \mathcal{C} = \{1\}$. The pair $\underset{0}{!}, \underset{0}{?}$ will play the role of the **S4**-modalities $\Box, \Diamond$; the pair $\underset{1}{!}, \underset{1}{?}$ is the usual **LL**-pair $!, ?$. (And, of course, we can and will just write $\Box, \Diamond$ for $\underset{0}{!}, \underset{0}{?}$, and $!, ?$ for $\underset{1}{!}, \underset{1}{?}$.) Let us call the corresponding (bi-coloured) linear logic: **2-LL**. We will refer to $\Box, \Diamond$ as the *0-indexed* exponentials in a **2-LL**-derivation, to $!, ?$ as the *1-indexed* exponentials.

Our second ingredient are the (dual) translations t and q of classical into linear logic that we mentioned above, translations having the special property of giving rise to *decorations* of derivations in sequent calculus: they extend to translations of proofs.

The mappings t and q correspond on the nose to two, again dual, ways to treat the 'structural phase' (neither of the cut formulas has been introduced in the last inference before the cut) when eliminating cuts from a classical proof: the t-translation corresponds to first permuting the cut upwards in the subderivation of the *right* premiss, while q corresponds to first permuting the cut upwards in the subderivation of the *left premiss* (cf. [5], [2]). Using the translations uniformly corresponds to persistently applying either the right- or the left-protocol.

Another option is to first apply a *marking* to a given derivation π: call occurrences of (sub)formulas in π either t or q, the only limitation being that this marking should respect *identity-classes* (i.e. corresponding occurrences of formulas in premiss and conclusion of a rule get the same marking, corresponding occurrences of subformulas in the two formulas of an axiom or in the two cut formulas are marked identically, corresponding occurrences of subformulas in contracted formulas are given the same sign, etcetera[5]). We will note the conclusion of a marked derivation π as $\Gamma_1^{\mathsf{t}}, \Gamma_2^{\mathsf{q}} \Rightarrow \Delta_1^{\mathsf{t}}, \Delta_2^{\mathsf{q}}$, where Γ_1^{t} denotes the multiset of all formulas in the antecedent of the conclusion that are marked t, etcetera. One then defines a translation of marked formulas, which once more, when applied inductively to a (marked) derivation results in a decoration of π. (Of course the uniform translations are the two limit-cases ('all q, all t') of the 'mixed' translation, which, obviously, corresponds to a cut-elimination that 'mixes' protocols (i.e. left or right depends on the marking of the cut-formula).

The **LL**-decorations of classical derivations thus obtained are decorations in **2-LL** as well: an **LL**-proof is nothing but a proof in **2-LL** that does not use 0-indexed exponentials.

Hence the idea will be clear. We are going to extend our translations to **S4**, by mapping the **S4**-modalities to 0-indexed exponentials in **2-LL**. The principle, all tricks and traps will already be sufficiently illustrated by considering the implication-only fragment of **S4**. (By herself or with a little help from [2] the reader easily finds the extension to the full system.)

The translation is defined as follows: for atomic formulas p, whether they are marked t or q, it is the identity, i.e. p; for compound formulas one translates according to the following table:

A/B	t	q	A	t	q
t	$!?A \multimap !?B$	$!?A \multimap ?!B$	$\Box A$	$\Box !?A$	$\Box ?!A$
q	$!?!A \multimap ?!?B$	$?!A \multimap ?!B$	$\Diamond A$	$\Diamond !?A$	$\Diamond ?!A$

Given a marked formula ϕ, we will denote by ϕ^* a **2-LL**-translation of ϕ according to this table. So ϕ^* will depend on the markings of ϕ's (proper) subformulas. To give an example, if $\phi \equiv (\Diamond((\Box A^{\mathsf{t}})^{\mathsf{q}} \rightarrow B^{\mathsf{t}})^{\mathsf{q}})^{\mathsf{t}}$ then ϕ^* is $\Diamond ?!(!?!\Box !?A \multimap ?!?B)$.

Here is our first proposition.

[5] A formal definition can be found in [1] and [5]

PROPOSITION 1 *Suppose that π is a marked* **S4***-derivation with conclusion* Γ_1^{t}, $\Gamma_2^{\mathrm{q}} \Rightarrow \Delta_1^{\mathrm{t}}, \Delta_2^{\mathrm{q}}$. *Then we can construct a* **2-LL***-derivation* π^* *of* $!?\Gamma_1^*, !\Gamma_2^* \Rightarrow ?\Delta_1^*, ?!\Delta_2^*$, *which moreover has the property that, by deleting all 1-indexed exponentials, one recovers (modulo possible repetitions of sequents) the original proof* π.

Proof. Of course our claim is shown to hold by induction on the length of marked **S4**-derivations.

- It is easily verified for axioms, and if our last inference was one of the structural rules, then note that by induction hypothesis the contracted, resp. weakened, formula(s) are/is prefixed by the 'right' exponential, and thus can be contracted/ weakened in **2-LL**.

- Suppose the last inference was a cut between $\Gamma_1^{\mathrm{t}}, \Gamma_2^{\mathrm{q}} \Rightarrow \Delta_1^{\mathrm{t}}, \Delta_2^{\mathrm{q}}, A^{\mathrm{q}}$ and $A^{\mathrm{q}}, \Gamma_3^{\mathrm{t}}, \Gamma_4^{\mathrm{q}} \Rightarrow \Delta_3^{\mathrm{t}}, \Delta_4^{\mathrm{q}}$, then, by induction hypothesis, we have **2-LL**-derivations of $\ldots \Rightarrow \ldots, ?!A^*$ and $!A^*, \ldots \Rightarrow \ldots$. We proceed by applying an exponential contextual rule to this latter derivation, 'promoting' $!A^*$ to $?!A^*$; and we are done by a cut in **2-LL**. (As for the structural rules, also the application of exponential contextual rules is *always* possible: all formulas are prefixed by the 'right' exponentials.)

In case of a cut between $\Gamma_1^{\mathrm{t}}, \Gamma_2^{\mathrm{q}} \Rightarrow \Delta_1^{\mathrm{t}}, \Delta_2^{\mathrm{q}}, A^{\mathrm{t}}$ and $A^{\mathrm{t}}, \Gamma_3^{\mathrm{t}}, \Gamma_4^{\mathrm{q}} \Rightarrow \Delta_3^{\mathrm{t}}, \Delta_4^{\mathrm{q}}$ we have, by induction hypothesis, **2-LL**-derivations of $\ldots \Rightarrow \ldots, ?A^*$ and $!?A^*, \ldots \Rightarrow \ldots$. Now we proceed by applying an exponential contextual rule to the former derivation, promoting $?A^*$ to $!?A^*$.[6]

- In case our last inference is a left/right implication introduction rule we reason similarly. We leave the details to the reader. It comes to the fore in case antecedent, marked

- What is left to verify is the case of the introduction of the modalities. Let us start by considering the 'dereliction'-rules, e.g. L$\Box$.

Suppose we got $\Gamma_1^{\mathrm{t}}, \Gamma_2^{\mathrm{q}}, (\Box A)^{\mathrm{q}} \Rightarrow \Delta_1^{\mathrm{t}}, \Delta_2^{\mathrm{q}}$ from $\Gamma_1^{\mathrm{t}}, \Gamma_2^{\mathrm{q}}, A^{\mathrm{t}} \Rightarrow \Delta_1^{\mathrm{t}}, \Delta_2^{\mathrm{q}}$. By induction hypothesis we have a **2-LL**-derivation of $!?\Gamma_1^*, !\Gamma_2^*, !?A^* \Rightarrow ?\Delta_1^*, ?!\Delta_2^*$. We continue as in:

$$\dfrac{\dfrac{\vdots \atop !?\Gamma_1^*, !\Gamma_2^*, !?A^* \Rightarrow ?\Delta_1^*, ?!\Delta_2^*}{!?\Gamma_1^*, !\Gamma_2^*, \Box !?A^* \Rightarrow ?\Delta_1^*, ?!\Delta_2^*}}{!?\Gamma_1^*, !\Gamma_2^*, !\Box !?A^* \Rightarrow ?\Delta_1^*, ?!\Delta_2^*}$$

We leave other markings of the active formula in case of L$\Box$ and all instances of the dual rule R$\Diamond$ to the reader. So what remains are the contextual rules, e.g. L$\Diamond$.

[6] If the cut is on a formula marked q, then, in 'linear terminology', in the decoration the derivation of the *right* premiss is a box, if it is on a formula marked t, the derivation of the *left* premiss is a box.

Let us suppose we derived $(\Box\Gamma_1)^{\mathsf{t}}, (\Box\Gamma_2)^{\mathsf{q}}, (\Diamond A)^{\mathsf{t}} \Rightarrow (\Diamond\Delta_1)^{\mathsf{t}}, (\Diamond\Delta_2)^{\mathsf{q}}$ from $(\Box\Gamma_1)^{\mathsf{t}}, (\Box\Gamma_2)^{\mathsf{q}}, A^{\mathsf{q}} \Rightarrow (\Diamond\Delta_1)^{\mathsf{t}}, (\Diamond\Delta_2)^{\mathsf{q}}$. We find:

$$\dfrac{\dfrac{\dfrac{\dfrac{\dfrac{\vdots}{!?(\Box\Gamma_1)^*, !(\Box\Gamma_2)^*, !A^* \Rightarrow ?(\Diamond\Delta_1)^*, ?!(\Diamond\Delta_2)^*}}{!?(\Box\Gamma_1)^*, !(\Box\Gamma_2)^*, ?!A^* \Rightarrow ?(\Diamond\Delta_1)^*, ?!(\Diamond\Delta_2)^*}}{!?(\Box\Gamma_1)^*, !(\Box\Gamma_2)^*, \Diamond ?!A^* \Rightarrow ?(\Diamond\Delta_1)^*, ?!(\Diamond\Delta_2)^*}}{!?(\Box\Gamma_1)^*, !(\Box\Gamma_2)^*, ?\Diamond ?!A^* \Rightarrow ?(\Diamond\Delta_1)^*, ?!(\Diamond\Delta_2)^*}}{!?(\Box\Gamma_1)^*, !(\Box\Gamma_2)^*, !?\Diamond ?!A^* \Rightarrow ?(\Diamond\Delta_1)^*, ?!(\Diamond\Delta_2)^*}$$

Note that it is here (as in the other instances of L$\Diamond$ and those of R$\Box$, all left to the reader) that we (and essentially so!) benefit from the special form of the **2-LL** exponential contextual rules, which allows 'promotion' to an exponential having an index *below* the indices of the contextual exponentials: we may introduce $\Box/\Diamond$ in contexts where all formulas are prefixed by !/?. ■

So when inductively applying our translations, an **S4**-derivation π becomes 'wrapped up' in thick layers of linear exponentials. (Many of these in fact will be 'superfluous'. They can be identified and removed using the techniques developed in [1]. This will lead to *optimal* (i.e. using a minimal number of exponentials) linear interpretations of our modal proof. But for the moment that's beside the point.)

REMARK 2 *Whereas for classical logic the converse of Proposition 1 trivially holds, this is no longer true in the present case: note e.g. that* $!A \Rightarrow ?!\Box?!A$ *(the translation of* $A^{\mathsf{q}} \Rightarrow (\Box A^{\mathsf{q}})^{\mathsf{q}}$*) is derivable in* **2-LL**.

It is not unlikely that a more complex 'multicolour' embedding can be constructed that does *have both properties. Let us, however, for the moment stick to the 'simple' one.*

DEFINITION 3 *A marking of an* **S4***-derivation that assigns* t *to all occurrences of (sub)formulas of the form* $\Diamond A$*, and* q *to all occurrences of the form* $\Box A$*, is called a* sound marking.

And now for the upshot.

THEOREM 4 *Let* π *be an* **S4***-derivation. If we reduce in linear logic a decoration* π^* *corresponding to a sound marking of* π*, and delete all 1-indexed exponentials from the reduct, then (modulo possible repetitions of sequents) the result is an* **S4***-derivation* π' *which is a reduct of* π *(in other words, the diagram of page 34 applies to* **S4***).*

Proof. Due to the fact that all formulas $\Diamond A$ are marked t and all formulas $\Box A$ are marked q, the reductions of the decoration can not induce permutations that are impossible in **S4**. ■

REMARK 5 *The property expressed by Theorem 4 does, in general,* not *hold for decorations π^* obtained from markings of an* **S4**-*derivation that are not 'sound'. It will be most instructive to show* why. *(This moreover should clarify the proof of the theorem.)*

Consider the following possible occurrence of a cut in an **S4**-*proof:*

$$\dfrac{\dfrac{\begin{matrix}\vdots\\ C, \Box\Gamma_1 \Rightarrow \Diamond A, \Diamond\Delta_1\end{matrix}}{\Diamond C, \Box\Gamma_1 \Rightarrow \Diamond A, \Diamond\Delta_1} \qquad \dfrac{\begin{matrix}\vdots\\ \Gamma_2, \Diamond A \Rightarrow \Delta_2\end{matrix}}{\Gamma_2, \Diamond A \Rightarrow \Delta_2, B}}{\Diamond C, \Box\Gamma_1, \Gamma_2 \Rightarrow \Diamond\Delta_1, \Delta_2, B}$$

(The exact nature of the last inference in the right premiss in fact is irrelevant. What counts is that $\Diamond A$ is not main formula of the rule.)

Suppose we mark $\Diamond A$ by q, *and all other occurring formulas as well (this however, as the reader will see, is of no consequence for the argument). Then in a corresponding decoration π^* the instance of cut becomes*

$$\dfrac{\dfrac{\dfrac{\dfrac{\begin{matrix}\vdots\\ !C^*, !(\Box\Gamma_1)^* \Rightarrow ?!(\Diamond A)^*, ?!(\Diamond\Delta_1)^*\end{matrix}}{?!C^*, !(\Box\Gamma_1)^* \Rightarrow ?!(\Diamond A)^*, ?!(\Diamond\Delta_1)^*}}{\Diamond ?!C^*, !(\Box\Gamma_1)^* \Rightarrow ?!(\Diamond A)^*, ?!(\Diamond\Delta_1)^*}}{!\Diamond ?!C^*, !(\Box\Gamma_1)^* \Rightarrow ?!(\Diamond A)^*, ?!(\Diamond\Delta_1)^*} \quad \dfrac{\dfrac{\begin{matrix}\vdots\\ !\Gamma_2^*, !(\Diamond A)^* \Rightarrow ?!\Delta_2^*\end{matrix}}{!\Gamma_2^*, !(\Diamond A)^* \Rightarrow ?!\Delta_2^*, ?!B^*}}{!\Gamma_2^*, ?!(\Diamond A)^* \Rightarrow ?!\Delta_2^*, ?!B^*}}{!\Diamond ?!C^*, !(\Box\Gamma_1)^*, !\Gamma_2^* \Rightarrow ?!(\Diamond\Delta_1)^*, ?!\Delta_2^*, ?!B^*}$$

As, in the linear decoration, the right premiss now is a box, we (have to!) reduce the decorated cut in linear logic by permuting to the left. This is perfectly allright from the point of view of **2-LL**, *but not from that of* **S4**: *were we to perform this reduction in* **2-LL**, *then, after two permutations, when deleting all 1-indexed exponentials we are likely to find that we left the set of* **S4**-*derivations, as the result would be*

$$\dfrac{\dfrac{\begin{array}{c}\vdots\\ C, \Box\Gamma_1 \Rightarrow \Diamond A, \Diamond\Delta_1\end{array} \quad \dfrac{\begin{array}{c}\vdots\\ \Gamma_2, \Diamond A \Rightarrow \Delta_2\end{array}}{\Gamma_2, \Diamond A \Rightarrow \Delta_2, B}}{C, \Box\Gamma_1, \Gamma_2 \Rightarrow \Diamond\Delta_1, \Delta_2, B}}{\Diamond C, \Box\Gamma_1, \Gamma_2 \Rightarrow \Diamond\Delta_1, \Delta_2, B}$$

In order for this to be a correct derivation all formulas in Γ_2 *need to be prefixed by* $\Box$, *all formulas in* $\Delta_2 \cup \{B\}$ *by* $\Diamond$. *But this we cannot guarantee.*

Precisely this situation and its duals force us to consider decorations obtained via sound *markings.*

The immediate fruits of our efforts are gathered in the following two corollaries to Theorem 4.

COROLLARY 6 *All instances of the cut rule are eliminable from* **S4**-*derivations.*

COROLLARY 7 *Reductions of* **S4**-*derivations* π *defined as the reflection of the normalization in linear logic of decorations* π^*, *corresponding to sound markings of* π, *are strongly normalizing and confluent (cf. [2]).*

3 PROSPECTS

We had to be sketchy, and the above is far from being a complete 'linear analysis' of **S4**-derivations. If, however, it did succeed in bringing across some of the strength of linear logic as a proof theoretical tool, it will have served its purpose.

The same techniques can be applied directly to several modal logics weaker than **S4**. Also, the above results in fact apply to *second order* extensions of **S4** (whatever the interest of these might be). Yes, they will indeed apply in all those cases where the logic under consideration can be mapped to linear logic by a 'decoration-type' embedding. Hence, however far from universal, the approach sketched merits serious consideration as a *method*, providing e.g. a uniform alternative to several 'one-shot' solutions. Exploring its limits should be the subject of further research.

Utrecht University, The Netherlands

REFERENCES

[1] V. Danos, J.-B. Joinet, H. Schellinx. The structure of exponentials: uncovering the dynamics of linear logic proofs. In: G. Gottlob, A. Leitsch, D. Mundici, (eds.), *Computational Logic and Proof Theory*, Lecture Notes in Computer Sci. 713, Proceedings of the Third Kurt Gödel Colloquium, Brno, Czech Republic, pp. 159–171. Springer-Verlag, Berlin, 1993.

[2] V. Danos, J.-B. Joinet, H. Schellinx. A new deconstructive logic: linear logic. Prépublication 52, Équipe de Logique Mathématique, Université Paris VII, octobre 1994. To appear in: *Journal of Symbolic Logic*.
[3] J.-Y. Girard. Linear logic. *Theoretical Computer Science*, **50**, 1–102, 1987.
[4] S. Martini, A. Masini. A modal view of linear logic. *Journal of Symbolic Logic*, **59**, 888–899, 1994.
[5] H. Schellinx. *The Noble Art of Linear Decorating*. ILLC Dissertation Series, 1994-1. Institute for Language, Logic and Computation, University of Amsterdam, 1994.
[6] A. S. Troelstra. *Lectures on Linear Logic*. CSLI Lecture Notes 29, Center for the Study of Language and Information, Stanford, 1992.

TOMASZ SKURA

REFUTATIONS AND PROOFS IN S4

1 INTRODUCTION

J. Łukasiewicz introduced the concept of a refutation calculus that axiomatizes the set of non-theorems of a logic (cf. [2]). For instance the classical propositional logic can be characterized by the following system of this kind:

Axiom: $\dashv \bot$
Rules:

(r_s) $\qquad \dfrac{\dashv e(\alpha)}{\dashv \alpha}$ $\qquad$ *where $e(\alpha)$ is a substitution instance of α*

(r_{mp}) $\qquad \dfrac{\vdash \alpha \rightarrow \beta \qquad \dashv \beta}{\dashv \alpha}$

Refutation calculi for non-classical logics may be obtained by adding rules of the form

$$\frac{\dashv \alpha_1, ..., \dashv \alpha_n}{\dashv \alpha}$$

to the above system. The interest of such formulations is mainly theoretical (cf. [5, 6]). In particular they provide a method of obtaining decidability results by enumerating non-theorems.

However it is possible to give more practical refutation rules that are applicable to formulas in normal form and that are justified by theorems of the following form

$$\vdash \alpha \textit{ iff for some } 1 \leq i \leq n, \vdash \alpha_i$$

where α is a formula in normal form and all α_i are simpler formulas. Using such rules for every formula we can construct either a proof or a disproof of it. In fact such a system for the intuitionistic propositional logic was given by D. Scott in [4].

In this paper we present a system of this kind for the modal logic $S4$.

2 PRELIMINARIES

Let FOR be the set of all formulas generated from the set VAR of propositional variables by the connectives $\wedge, \vee, \neg, \Box$, and the constants $\top$ (truth), $\bot$ (falsity). The connectives $\rightarrow, \equiv, \Diamond$ are defined in the usual way. Formulas with no occurrences of $\Box$ are called modal-free. For any $\alpha, \beta \in FOR$, $p \in VAR$ the symbol

H. Wansing (ed.), Proof Theory of Modal Logic, 45–51.

$\alpha(p/\beta)$ denotes the result of substituting β for p in α. We say that α is equivalent to β if the equivalence can be shown using theorems of $S4$, the modus ponens rule, and the substitution rule. The symbol $S4$ denotes also the set of all theorems of $S4$.

By an $S4$ algebra we mean an algebra $A = (B, I)$, where B is a Boolean algebra and I is an interior operation (cf. [1]). The Boolean operations, the unit element, the zero element of A will be denoted by $\wedge, \vee, \neg, 1, 0$ respectively. An element x is said to be open iff $I(x) = x$. The symbol B_1 will denote the 2-element Boolean algebra. For any $S4$ algebras A_1, A_2, $A_1 \times A_2$ is the direct product of A_1 and A_2. We also write A^n instead of $A_1 \times ... \times A_n$ if $A_1 = ... = A_n = A$.

3 RESULTS

First observe the following simple facts:

LEMMA 1 *Let $p \in VAR$, $\alpha \in FOR$. Then the formula $\Box p \to \alpha$ is equivalent to the formula $\alpha(p/\top)$.*

LEMMA 2 (cf. [3]) *Let $\beta \in FOR$. Then β is equivalent to a formula of the form*

$$\bigwedge\{\Box(\Box r_j \equiv q_j) : 1 \leq j \leq n\} \wedge \Box\alpha \to p_1$$

where all $r_j, q_j, p_1 \in VAR$, $q_j \neq q_k$ for $1 \leq j \neq k \leq n$, and α is a modal-free formula.

LEMMA 3 *Let $\beta \in FOR$. Then β is equivalent to a formula of the form*

$$\bigwedge\{\Box(\Box r_j \equiv q_j) : 1 \leq j \leq n\} \wedge \bigwedge\{\Box(\Box p_i \equiv q_0) : 1 \leq i \leq m\} \wedge \Box\alpha \to p_1$$

where $m \geq 1$, all $r_j, q_j, p_i, q_0 \in VAR$, $q_j \neq q_k$ for $0 \leq j \neq k \leq n$, and α is modal-free.

LEMMA 4 *Let A=(B,I) be an S4 algebra and let d be an open element in it. Then the function $I' : B \to B$ defined thus*

$$I'(x) = \begin{cases} I(x) \wedge d & if \quad x \neq 1 \\ 1 & otherwise \end{cases} \qquad (x \in B)$$

is an interior operation.

Now we prove the following:

THEOREM 5 *Let* $\varphi = \bigwedge\{\Box(\Box\delta_j \equiv \gamma_j) : 1 \leq j \leq n\} \wedge \bigwedge\{\Box(\Box\beta_i \equiv \gamma_0) : 1 \leq i \leq m\} \wedge \Box\alpha$, *where all* $\delta_j, \gamma_j, \beta_i, \gamma_0, \alpha$ *are modal-free formulas,* $m \geq 1$. *Then*

$$\varphi \to \beta_1 \in S4$$

iff for some $1 \leq i \leq m$

$$\alpha \to \beta_i \vee \gamma_0 \vee ... \vee \gamma_n \in S4, \qquad \Box\delta_j \wedge \varphi \to \beta_1 \in S4 \quad (1 \leq j \leq n)$$

or for some $1 \leq j \leq n$

$$\varphi \to (\Box\beta_1 \equiv \Box\delta_j) \in S4, \qquad \varphi \wedge \Box(\Box\delta_j \equiv \gamma_0) \to \beta_1 \in S4.$$

Proof. ($\Leftarrow$)
Case 1. $\alpha \to \beta_i \vee \gamma_0 \vee ... \vee \gamma_n \in S4$ and $\Box\delta_j \wedge \varphi \to \beta_1 \in S4$ $(1 \leq j \leq n)$. Then

$$\varphi \to (\Box\delta_j \to \beta_1) \in S4 \qquad (1 \leq j \leq n)$$

$$\varphi \to (\Box\delta_j \to \Box\beta_1) \in S4 \qquad (1 \leq j \leq n)$$

$$\varphi \to (\Box\beta_1 \equiv \Box\beta_i) \in S4$$

$$\varphi \to (\Box\delta_j \to \Box\beta_i) \in S4 \qquad (1 \leq j \leq n)$$

$$\varphi \to (\Box\delta_j \to \beta_i) \in S4 \qquad (1 \leq j \leq n)$$

Also

$$\varphi \to \beta_i \vee \gamma_0 \vee ... \vee \gamma_n \in S4$$

$$\varphi \to (\gamma_j \to \Box\delta_j) \in S4 \qquad (1 \leq j \leq n)$$

$$\varphi \to (\gamma_0 \to \Box\beta_i) \in S4$$

$$\varphi \to (\gamma_0 \to \beta_i) \in S4$$

Therefore

$$\varphi \to \beta_i \in S4$$

$$\varphi \to \Box\beta_i \in S4$$

$$\varphi \to \Box\beta_1 \in S4$$

$$\varphi \to \beta_1 \in S4$$

Case 2. $\varphi \to (\Box\beta_1 \equiv \Box\delta_j) \in S4$ and $\varphi \wedge \Box(\Box\delta_j \equiv \gamma_0) \to \beta_1 \in S4$.
Then

$$\varphi \to (\Box\delta_j \equiv \gamma_0) \in S4$$

$$\varphi \to \Box(\Box\delta_j \equiv \gamma_0) \in S4$$

$$\varphi \to \beta_1 \in S4$$

($\Rightarrow$) Assume that for every $1 \leq i \leq m$
$\alpha \to \beta_i \vee \gamma_0 \vee ... \vee \gamma_n \notin S4$ or $\Box\delta_j \wedge \varphi \to \beta_1 \notin S4$ for some $1 \leq j \leq n$
and for every $1 \leq j \leq n$
$\varphi \to (\Box\beta_1 \equiv \Box\delta_j) \notin S4$ or $\varphi \wedge \Box(\Box\delta_j \equiv \gamma_0) \to \beta_1 \notin S4$.
Now if

$$\Box\delta_j \wedge \varphi \to \beta_1 \notin S4$$

or

$$\varphi \to (\Box\delta_j \to \beta_1) \notin S4$$

or

$$\varphi \wedge \Box(\Box\delta_j \equiv \gamma_0) \to \beta_1 \notin S4$$

then

$$\varphi \to \beta_1 \notin S4.$$

So we assume that for all $1 \leq i \leq m$

$$\alpha \to \beta_i \vee \gamma_0 \vee ... \vee \gamma_n \notin S4$$

and for all $1 \leq j \leq n$

$$\varphi \wedge \Box\beta_1 \to \delta_j \notin S4.$$

It follows that for every $1 \leq i \leq m$ there is a valuation v_i in B_1 such that $v_i(\alpha) = 1$, $v_i(\beta_i) = v_i(\gamma_j) = 0$ $(0 \leq j \leq n)$ and for every $1 \leq j \leq n$ there is a valuation w_j in some $S4$ algebra A_j such that $w_j(\varphi) = w_j(\beta_1) = 1$, $w_j(\delta_j) \neq 1$. Hence $w_j(\beta_i) = 1$ $(1 \leq i \leq m)$ and $w_j(\Box\beta) = 1$ $(1 \leq j \leq n)$, where $\beta = \beta_1 \wedge ... \wedge \beta_m$. Moreover

$$v_i(\Box\delta_j \wedge \Box\beta) = 0 = v_i(\gamma_j) \qquad (1 \leq j \leq n)$$

$$v_i(\Box\beta_l \wedge \Box\beta) = 0 = v_i(\gamma_0) \qquad (1 \leq l \leq m)$$

$$w_j(\Box\delta_k \wedge \Box\beta) = w_j(\Box\delta_k) = w_j(\gamma_k) \qquad (1 \leq k \leq n)$$

$$w_j(\Box\beta_i \wedge \Box\beta) = w_j(\Box\beta_i) = w_j(\gamma_0) \qquad (1 \leq i \leq m)$$

Consider the $S4$ algebra $A = B_1^m \times A_1 \times ... \times A_n$. Let v be the valuation in A defined thus: $v(p) = (v_1(p), ..., v_m(p), w_1(p), ..., w_n(p))$ $(p \in VAR)$.
Then

$$v(\delta_j) \neq 1 \qquad (1 \leq j \leq n)$$

$$v(\beta_i) \neq 1 \qquad (1 \leq i \leq m)$$

Also

$$v(\Box\delta_j \wedge \Box\beta) = v(\gamma_j) \qquad (1 \leq j \leq n)$$

$$v(\Box\beta_i \wedge \Box\beta) = v(\gamma_0) \qquad (1 \leq i \leq m)$$

$$v(\alpha) = 1$$

Now using Lemma 4 we define the $S4$ algebra A' obtained from A by replacing the interior operation I of A with the function I' defined thus

$$I'(x) = \begin{cases} I(x) \wedge v(\Box\beta) & if \quad x \neq 1 \\ 1 & otherwise \end{cases} \qquad (x \in A)$$

Let v' be the valuation in A' defined as follows: $v'(p) = v(p)$ $(p \in VAR)$.
Then

$$v'(\Box\delta_j) = v'(\gamma_j) \qquad (1 \leq j \leq n)$$

$$v'(\Box\beta_i) = v'(\gamma_0) \qquad (1 \leq i \leq m)$$

$$v'(\Box\alpha) = 1$$

$$v'(\beta_1) \neq 1$$

Therefore $v'(\varphi \to \beta_1) \neq 1$, whence $\varphi \to \beta_1 \notin S4$. ■

Next we define the following refutation system for $S4$:

Axioms: $\dashv \alpha$, *where α is a modal-free formula that is not a theorem of Classical Logic*

Rules: $r_{mp}, r_s,$

$$(r)\ \frac{\dashv \alpha \to \beta_i \vee \gamma_0 \vee ... \vee \gamma_n \quad (1 \leq i \leq m) \qquad \dashv \varphi \wedge \Box\beta_1 \to \delta_j \quad (1 \leq j \leq n)}{\dashv \varphi \to \beta_1}$$

where $\varphi = \bigwedge\{\Box(\Box\delta_j \equiv \gamma_j) : 1 \leq j \leq n\} \wedge \bigwedge\{\Box(\Box\beta_i \equiv \gamma_0) : 1 \leq i \leq m\} \wedge \Box\alpha$, $m \geq 1$, *and all* $\alpha, \beta_i, \gamma_j, \delta_j, \gamma_0$ *are modal-free.*

REMARKS 6 *1. If $n = 0$ then the rule r has the form*

$$\frac{\dashv \alpha \rightarrow \beta_i \vee \gamma_0 \qquad (1 \leq i \leq m)}{\dashv \bigwedge\{\Box(\Box\beta_i \equiv \gamma_0) : 1 \leq i \leq m\} \wedge \Box\alpha \rightarrow \beta_1}$$

2. *From the proof of Theorem 5($\Rightarrow$) it is clear that the refutation rule r is valid in $S4$, i.e. the set $FOR - S4$ is closed under this rule. Hence the above refutation system is really a refutation system for $S4$, i.e. no theorem of $S4$ can be refuted in it.*

Finally we can prove the following completeness theorem for $S4$.

THEOREM 7 *Let ψ be a formula of the form $\varphi \rightarrow p_1$, where $\varphi = \bigwedge\{\Box(\Box r_j \equiv q_j) : 1 \leq j \leq n\} \wedge \bigwedge\{\Box(\Box p_i \equiv q_0) : 1 \leq i \leq m\} \wedge \Box\alpha$, $m \geq 1$, all $p_i, q_j, r_j, q_0 \in VAR$, $q_j \neq q_k$ for $0 \leq j \neq k \leq n$, and α is modal-free. Then either $\vdash \psi$ or $\dashv \psi$.*

Proof. By induction on the number of variables occurring in ψ.

1. ψ has just one variable. Then $\psi = \Box(\Box p_1 \equiv p_1) \wedge \Box\alpha \rightarrow p_1$. Consider the formula $\alpha \rightarrow p_1$. It is modal-free, so either $\vdash \alpha \rightarrow p_1$ or $\dashv \alpha \rightarrow p_1$. (We assume that every theorem of Classical Logic has a proof.) Hence either $\vdash \psi$ or $\dashv \psi$.

2. ψ has more variables than one. Then the formulas

$$(\varphi \rightarrow p_1)(r_j/\top)$$

$$(\varphi \rightarrow r_j)(p_1/\top)$$

$$(\varphi \wedge \Box(\Box r_j \equiv q_0) \rightarrow p_1)(q_j/q_0),$$

where $1 \leq j \leq n$, have less variables than ψ. Also after some simple reductions their form becomes that of ψ. Hence by the inductive hypothesis every one of them has either a proof or a disproof. Moreover the above formulas are equivalent to

$$A_j = \Box r_j \wedge \varphi \rightarrow p_1$$

$$B_j = \varphi \wedge \Box p_1 \rightarrow r_j$$

$$C_j = \varphi \wedge \Box(\Box r_j \equiv q_0) \rightarrow p_1$$

respectively, so that all A_j, B_j, C_j are either provable or refutable.

Also for every $1 \leq i \leq m$ either $\vdash D_i$ or $\dashv D_i$, where $D_i = \alpha \to p_i \vee q_0 \vee \ldots \vee q_n$, because each D_i is modal-free.

Now we have the following cases:

1. $\dashv D_i$ $(1 \leq i \leq m)$ and $\dashv B_j$ $(1 \leq j \leq n)$. Then $\dashv \varphi \to p_1$.

2. $\vdash D_i$ for some $1 \leq i \leq m$ or $\vdash B_j$ for some $1 \leq j \leq n$.

If $\dashv A_j$ or $\dashv C_j$ then $\dashv \varphi \to p_1$, so we assume that $\vdash A_j$ and $\vdash C_j$ $(1 \leq j \leq n)$. Hence by the argument in the proof of Theorem 5($\Leftarrow$) we obtain $\vdash \varphi \to p_1$. Therefore either $\vdash \psi$ or $\dashv \psi$. ■

ACKNOWLEDGEMENTS

The author wishes to thank the Alexander-von-Humboldt Foundation for supporting this research and André Fuhrmann for his hospitality.

University of Wrocław, Poland.

REFERENCES

[1] R. Bull, K. Segerberg. Basic modal logic. In: D. M. Gabbay, F. Guenthner (eds.), *Handbook of Philosophical Logic, Vol. II*, pp. 1–88, Reidel, Dordrecht, 1984.

[2] J. Łukasiewicz. *Aristotle's Syllogistic from the Standpoint of Modern Formal Logic*. Oxford, 1951.

[3] G. Mints. Gentzen-type systems and resolution rules. Part I. In: P. Martin-Löf, G. Mints (eds.), *COLOG-88*, Lecture Notes in Computer Science, 417, pp. 198–231, Springer-Verlag, Berlin, 1990.

[4] D. Scott. Completeness proofs for the intuitionistic sentential calculus, *Summaries of talks presented at the Summer Institute for Symbolic Logic*, Cornell University, 1957.

[5] T. Skura. Refutation calculi for certain intermediate propositional logics, *Notre Dame Journal of Formal Logic*, **33**, 552–560, 1992.

[6] T. Skura. *A Łukasiewicz-style refutation system for the modal logic S4*. Forthcoming.

PART II

EXTENDED FORMALISMS

EWA ORLOWSKA

RELATIONAL PROOF SYSTEMS FOR MODAL LOGICS

1 INTRODUCTION

The purpose of this paper is to give a survey of the relational formalization of modal logics. The paradigm 'formulas are relations' leads to the development of a relational logic based on algebras of relations. The logic can be viewed as a generic logic for the representation of nonclassical logics; in particular a broad class of multimodal logics can be specified within its framework. As a consequence, proof systems for the relational logic become a convenient tool for the development of a proof theory for nonclassical logics. The relational logic enables us to represent within a uniform formalism the three basic components of any propositional logical system: syntax, semantics and deduction apparatus. The essential observation, leading to a relational formalization of logical systems, is that a standard relational structure (a Boolean algebra with a monoid) constitutes a common core of a great variety of nonclassical logics. Exhibiting this common core on all the three levels of syntax, semantics and deduction, enables us to create a general framework for representation, investigation and implementation of nonclassical logics.

The relational formalization of nonclassical logics is realized on the following methodological levels:

Syntax: With the formal language of a logic L there is associated a language of relational terms.

Semantics and model theory: With logic L there is associated a class of relational models for L, and in these models the formulas from L are interpreted as relations. Each relational model determines an algebra of relations. The class of algebras derived from the underlying class of relational models for L provides a relation-algebraic semantics for L.

Proof theory: With logic L there is associated a relational logic for L such that its proof system provides a deduction method for L.

The semantics of nonclassical logics usually carries on twofold information: information about extensional components of the logical system and information about its intensional components. Relational formalisms are rich enough to represent uniformly these two types of information. Extensional information is modeled by means of a Boolean part of algebras of relations and intensional information by

H. Wansing (ed.), Proof Theory of Modal Logic, 55–78.

means of its monoid part. The crucial element of this construction is a decomposition of both extensional and intensional elements of a logical system under consideration into primitive ingredients and next the explication of those of them that can be treated as elements of a Boolean algebra or a monoid.

In the relational representation of logical systems we are going to articulate explicitly information about both their syntax and semantics. Generally speaking, formulas are represented as terms over algebras of relations. Each of the propositional connectives becomes a 'logical' relational operation, and in this way an original syntactic form of formulas is preserved. Semantic information about a formula, which normally is included in a satisfiability condition, consists of the two basic parts: first, we say which states satisfy the subformulas of the given formula, and second, how those states are related to each other via the accessibility relation. Those two ingredients of semantic information are of course interrelated and unseparable. In the relational representation of formulas, terms representing accessibility relations are included explicitly in the respective relational terms corresponding to the formulas. They become the arguments of the relational operations in a term in the same way as the other of its subterms, obtained from subformulas of the given formula. In this way semantic information is provided explicitly on the same level as syntactic information, and the traditional distinction between syntax and semantics disappears, in a sense. In the relational term corresponding to a formula both syntactic and semantic information about the formula are integrated into a single information item.

One of the advantages of the relational representation of formulas is that we gain compositionality. In most of the nonclassical logics their formulas that are built with intensional propositional connectives, for example with modal or temporal operators, are not compositional. The meaning of a compound formula is not necessarily a function of meanings of its subformulas. In the relational formalism the counterparts of these connectives become compositional. The classical opposition between compositional and noncompositional is eliminated when we pass to the relational counterpart of a given nonclassical logic. In several applications of modal logics there is a need to distinguish between information about static facts and dynamic transitions between states in a domain which a given logic is intended to model. The relational representation of modal formulas enables us to express an interaction of these two components of information in a single, uniform formalism. In relational logics an opposition between static and dynamic or between declarative and procedural is transformed into a coexistence and interaction of these two types of information in a uniform framework.

Relational proof theory enables us to build proof systems for nonclassical logics in a systematic modular way. First, deduction rules are defined for the common relational core of the logics. These rules constitute a core of all the relational proof systems. Next, for any particular logic some specific rules are designed and adjoined to the core set of rules. Hence, we need not implement each deduction system from the scratch, we should only extend the core system with a module corresponding to a specific part of a logic under consideration.

From the algebraic perspective the relational formalization of modal logics leads to what might be called 'nonclassical' algebras of relations. In these algebras the 'logical' relational operations are admitted which are relational counterparts of intensional propositional connectives. They are not always expressible in terms of the standard relational operations.

The standard semantics of modal logics is usually defined in terms of frames ([5, 6]), that is relational systems consisting of a set W of states and a family of accessibility relations in W. Then the meaning of a formula is defined, first, by means of an assignment m of subsets of W to propositional variables, and second, by extending m to all formulas of the language under consideration. In this way every formula A is interpreted as a subset of states, with the intuition that $m(A)$ consists of those states in which A is true. The meaning $m(A)$ of formulas built with the classical propositional connectives of negation ($\neg$), disjunction ($\vee$), conjunction ($\wedge$), implication ($\rightarrow$), and equivalence ($\leftrightarrow$) is defined from the meanings of the subformulas of A by the well known interpretation of these connectives in terms of Boolean operations of complement ($-$), join ($\cup$), and meet ($\cap$). The meaning $m(A)$ of formulas built with intensional operators, such as modal operators of possibility and necessity, dynamic logic operators determined by programs ([2, 16]), information operators determined by an indiscernibility or similarity relation ([11, 21]), is usually defined in terms of the values of m for the subformulas of A and in terms of an accessibility relation, which is most often a binary or ternary relation in W.

The idea of the relational formalization of nonclassical logics comes from the interpretability of logics in a relational logic ([12, 13, 14]). From the logical perspective the interpretability is established by means of a validity preserving translation of formulas of logical systems into relational terms. Under that translation both formulas, formerly understood as sets of states and accessibility relations, receive a uniform representation as relations of the same finite rank, and propositional connectives are transformed into logical relational operations. A survey of the relational semantics for nonclassical logics can be found in [15]. From the algebraic perspective the translation leads to the algebras of right ideal relations.

Let W be a nonempty set. The full algebra of binary relations over W is the algebra:

$$Re(W) = (Sb(W \times W), -, \cup, \cap, 1, 0, ;, ^{-1}, I)$$

where $(Sb(W \times W), -, \cup, \cap, 1, 0)$ is the Boolean algebra of subsets $W \times W$, $1 = W \times W$ is the universal relation, 0 is the empty relation, $I = \{(w, w) : w \in W\}$ is the identity in W, operations ; and $^{-1}$ are relational composition and conversion, respectively. The class **RRA** of representable relation algebras consists of isomorphic copies of subalgebras and direct products of full algebras of relations. Every algebra from **RRA** is isomorphic to an algebra whose elements are binary relations, 1 is an equivalence relation, and I is an identity on the field of 1. If 1 has exactly one equivalence class, then such an algebra is simple. The class **RRA** is a variety (Tarski [19, 20]), but its equational theory is not finitely axiomatizable, and moreover, an infinite axiomatization of **RRA** requires infinitely many relation variables (Monk [9, 10]). For any relations P, Q in $Re(W)$ the following conditions are satisfied:

PROPOSITION 1

(a) $P \subseteq Q$ *iff* $-P \cup Q = 1$

(b) $P = Q$ *iff* $(-P \cup Q) \cap (-Q \cap P) = 1$

(c) $P \neq 1$ *iff* $1; (-P); 1 = 1$

(d) $P = 1$ *and* $Q = 1$ *iff* $(P \cap Q) = 1$

(e) $P = 1$ *or* $Q = 1$ *iff* $1; -(1; (-P); 1); 1 \cup Q = 1$

(f) $P = 1$ *implies* $Q = 1$ *iff* $1; (-P); 1 \cup Q = 1$

Conditions (a)–(f) show that any open formula of the equational theory of algebras of relations is expressible as an equation of the form $P = 1$, where P is a relational term.

In the relational representation of logical formulas the operations of right (/) and left (\) residuation play an important role:

$$R/P = \{(y, z) : (x, y) \in P \text{ implies } (x, z) \in R \text{ for all } x\}$$

$$Q \setminus R = \{(x, y) : (y, z) \in Q \text{ implies } (x, z) \in R \text{ for all } z\}.$$

Residuations are definable in terms of the standard relational operations:

$$R/P = -(P^{-1}; -R), \quad Q \setminus R = -(-R; Q^{-1}).$$

A binary relation P is said to be a (right) ideal relation whenever $P; 1 = P$. If P, Q are ideal relations, then so are $-P$, $P \cup Q$, $P \cap Q$. Moreover, if Q is an ideal relation, then for any P relations $P; Q$ and Q/P are ideal as well. Ideal relations can be viewed as unary relations (or sets) which are dummy embedded into relations of a higher rank. It follows that the class **RRA** is a generalized reduct of a class of subalgebras of algebras of relations of any rank higher than 2. A comprehensive survey of the theory of algebras of relations can be found in Maddux [7].

2 RELATIONAL SEMANTICS

The first step toward a relational formalization of a nonclassical logic L is a transformation of the language of L into the language of relational terms. In the present section we give a number of examples of how formulas from various nonclassical logics are represented as relational terms. We define inductively a function t

from formulas of a nonclassical logic into relational terms. The function can also be interpreted in semantic terms as a meaning function that assigns relations to formulas, and therefore we often refer to t as to the mapping that provides a relational semantics. Assume that every propositional variable p is translated into a relational variable $t'(p)$. From the semantic point of view it means that a relation is assigned to every propositional variable. Next we define:

$$t(p) = t'(p); 1$$

This condition assures that propositional variables are interpreted as right ideal relations. Furthermore, classical propositional connectives are interpreted as Boolean operations:

$$t(\neg A) = -t(A),\ t(A \vee B) = t(A) \cup t(B),\ t(A \wedge B) = t(A) \cap t(B)$$

The translation of modal formulas built with the possibility operator $\langle R \rangle$ and necessity operator $[R]$ determined by a binary accessibility relation R is:

$$t(\langle R \rangle A) = R; t(A), \quad t([R]A) = -(R; -t(A))$$

Observe, that the necessity operator can be equivalently expressed by means of right residuation:

$$t([R]A) = t(A)/R^{-1}$$

In particular, for dynamic logic modal operators we have:

$$t(\langle R \cup P \rangle A) = (R \cup P); t(A) = R; t(A) \cup P; t(A),$$

$$t(\langle R; P \rangle A) = R; P; t(A), \quad t(\langle R^* \rangle A) = R^*; t(A)$$

In dynamic logic with specification operators that have been introduced in [4] we have in addition:

$$t(\langle R/P \rangle A) = (R/P); t(A), \quad t(\langle Q \setminus R \rangle A) = (Q \setminus R); t(A)$$

The relational semantics for the intuitionistic operations of negation ($\neg_{int}$) and implication ($\rightarrow_{int}$) is:

$$t(A \rightarrow_{int} B) = -(R; (t(A) \cap -t(B)))$$

$$t(\neg_{int} A) = -(R; t(A))$$

where R is a reflexive and transitive binary accessibility relation such that for all propositional variables p we have the following counterpart of the atomic heredity condition:

$$R^{-1}; t(p) \subseteq t(p).$$

The relational semantics for the temporal operator $Next$ is:

$$t(NextA) = (R; t(A)) \cap -(R; P)$$

where R is a reflexive, transitive, and weakly connected accessibility relation.

Post operations D_i, $i = 1, 2, \ldots$, and Post constants E_j, $j = 0, \ldots, \omega$ in ω^+-valued logic ([17]) have the following relational semantics:

$$t(D_i A) = d_i; t(A), \quad t(E_j A) = (I \cap (d_j; R)); 1$$

where I is the identity relation, R is an accessibility relation satisfying the same conditions as the intuitionistic accessibility relation, and d_i, $i = 1, 2, \ldots$ are the functions that satisfy the following conditions:

$d_j; d_i \subseteq d_i$

$I \subseteq d_1; R^{-1}$

$I \subseteq d_{i+1}; R; d_i^{-1}$

$R \subseteq d_i; R; d_i$

$I \subseteq d_i; R$ iff not $I \subseteq d_{i+1}; R^{-1}$

There is $i = 1, 2, \ldots$ such that $I \subseteq d_i; R$

The above conditions are the counterparts of the respective conditions assumed in Post frames ([8]).

The relational semantics for relevant logics can be found in Orlowska [13]. It is easy to see that in all the above examples for any formula A its relational counterpart $t(A)$ is a right ideal relation.

The basic idea behind the relational semantics is that we assign right ideal relations to formulas. This interpretation has a natural motivation. To define a concept, for example the concept 'prime number', we normally say that a prime number is a natural number whose divisors are 1 and the number itself. This definition consists of two parts. The first part 'natural number' refers to a broader concept whose instances form a universe from which the instances of our concept 'prime number' come from. The second part of the definition describes the positive instances of our concept. The relational semantics of formulas is designed just according to that pattern. The domain of the relation assigned to a given formula A consists of the positive instances of a concept defined by A, and the range of that relation coincides with the admitted universe of objects to which the concept applies. The important fact is that under the relational semantics the accessibility relations are 'taken out' of intensional operators and they are treated in the same way as formulas. This enables us to formalize multimodal logical systems such that an algebraic structure is assumed in the family of accessibility relations. For example, in an extension of dynamic logic obtained by adjoining operations of complement and intersection to program constructors ([14]) we have $t(\langle -R\rangle A) = -R; t(A)$ and $t(\langle R \cap P\rangle) = (R \cap P); t(A)$, and hence the logic can be formalized as a suitable fragment of the relational calculus and treated in the relational framework in the same way as the standard dynamic logic.

3 RELATIONAL LOGICS

In Sections 3 and 4 we show how a relational framework is constructed for propositional nonclassical logics. Given a propositional logic L, whose frames consist of a set W and an accessibility relation in this set or a family of accessibility relations, we define a relational logic $RelL$ for L. Let VAR be a set of individual variables and let CON_L be a set of relational constants representing accessibility relations from frames of L. For example, if L is a modal or standard temporal logic, then $CON_L = \{R\}$ consists of a single constant R. If L is a dynamic logic, then CON_L consists of atomic programs. The set $EXPREL_L$ of relational expressions is generated from $VARREL \cup \{1, I\} \cup CON_L$ with the standard relational operations $(-, \cup, \cap, ;, /, \backslash, ^{-1})$ and possibly with new relational operations which might not be definable in terms of the standard ones, that correspond to some of the intensional propositional operations of L. Formulas of the relational logic are of the form xAy, where $x, y \in VAR$ and A is a relational expression from set $EXPREL_L$. If in frames for L an algebraic structure is assumed in the universes of those frames, then we define a set TMS_L of terms generated by VAR with all the underlying algebraic operations. For example, if in models of a temporal logic a unary successor operation s is admitted in the set of moments of time, then the respective set TMS_L contains terms of the form $s(x)$ for all $x \in TMS_L$. Next, formulas of the relational logic for L are defined as expressions of the form xAy, where $x, y \in TMS_L$ and $A \in EXPREL_L$. Semantical structures for relational logics of the form $RelL$ are models of the form:

$$M = (W, m)$$

where W is a nonempty set and meaning function $m : VARREL \cup \{1, I\} \cup CON_L \longrightarrow Sb(W \times W)$ assigns binary relations in W to relational variables and constants, in such a way that 1, I, and constants from CON_L receive their intended interpretation as the universal relation, identity and the accessibility relations from models of L, respectively, and the relation variables are mapped into right ideal relations. Function m extends to all the relational expressions from $EXPREL_L$ in a homomorphic way. Resuming, the meaning function in relational models is defined as follows:

$$m(P) \subseteq W \text{ and } m(P); m(1) = m(P), \text{ for } P \in VARREL$$

$$m(R) \subseteq W \times W \text{ for every } R \in CON_L$$

$$m(1) = W \times W, \quad m(I) = \{(w, w) : w \in W\}$$

$$m(-A) = -m(A), \quad m(A^{-1}) = m(A)^{-1}$$

$$m(A \cup B) = m(A) \cup m(B), \quad m(A \cap B) = m(A) \cap m(B)$$

$$m(A; B) = m(A); m(B)$$

and, moreover, if a relation represented by $R \in CON_L$ satisfies some constraints in logic L, then we assume that in the models of the relational logic for L relation $m(R)$ satisfies the same constraints. For example, in temporal logic of linear time we assume that in every model relation $m(R)$ is weakly connected.

By a valuation in M we mean a function $v : VAR \longrightarrow W$ assigning elements of W to the individual variables. A relational formula xAy is satisfied by v in M whenever the following holds:

$$M, v\ sat\ xAy \text{ iff } (v(x), v(y)) \in m(A).$$

Formula xAy is true in M iff $M, v\ \ sat\ \ xAy$ for all the valuations v in M, and xAy is valid iff it is true in all models. In other words, formula xAy is true in a model M whenever $m(A) = m(1)$ holds in M. Relational logics $RelL$ defined according to the above scheme and such that set CON_L is nonempty, or the set of relational operations includes some nonstandard operations, are called nonclassical relational logics. The classical relational logic is the logic with formulas xAy such that A is a term built with the standard relational operations as defined in full algebras of relations. The semantics of relational logics defined above provides an interpretation of relational variables as right ideal relations. The respective models determine classes of algebras of relations generated by right ideal relations. These models are referred to as RI-models, and the corresponding relational semantics is called RI-semantics.

A semantic relationship between a nonclassical logic L and the relational logic $RelL$ for L is established in the lemmas of the following scheme.

PROPOSITION 2 *For every model of L of the form $M = (W,$ family of relations, $m)$, there is a model $M' = (W, m')$ of the relational logic for L such that for any formula A of L and for any $w \in W$ we have:*

(i) $M, w\ sat\ A$ *iff* $(w, z) \in m'(t(A))$ *for all* $z \in W$.

Proof. We define model M' as follows. Its universe coincides with the universe W of M. If $P \in VARREL$ and $P = t'(p)$ for a propositional variable p, then we put

$m'(P) = m(p) \times W$. If $R \in CON_L$, then we put $m'(R) = R$, that is the meaning of constant R in the relational model is the relation from model M denoted by that constant, as usual we use the same symbols for both of them. The proof of the required condition is by induction with respect to the complexity of A. We show the induction step for a modal logic L and for a formula of the form $\langle R \rangle A$. We have $M, w\ sat\ \langle R \rangle A$ iff there is $t \in W$ such that $(w, t) \in R$ and $M, t\ sat\ A$. By the induction hypothesis there is t such that $(w, t) \in m'(R)$ and $(t, z) \in m'(t(A))$ for all $z \in W$, which yields $(w, z) \in m'(R); m'(t(A)) = m'(R; t(A)) = m'(t(\langle R \rangle A))$. ■

PROPOSITION 3 *For every model $M' = (W, m')$ of the relational logic for L there is a model M of L such that condition (i) is satisfied.*

Proof. We define model M as follows. Its universe coincides with the universe W of M'. Accessibility relations in M are all the relations $m'(R) for R \in CON_L$. For any propositional variable p we put $m(p)$ = domain of $m'(P)$ where $P = t'(p)$. By induction on the complexity of a formula A one can show that condition (i) is satisfied. ■

PROPOSITION 4 *(a) A formula A of logic L is valid in L iff $xt(A)y$ is valid in RelL.(b) Formulas $A_1, \ldots, A_n$ imply a formula B in L iff $x1; -(t(A_1) \cap \ldots \cap t(A_n)); 1 \cup t(B)y$ is valid in RelL.*

Proof. To prove (a) assume that a formula A is valid in logic L. Suppose that there is a model $M' = (W', m')$ of the relational logic for L and there are x, $y \in W'$ such that $(x, y) \notin m'(t(A))$. By Proposition 3 there is a model M of L with the universe W' such that not $M, x\ \ sat\ \ A$, which contradicts the assumption. Now assume that $xt(A)y$ is valid in the relational logic for L and suppose that A is not valid in L. Hence there is a model $M = (W$, family of accessibility relations, $m)$ of L and there is $w \in W$such that not $M, w\ \ sat\ \ A$. By Proposition 2 there is a model M' of the relational logic for L with the universe W, and there is $z \in W$ such that $(w, z) \notin m'(t(A))$, which contradicts the assumption.We prove condition (b) for $n = 1$. Assume that for every model M of L, if A is true in M, then B is true in M. Suppose that there is a model $M' = (W, m)$ of logic $RelL$ such that $m(1; -t(A); 1 \cup t(B)) \neq m(1)$. By 1 (f) from Section 1 it yields $m(t(A)) = m(1)$ and $m(t(B)) \neq m(1)$. By Proposition 2 there is a model M of L having the same universe as M' such that A is true in M and B is not true in M, a contradiction. Now assume that $x1; -t(A); 1 \cup t(B)y$ is valid in $RelL$. Suppose that A does not imply B in L, that is, there is a model M of L such that A is true in M but B is not true in M. By Proposition 2 there is a model $M' = (W, m')$ of $RelL$ with the same universe as the universe of M such that $m'(t(A)) = m'(1)$ and $m'(t(B)) \neq m'(1)$. By 1 (f) we obtain a contradiction which completes the proof. ■

We conclude that, given a nonclassical logic L, a deduction system of the relational logic $RelL$ can serve as a theorem prover for L. In the following sections relational proof systems will be outlined for some nonclassical logics.

4 PROOF SYSTEM FOR THE CLASSICAL RELATIONAL LOGIC

Proof systems for relational logics are Rasiowa-Sikorski style systems ([18]). In the present section we recall the deduction rules for the classical relational logic ([12]), that is the logic whose formulas xAy are built from terms A generated by $VARREL \cup \{1, I\}$ with the standard relational operations. The rules apply to finite sequences of relational formulas. There are two groups of rules: decomposition rules and specific rules. Decomposition rules enable us to decompose formulas in a sequence into some simpler formulas. Decomposition depends on relational operations occurring in a formula. As a result of decomposition we usually obtain finitely many new sequences of formulas, and sometime, as in the case of operation $-*$ in dynamic logic, infinitely many sequences. The specific rules enable us to modify a sequence to which they are applied, they have a status of structural rules. The role of axioms is played by what is called fundamental sequences. In what follows K and H denote finite, possibly empty, sequences of formulas of a relational logic. A variable is said to be restricted in a rule whenever it does not appear in any formula of the upper sequence in that rule.

(DEC) Decomposition rules for the standard relational operations:

$(\cup)$ $$\frac{K, xA \cup By, H}{K, xAy, xBy, H}$$ $(-\cup)$ $$\frac{K, x-(A \cup B)y, H}{K, x-Ay, H \quad K, x-By, H}$$

$(\cap)$ $$\frac{K, xA \cap By, H}{K, xAy, H \quad K, xBy, H}$$ $(-\cap)$ $$\frac{K, x-(A \cap B)y, H}{K, x-Ay, x-By, H}$$

$(--)$ $$\frac{K, x--Ay, H}{K, xAy, H}$$

$(^{-1})$ $$\frac{K, xA^{-1}y, H}{K, yAx, H}$$ $(-^{-1})$ $$\frac{K, x-(A^{-1})y, H}{K, y-Ax, H}$$

(;)
$$\frac{K, xA;By, H}{K, xAz, H, xA;By \quad K, zBy, H, xA;By}$$

z is a variable

(−;)
$$\frac{K, x-(A;B)y, H}{K, x-Az, z-By, H}$$

z is a restricted variable

(\)
$$\frac{K, xA\setminus By, H}{K, y-Az, xBz, H}$$

z is a restricted variable

(−\)
$$\frac{K, x-(A\setminus B)y, H}{K, yAz, H, x-(A\setminus B)y \quad K, x-Bz, H, x-(A\setminus B)y}$$

z is a variable

(/)
$$\frac{K, xA/By, H}{K, z-Bx, zAy, H}$$

z is a restricted variable

(−/)
$$\frac{K, x-(A/B)y, H}{K, zBx, H, x-(A/B)y \quad K, z-Ay, H, x-(A/B)y}$$

z is a variable

(SPE) Specific rules:

($I1$)
$$\frac{K, xAy, H}{K, xIz, xAy, H \quad K, zAy, xAy, H}$$

z is a variable, $A \in VARREL$

($I2$)
$$\frac{K, xAy, H}{K, xAz, xAy, H \quad K, zIy, xAy, H}$$

z is a variable, $A \in VARREL$

$(I3)$ $$\frac{K, xIy, H}{K, yIx, H, xIy}$$

$(I4)$ $$\frac{K, xIy, H}{K, xIz, H, xIy \quad K, zIy, H, xIy}$$

z is a variable

$(ideal)$ $$\frac{K, xAy, H}{K, xAz, H, xAy}$$

z is a variable, $A \in VARREL$

(FND) Fundamental sequences:

A sequence of formulas is said to be fundamental whenever it contains a subsequence of either of the following forms:
(f1) $xAy, x - Ay$ where A is a relational expression
(f2) $x1y$
(f3) xIx

A sequence K of relational formulas is valid iff for every model M of the relational logic and for every valuation v over M there is a formula in K which is satisfied by v in M. It follows that sequences of formulas are interpreted as (metalevel) disjunctions of their elements. A relational rule of the form $K/\{H_t : t \in T\}$ is admissible whenever sequence K is valid iff for all $t \in T$ sequence H_t is valid. It is easy to see that the fundamental sequences are valid and all the rules given above are admissible.

PROPOSITION 5 *(a) All the rules in (DEC) and (SPE) are admissible.*
(b) All the sequences in (FND) are valid.

Admissibility of the decomposition rules follows from the definitions of the respective relational operations, and admissibility of the specific rules follows from the properties of the relational constants or (in the case of rule $(ideal)$) variables reflected by those rules. For example, rules $(I1)$ and $(I2)$ are admissible because we have $I; A = A = A; I$ for any relation A. Rules $(I3)$ and $(I4)$ are admissible because of symmetry and transitivity of I, respectively. Rule $(ideal)$ is admissible because every relational variable is interpreted as a right ideal relation.

Relational proofs have the form of trees. Given a relational formula xAy, where A might be a compound relational expression, we successively apply decomposition or specific rules. In this way we form a tree whose root consists of xAy and whose nodes consist of finite sequences of relational formulas. We stop applying rules to formulas in a node after obtaining a fundamental sequence, or when none of the rules is applicable to the formulas in this node. A branch of a proof tree is said to be closed whenever it contains a node with a fundamental sequence of formulas. A tree is closed iff all of its branches are closed.

In the following a completeness theorem is given for the classical relational logic with respect to the class of its RI-models.

PROPOSITION 6 (Completeness theorem) *A relational formula xAy is valid iff there is a closed proof tree with the root xAy.*

Proof. ($\Rightarrow$) Suppose that there is no closed proof tree for xAy and consider a tree satisfying the following conditions for every non-closed branch b. We write $G \in b$ whenever formula G is a member of a sequence of formulas in a certain node of b.
(b1) $xAy \in b$
(b2) If $x(B \cup C)y$ $(x - (B \cap C)y) \in b$ then both xBy $(x - By) \in b$ and xCy $(x - Cy) \in b$ obtained by application of rule $(\cup)$ (resp. $(-\cap)$)
(b3) If $x - (B \cup C)y$ $(x(B \cap C)y) \in b$ then either $x - By$ $(xBy) \in b$ or $x - Cy$ $(xCy) \in b$ obtained by application of rule $(-\cup)$ (resp. $(\cap)$)
(b4) If $x(B;C)y \in b$ then for every $z \in VAR$ either $xBz \in b$ or $zCy \in b$ obtained by application of rule $(;)$
(b5) If $x - (B;C)y \in b$ then for some $z \in VAR$ both $x - Bz \in b$ and $z - Cy \in b$ obtained by application of rule $(-;)$
(b6) If $x - -By \in b$ then $xBy \in b$ obtained by application of rule $(--)$
The similar closure conditions are postulated for formulas built with operations of conversion and residuations.
(b7) If $xBy \in b$ with $B \in VARREL$ then for every $z \in VAR$ either $xIz \in b$ or $zBy \in b$ obtained by application of rule $(I1)$
(b8) If $xBy \in b$ with $B \in VARREL$ then for every $z \in VAR$ either $xBz \in b$ or $zIy \in b$ obtained by application of rule $(I2)$
(b9) If $xIy \in b$, then $yIx \in b$ obtained by application of rule $(I3)$
(b10) If $xIy \in b$, then for every $z \in VAR$ either $xIz \in b$ or $zIy \in b$ obtained by application of rule $(I4)$
(b11) If $xBy \in b$ with $B \in VARREL$ then for every $z \in VAR$ we have $xBz \in b$ obtained by application of rule $(ideal)$.
Any tree satisfying conditions (b1), ..., (b11) is referred to as a complete proof tree. The standard proof-theoretic construction shows that for every formula there is a complete proof tree with this formula in a root.

Let b be a non-closed branch of a complete proof tree. We define system:

$M^b = (W^b, m^b)$ such that

$W^b = VAR$

$m^b(P) = \{(x, y) \in W^b \times W^b : xPy \notin b\}$ for $P \in VARREL \cup \{1, I\}$

We extend m^b in a homomorphic way to all the relational expressions.
Observe that:

(i) $m^b(I)$ is an equivalence relation in set W^b.
For suppose that for some x we have $(x,x) \notin m^b(I)$. It follows that $xIx \in b$, and then branch b would be closed, a contradiction. Now let $(x,y) \in m^b(I)$, and hence $xIy \notin b$. If $(y,x) \notin m^b(I)$, then $yIx \in b$, and by (b9) we have $xIy \in b$, a contradiction. Similarly, (b10) enables us to prove transitivity of I.
(ii) $m^b(I); m^b(B) = m^b(B) = m^b(B); m^b(I)$ for any relational expression B.
If B is a relational variable, then the condition holds by (b7) and (b8). For the compound relational expressions the proof is by induction with respect to the complexity of the expression.
(iii) $m^b(1)$ is the universal relation on W^b.
For otherwise $x1y \in b$ for some x, y which would make branch b closed.
(iv) $m^b(P)$ is an ideal relation for any $P \in VARREL$.
This condition follows from (b11).
Let v^b be a valuation in M^b such that $v^b(x) = x$ for every object variable x.
We say that formula xBy is indecomposable whenever $B \in VARREL \cup \{1, I\}$.
Let IND^b be the set of all the indecomposable formulas occurring in the nodes of branch b. From the definition of m^b we have:
(v) For every $zBt \in IND^b$ we have not M^b, v^b *sat* zBt.
We define an ordering of relational terms:
if P is a relational variable then $ord(P) = ord(1) = ord(I) = 1$
if $ord(B) = n$ then for any unary relational operation * we define $ord(^*B) = n+1$
if $ord(B) \leq n$ and $ord(C) \leq n$ and at least one of the inequalities is $=$, then for every binary relational operation $\sharp$ we define $ord(B\sharp C) = n + 1$.
We will show that:
(vi) not M^b, v^b *sat* xAy.
For suppose conversely, and let X^b be the set of formulas zBt on b such that M^b, v^b *sat* zBt. X^b is nonempty since xAy is its member. Let C be a term of a minimal order such that uCw is in X^b for some variables u, w. We show that C must be either a relational variable or 1 or I. C cannot be of the form $u - Pw$ for a relational variable P, for otherwise we would have $u - Pw \in b$ and M^b, v^b *sat* $u - Pw$, and from the definition of m^b the latter is equivalent to $uPw \in b$. A similar argument shows that C is neither of the form -1 nor $-I$. Suppose that C is of the form $C_1; C_2$. Hence conditions (a1) $uC_1; C_2w \in b$ and (a2) M^b, v^b *sat* $uC_1; C_2w$ hold. From (a1) and (b4) we conclude that for all z either $uC_1z \in b$ or $zC_2w \in b$. From (a2) we have that there is t such that M^b, v^b *sat* uC_1t and M^b, v^b *sat* tC_2w. Hence either $uC_1t \in X^b$ or $tC_2w \in X^b$, and C_1, C_2 have a smaller value of ord than C, a contradiction. In a similar way we show that C is neither an expression built with any other relational operator nor a complemented compound expression. In view of the above, $uCw \in IND^b$, and hence by (v) we have not M^b, v^b *sat* uCw, a contradiction that completes the proof of (vi).
To complete the proof of the $(\Rightarrow)$ part of the theorem we define a quotient structure:

$M^q = (W^q, m^q)$ such that

$W^q = \{/x/ : x \in VAR\}$ and $/x/$ is the equivalence class of $m^b(I)$ generated by x,

$m^q(P) = \{(/x/, /y/) \in W^q \times W^q : (x, y) \in m^b(P)\}$ for $P \in VARREL$,

and m^q is extended in a homomorphic way to all the relational expressions. It follows from (ii) that the definition of $m^q(P)$ is correct, that is, it does not depend on the choice of representatives from the equivalence classes. It is easy to see that in M^q the relation $m^q(I)$ is the indentity in W^q, since we have $/x/ = /y/$ iff $(x, y) \in m^b(I)$ iff $(/x/, /y/) \in m^q(I)$. Moreover, a standard inductive argument shows that we have $(/x/, /y/) \in m^q(B)$ iff $(x, y) \in m^b(B)$ for any relational expression B. It follows that M^q is a model of the relational logic such that xAy is not true in M^q, a contradiction. The proof of part ($\Leftarrow$) of the theorem is based on Proposition 5. ∎

Clearly, a similar completeness theorem holds for a relational logic whose language coincides with the language of the classical relational logic and whose semantics is less restrictive, that is the meaning function in the respective models assigns arbitrary binary relations to relational variables, not necessarily right ideal relations. The deduction system for this logic can be obtained from the deduction system defined above by deleting rule (ideal).

EXAMPLE 7 We show that in the classical relational logic with RI-semantics formula $x(A \cup -(A;1))y$ is valid.

$$x(A \cup -(A;1))y$$

$$(\cup)$$

$$xAy,\ x - (A;1)y$$

$$(-;)\text{ restricted variable } := z$$

$$xAy,\ x - Az,\ z - 1y$$

$$\text{(ideal) new variable } := z$$

$$xAz,\ x - Az,\ z - 1y$$

$$\text{fundamental}$$

EXAMPLE 8 We show that $-(A \setminus B) = (-B \setminus -A) \setminus -I$ holds for any binary relations A, B. In view of fact 1 (b) it is sufficient to show that $((A \setminus B) \cup ((-B \setminus$

$-A) \setminus -I)) \cap (-(A \setminus B) \cup -((-B \setminus -A) \setminus -I)) = 1$:

$$x((A \setminus B) \cup ((-B \setminus -A) \setminus -I)) \cap (-(A \setminus B) \cup -((-B \setminus -A) \setminus -I))y$$

$(\cap)$

$x(A \setminus B) \cup ((-B \setminus -A) \setminus -I)y$ $\qquad$ $x - (A \setminus B) \cup -((-B \setminus -A) \setminus -I)y$

$(\cup)$ $\qquad$ $(\cup)$

$xA \setminus By,\ x(-B \setminus -A) \setminus -Iy$ $\qquad$ $x - (A \setminus B)y,\ F = x - ((-B \setminus -A) \setminus -I)y$

$(\setminus)$ *twice* $\qquad$ $(-\setminus)$ *new variable* $:= x$

new variables $:= z, t$ $\qquad$ $x - (A \setminus B)y,\ y - B \setminus -Ax,\ F$ $\qquad$ $\ldots,\ xIx,\ \ldots$

$y - Az,\ xBz,\ y - (-B \setminus -A)t,\ x - It$ $\qquad$ $(\setminus)$ *new variable* $:= z$ $\qquad$ *fund.*

$(-\setminus)$ *new variable* $:= z$ $\qquad$ $x - (A \setminus B)y,\ xBz,\ y - Az$

$y - Az,\ xBz$ $\qquad$ $y - Az, xBz$ $\qquad$ $(-\setminus)$ *new variable* $:= z$

$t - Bz,\ x - It$ $\qquad$ $yAz, x - It$ $\qquad$ $yAz,\ \ldots$ $\qquad$ $x - Bz,\ \ldots$

$(I1)$ *to* xBz $\qquad$ *fund.* $\qquad$ *fund.* $\qquad$ *fund.*

new variable $:= t$

$\ldots, xIt, \ldots$ $\qquad$ $\ldots,\ tBz,\ \ldots$

fund. $\qquad$ *fund.*

5 PROOF SYSTEMS FOR NONCLASSICAL RELATIONAL LOGICS

Relational proof systems are fully modular. Given a reduct of the relational language, we obtain a deduction system for it by restricting the set of rules and the set of fundamental sequences to those which refer to the respective operators and constants from this reduct. For example, in the relational logic corresponding to modal logic K we do not need constant I, and hence we do not need the rules and fundamental sequences referring to it. Similarly, given an extension of the basic relational language, to obtain a deduction system for the underlying logic, we have to define decomposition rules for all the new operators and their complements, and the specific rules reflecting the properties of the new constants.

Let CT be a set of constraints which a binary relation is supposed to satisfy. We do not specify here any particular language to express these constraints, it can be for example a first order language or a language of relational terms. For a constraint $C \in CT$ and a relation R, we shall write $C(R)$ to denote a sentence obtained from formula C by interpreting the relational constant in C as R. Let $REL(CT)$ be the class of relations satisfying constraints from CT:

$$REL(CT) = \{R \in Re(W) : W \text{ is a set, and } C(R) \text{ is true for every } C \in CT\}.$$

Let a nonclassical logic $L(CT)$ be given such that the accessibility relation from its models belongs to $REL(CT)$. The formulas of relational logic $RelL(CT)$ for $L(CT)$ are built from terms generated by $VARREL \cup \{1, I, R\}$, where R is a constant representing the accessibility relation, with the standard relational operations, and possibly with some nonclassical relational operations. We have to provide in the language of $RelL(CT)$ a relational counterpart of every propositional operation from the language of logic $L(CT)$. Models of this relational language are systems $M = (W, m)$ such that meaning function m satisfies the conditions given in Section 3, and moreover $m(R) \in REL(CT)$. The relational proof system for $RelL(CT)$ consists of the rules and fundamental sequences given in Section 4, decomposition rules for all the nonclassical relational operations admitted in the language of $RelL(CT)$, and specific rules for the accessibility relation, reflecting the fact that it belongs to $REL(CT)$. An example of decomposition rules for nonclassical relational operations is given below.

Decomposition rules for the temporal operators *Until* and *Since*:

(Until)

$$\frac{K,\ xA\,Until B y,\ H}{K,\ xRt,\ H,\ xA\,Until By \quad K,\ tBy,\ H,\ xA\,Until By \quad K,\ x-Ru,\ u-Rt,\ uAy,\ H,\ xA\,Until By}$$

t is a variable,
u is a restricted variable

(–Until)

$$\frac{K,\ x-(A\,Until B)y,\ H}{H_1 \quad H_2 \quad H_3}$$

where $H_1 = K,\ x-Rt,\ t-By,\ xRu,\ H, x-(A\,Until B)y$

$H_2 = K,\ x-Rt,\ t-By,\ uRt,\ H,\ x-(A\,Until B)y$

$H_3 = K,\ x-Rt,\ t-By,\ u-Ay,\ H, x-(A\,Until B)y$

u is a variable, t is a restricted variable

(Since)

$$\frac{K,\ xASince By}{K,\ tRx,\ H,\ xASince By \quad K,\ tBy,\ H,\ xASince By \quad K,\ t-Ru,\ u-Rx,\ uAy,\ H,\ xASince By}$$

t is a variable,
u is a restricted variable

(–Since)

$$\frac{K,\ x-(ASince B)y,\ H}{H_1 \quad H_2 \quad H_3}$$

where $H_1 = K,\ t-Rx,\ t-By,\ tRu,\ H,\ x-(ASince B)y$

$H_2 = K,\ t-Rx,\ t-By,\ uRx,\ H,\ x-(ASince B)y$

$H_3 = K,\ t-Rx,\ t-By,\ u-Ay,\ H,\ x-(ASince B)y$

u is a variable, t is a restricted variable

Let $r(R)$ denote a relational rule of the form $K/\{H_t : t \in T\}$ in which relational constant R occurs, and let $RL(R)$ be a set of these rules. They are called specific rules for R. In a way similar to that developed in Section 4 we define the admissibility of a rule in a nonclassical relational logic. Sequence K of relational formulas is valid in $RelL(CT)$ iff for every model M of $RelL(CT)$ and every valuation v in M there is a formula F in K such that $M, v\ sat\ F$. We say that $r(R)$ is admissible in $RelL(CT)$ whenever sequence K is valid in $RelL(CT)$ iff for every $t \in T$ sequence H_t is valid in $RelL(CT)$. We say that:

A set $RL(R)$ of rules defines class $REL(CT)$ of relations whenever $R \in REL(CT)$ iff every rule from $RL(R)$ is admissible.

As usual, we use the same symbols for relations and the respective constants.

PROPOSITION 9 *If sets $RL_1(R)$, $RL_2(R)$ of specific rules for R define classes $REL(CT_1)$, $REL(CT_2)$ of relations, respectively, then class $REL(CT_1 \cup CT_2)$ is definable by $RL_1(R) \cup RL_2(R)$.*

In the following we give an example of a class of relations definable by means of a rule. A relation R is euclidean if it satisfies:

For all x, y, z if $(x, y) \in R$ and $(x, z) \in R$, then $(y, z) \in R$.

The corresponding relational rule is: (euc R)

$$\frac{K,\ xRy,\ H}{K,\ zRx,\ H,\ xRy \qquad K,\ zRy,\ H,\ xRy} \qquad z \text{ is a variable}$$

PROPOSITION 10 *Rule (euc R) defines $REL(Euclidean)$.*

Proof. Assume that rule (euc R) is admissible, and suppose that there are a, b, c such that $(a, b) \in R$, $(a, c) \in R$ and $(b, c) \notin R$. Consider an instance of rule (euc R) such that $K = z - Rx$, $z - Ry$ and H is empty. It follows that sequence $S = xRy, z-Rx, z-Ry$ is valid. However, a valuation v such that $v(x) = b$, $v(y) = c$, and $v(z) = a$ satisfies none of the formulas in S, a contradiction. Now assume that $R \in REL(Euclidean)$. If the upper sequence $S = K, xRy, H$ in rule (euc R) is valid, then so are the lower sequences $S_1 = K, zRx, H, xRy$ and $S_2 = K, zRy, H, xRy$. Now assume that for every model M and every valuation v in M there are formulas F_1 in S_1 and F_2 in S_2 such that $M, v\ sat\ F_1$ and $M, v\ sat\ F_2$. If $F_1 = zRx$ and $F_2 = zRy$, then since R is euclidean, we have $M, v\ sat\ xRy$. In all the remaining cases F_1 or F_2 occurs in S. We conclude that S is valid. ∎

EXAMPLE 11 We give a relational proof of the modal formula $\langle R\rangle A \to [R]\langle R\rangle A$ which is valid in the class of Kripke models with euclidean accessibility relations.

To simplify notation we assume that $t(A) = A$.

$$x(-(R;A) \cup (R;A)/R^{-1})y$$
$$(\cup)$$
$$x-(R;A)y,\ x(R;A)/R^{-1}y$$
$$(-;)\ \text{new variable}\ := z$$
$$x-Rz,\ z-Ay,\ x(R;A)/R^{-1}y$$
$$(/)\ \text{new variable};:= v(-^{-1})$$
$$x-Rz,\ z-Ay,\ x-Rv,\ vR;Ay$$
$$(;)\ \text{new variable}\ := z$$
$$x-Rz,\ z-Ay,\ x-Rv,\ vRz,\ vR;Ay \qquad zAy,\ \ldots$$
$$\text{(euc R) new variable} := x \qquad \text{fundamental}$$
$$xRv,\ \ldots \qquad xRz, \ldots$$
$$\text{fund.} \qquad \text{fund.}$$

Observe, that defining the formalization of modal logics within the framework of relational logics with RI-models we can use a simpler version of translation of propositional variables: $t(p) = t'(p)$, the property of being right ideal relation is guaranteed directly in the RI-semantics.

6 RESTRICTED CUT RULES

In relational proof systems the cut rule has the following form:

(cut)

$$K$$
$$K,\ xAy \qquad K,\ x-Ay \quad x,y\ \text{arbitrary variables}$$

Admissibility of this rule follows from the fact that a sequence K of formulas is valid iff the sequence $K,\ xA \cap -Ay$ is valid. In other words, we can extend any sequence of formulas in a node of a proof tree with an unsatisfiable formula.

In relational proof systems for modal logics we often admit a restricted cut rule such that the respective unsatisfiable formula is of a particular form. For example, in the proof system for the logic with a serial accessibility relation ([2]) we have the following restricted cut rule:

(serial)

$$K$$
$$K,\ x-Ry \quad x\ \text{is a variable},\ y\ \text{is a restricted variable.}$$

Clearly, if in every model the relation R is serial, then the rule (serial) is admissible. In the completeness proof of the system with this rule the following condition is added in the definition of a complete proof tree:
(b12) For every x there is a y such that $x - Ry \in b$ obtained by application of the rule (serial).

Next, in the definition of system M^b we extend the meaning function m^b by putting:

$$m^b(R) = \{(x, y) : xRy \notin b\}$$

Condition (b12) enables us to prove that $m^b(R)$ is serial. Furthermore, we extend the family of indecomposable formulas postulating that xRy is an indecomposable formula, and then the proof can be completed as in Section 4.

7 INFINITARY RULES

Infinitary rules in relational proof systems have the form:

$$K/\{Hi\}_{i \in \omega}$$

A rule of that form is admissible whenever sequence K is valid iff for every $i \in \omega$ sequence H_i is valid. An example of a relational proof system with an infinitary rule is the system for dynamic logic ([14]). The rules for the iteration operator * are as follows:

(*)

$$\frac{K,\ xA^*y,\ H}{K,\ xA^iy,\ H, xA^*y \text{ where } i \in \omega,\ A^0 = I,\ A^{i+1} = A; A^i}$$

(−*)

$$\frac{K,\ x - (A^*)y,\ H}{\{K,\ x - (A^i)y,\ H\}_{i \in \omega}}$$

The definition of a complete proof tree is extended with the following conditions:
(b12) If $xB^*y \in b$, then for every $i \in \omega$, $xB^iy \in b$ obtained by application of rule (*)
(b13) If $x - B^*y \in b$, then for some $i \in \omega$, $x - (B^i)y \in b$ obtained by application of rule (−*).

Moreover, we suitably modify the notion of ordering $<$ of relational terms:
(<1) Relational variables and constants are minimal with respect to $<$
(<2) $A < -A$, $A < A^{-1}$
(<3) $A < A\sharp B$ and $B < A\sharp B$ for $\sharp = \cup, \cap, ;$
(<4) $A^i < A^*$ for all $i \in \omega$
(<5) $A < B$ implies $-A < -B$

It follows that $-A < -(A\sharp B)$, $-B < -(A\sharp B)$, and $-A^i < -A^*$. Ordering $<$ is well founded, and in the proof of the completeness theorem analogous to 6 condition (v) is shown by induction with respect to $<$.

Other examples of infinitary rules in relational proof systems can be found in [1].

EXAMPLE 12 We give a relational proof of the induction axiom from dynamic logic. Let $F = p \wedge [R^*](p \to [R]p) \to [R^*]p$. In the relational representation of this formula we use the right residuation operator / for the translation of the necessity operation. For the sake of simplicity we assume that $t(p) = p$. We have $t(F) = -p \cup R^*; (p \cap (R; -p)) \cup p/R^{*-1}$.

$$xt(F)y$$

$$(\cup)$$

$$x - py,\ xR^*; (p \cap (R; -p))y,\ xp/R^{*-1}y$$

$$(/)\ (-^1)$$

$$\textit{new variable} := z$$

$$x - py,\ xR^*; (p \cap (R; -p))y,\ x - R^* z,\ zpy$$

$$(;)$$

$$\textit{new variable} := x$$

$$x - py,\ x - R*^z,\ zpy,\ xR^*x,\ xR^*; (p \cap (R; -p))y \qquad x - py,\ x - R^* z,\ zpy, xp \cap (R; -p)y,\ xR^*; (p \cap (R; -p))y$$

$$(*)\ i = 0 \qquad\qquad (\cap)$$

$$..., xIx, ... \qquad ..., xpy, ... \qquad x - py,\ x - R^* z,\ zpy,\ x(R; -p)y,\ xR^*; (p \cap (R; -p))y$$

$$\textit{fundamental} \qquad \textit{fundamental} \qquad (;)$$

$$\textit{new variable} := z$$

$$x - py,\ x - R^* z,\ zpy,\ xRz,\ xR^*; (p \cap (R; -p))y,\ x(R; -p)y \qquad ..., z - py, ...$$

$$(-^*) \qquad \textit{fundamental}$$

$$K_0\ K_1\ \ldots K_i \ldots$$

where $K_i(z) = x - py, x - R^i z, zpy, xRz, xR^*; (p \cap (R; -p))y, x(R; -p)y$ and z is any variable distinct from x and y.We show by induction that for any natural number i and for any $z \neq x, y$ there is a closed tree starting from $K_i(z)$.For $i = 0$

we have:

$$K_0 = \{x - py,\ x - Iz,\ zpy,\ xRz,\ xR^*;(p \cap (R;-p))y,\ x(R;-p)y\}$$

$$(I1) \text{ to } zpy$$

$$\text{new variable} := z$$

$$K_0,\ xIz \qquad K_0, xpy$$

$$\text{fundamental} \qquad \text{fundamental}$$

For $i = 1$ sequence K_1 contains $x - Rz$ and xRz, and hence it is fundamental. For $i = n$ we have:

$$x - py,\ x - (R^{n-1};R)z,\ zpy,\ xRz,\ xR^*;(p \cap (R;-p))y,\ x(R;-p)y$$

$$(-;)\ \textit{new variable} := t$$

$$x - py,\ x - R^{n-1}t,\ t - Rz,\ zpy,\ xRz,\ xR^*;(p \cap (R;-p))y,\ x(R;-p)y$$

$$(;)\ \textit{to}\ G = xR^*;(p \cap (R;-p))y$$

$$\textit{new variable} := t$$

$$\ldots, xR^*t, \ldots \qquad x - py,\ x - R^{n-1}t,\ t - Rz,\ zpy,\ xRz,\ t(p \cap (R;-p))y,\ x(R;-p)y\ G$$

$$(*)\ i = n - 1 \qquad (\cap)$$

$$\ldots, xR^{n-1}t, \ldots \qquad \textit{fundamental}$$

Left branch of $(\cap)$:

$$x - py,\ x - R^{n-1}t,$$
$$t - Rz,\ zpy,\ xRz,\ tpy,$$
$$G,\ x(R;-p)y$$
$$(;)\ \textit{to}\ x(R;-p)y$$
$$\textit{new variable} := t$$
$$\{x - py,\ x - R^{n-1}t,\ t - Rz,$$
$$zpy,\ xRz,\ tpy,\ G,\ xRt,$$
$$x(R;-p)y\} = S$$

Right branch of $(\cap)$:

$$x - py,\ x - R^{n-1}t,$$
$$t - Rz,\ zpy,\ xRz,$$
$$t(R;-p)y,\ G,\ x(R;-p)y$$
$$(;)\ \textit{to}\ t(R;-p)y$$
$$\textit{new variable} := z$$
$$\ldots, tpy, \ldots \qquad \ldots, tRz, \ldots \qquad \ldots, z - py, \ldots$$
$$\textit{fund.} \qquad \textit{fund.} \qquad \textit{fund.}$$

Sequence S includes sequence $K_{n-1}(t)$ and $t \neq x, y$. Hence by the induction hypothesis there is a closed tree starting from S. We conclude that there is a closed proof tree for the induction axiom.

8 CONCLUDING REMARKS

In the paper a method has been presented of transformation of nonclassical propositional logics into a relational logic. Our main concern has been in the development of relational proof systems for nonclassical logics. The basic steps leading to the construction of a relational proof system for a given logic L are:

Translate formulas of L into relational terms, introducing new constants and/or relational operations if needed.

Define the rules for these new constants and operators, and adjoin them to the proof system of the standard relational logic.

Relational proof systems consist of rules of the following types:

Decomposition rules: a compound relational term is replaced by its component terms. These rules are counterparts of operational rules in Gentzen style systems.

Specific rules: they reflect properties of relational constants. They are counterparts of structural Gentzen rules. Some of the specific rules have the form of:
Restricted cut rules: a cut formula has a particular fixed form.

A computational framework for relational proof systems is being creating within the ATINF inference system. The description of a module of graphical presentation of relational proofs can be found in [3].

Polish Academy of Sciences, Poland.

REFERENCES

[1] S. Demri, E. Orlowska. Logical analysis of demonic nondeterministic programs. Submitted, 1995.

[2] S. Demri, E. Orlowska, I. Rewitzky. Towards reasoning about Hoare relations, *Annals of Mathematics and Artificial Intelligence*, **12**, 265–289, 1994.

[3] M. Herment, E. Orlowska. Handling information logics in a graphical proof editor. *Computational Intelligence*, **11**, 297–322, 1995.

[4] C. A. R. Hoare, He Jifeng. The weakest prespecification, *Fundamenta Informaticae*, **9**, Part I: 51–84, Part II: 217–262, 1986.

[5] S. Kripke. Semantical analysis of modal logic I. *Zeitschrift für Mathematische Logik und Grundlagen der Mathematik*, **9**, 67–96, 1963.

[6] S. Kripke. Semantical analysis of intuitionistic logic. In: J. N. Crossley, M. A. Dummett, (eds.), *Formal Systems and Recursive Functions*, North Holland, Amsterdam, 1965.

[7] R. Maddux. The origin of relation algebras in the development and axiomatization of the calculus of relations. *Studia Logica*, **50**, 421–456, 1991.

[8] L. Maximova, D. Vakarelov. Semantics for ω+-valued predicate calculi. *Bulletin of the PAS, Ser. Math.*, **22**, 756–771, 1974.

[9] D. Monk. On representable relation algebras. *Michigan Mathematical Journal*, **11**, 207–210, 1964.

[10] D. Monk. Nonfinitizability of classes of representable cylindric algebras. *Journal of Symbolic Logic*, **34**, 331–343, 1969.

[11] E. Orlowska. Logic of nondeterministic information. *Studia Logica*, **44**, 93–102, 1985.

[12] E. Orlowska. Relational interpretation of modal logics. In: H. Andreka, D. Monk, I. Nemeti, (eds), *Algebraic Logic. Colloquia Mathematica Societatis Janos Bolyai* 54, North Holland, Amsterdam, 443–471, 1988.

[13] E. Orlowska. Relational proof system for relevant logics. *Journal of Symbolic Logic*, **57**, 1425–1440, 1992.

[14] E. Orlowska. Dynamic logic with program specifications and its relational proof system. *Journal of Applied Non-Classical Logics*, **3**, 147–171, 1993.

[15] E. Orlowska. Relational semantics for nonclassical logics: Formulas are relations. In: J. Wolenski, (ed), *Philosophical Logic in Poland*, pp. 167–186, Kluwer, Dordrecht, 1994.

[16] V. R. Pratt. Semantical considerations on Floyd–Hoare logic. In: *Proceedings of the 17th IEEE Symposium on Foundations of Computer Science*, pp. 109–121, 1976.

[17] H. Rasiowa. On generalized Post algebras of order ω+ and ω+- valued predicate calculi. *Bulletin od the PAS, Ser. Math.*, **21**, 209–219, 1973.

[18] H. Rasiowa, R. Sikorski. *The Mathematics of Metamathematics*, Polish Science Publishers, Warsaw, 1963.

[19] A. Tarski. On the calculus of relations. *Journal of Symbolic Logic*, **6**, 73–89, 1941.
[20] A. Tarski. Contributions to the theory of models. *Indagationes Mathematicae*, **17**, 56–64, 1955.
[21] D. Vakarelov. Modal logics for knoweldge representation systems. *Theoretical Computer Science*, **90**, 433–456, 1991.

NUEL BELNAP

THE DISPLAY PROBLEM

A Gentzen calculus has the 'display property' if every antecedent [consequent] constituent can be displayed as the antecedent [consequent] standing alone. It is explained why this property is interesting. The 'display problem' is the problem of designing a calculus with the display property. A perspective is suggested from which the solution of Wansing [19] can easily be seen to be incomparable with that of [4]. The perspective suggests some other solutions, which are briefly surveyed. Additional questions are raised.

1 THE DISPLAY PROPERTY

Display logic is a refinement of the sequent calculuses of Gentzen [11]–which for reasons given in Anderson and Belnap [2] we call 'consecution calculuses'[1]. These calculuses concern proof-theoretical implication statements expressed by Gentzen with the arrow and by Curry [6] with the turnstile '⊢'. Our usage follows Curry. Thus, the consecution

$$X \vdash Y$$

is taken as expressing the fundamental proof-theoretical implicative idea. In Gentzen, always X and sometimes Y can be sequences of constituents separated by commas. We call those in X 'antecedent' constituents, and those in Y 'consequent' constituents. Later this simple situation will be complicated by the presence of nested structuring, and one should then adjust the terminology so that e.g. antecedent constituents of antecedent constituents are themselves consequent constituents. In thinking this through it is sometimes helpful mentally to replace 'antecedent/consequent' by 'negative/positive'.

Central to display logic is the 'display property'. The property is in a way technical and accordingly difficult to state with a combination of precision and generality. We settle for an informal statement.

[1]There is no hope of exporting our usage; but it is so clear and understandable that there seems no reason why we ourselves should not consistently use it. Besides, there ought to be a controversy concerning whether 'consecution' contains 'cut' as a well-formed part.

H. Wansing (ed.), Proof Theory of Modal Logic, 79–92.

THE DISPLAY PROPERTY

> The idea is that if a constituent appears structured as a 'negative' or 'antecedent' constituent of a proof-theoretical statement, then there is a tightly equivalent way of expressing exactly the same thing in which that constituent is 'displayed' as the antecedent *standing alone*. And similarly for each constituent that appears structured as a 'positive' or 'consequent' constituent. 'Tightly equivalent' is to mean at least 'interderivable by the cut-free rules', not merely 'equi-provable using cut'.[2]

The systems of Gentzen do not have the display property. For example, start with

$$A, B \vdash C,$$

where A, B, and C are formulas in Gentzen's formulation of (even) classical logic. The display property asks for a 'tightly equivalent' consecution having the form

$$A \vdash X.$$

The natural candidate is

$$A \vdash \sim B, C.$$

That consecution is certainly equiprovable with the help of cut, but with equal certainty is not interderivable by means of the cut-free rules. In particular, the negation rules are not reversible–on pain of losing the subformula property.

The display property has technical point. In particular, the display property enables an 'essentials-only' proof of cut elimination relying on easily established and maximally general properties of structural and connective rules. Display logic is generally useful for becoming clearer on substructural logics just because–within wide and easily understood limits–proof of cut elimination for a display logic does not depend on which structural rules are postulated. In this way, display logics provide the only known cut-free formulations of several rich relevance logics (ones including boolean connectives and both positive and negative relevance connectives) and of the brouwerische modal logic. Display logic formulations of S5 are the only cut-free formulations that don't appear cooked up specially just to be cut-free. There have been display formulations of many other logics. For example, earlier proofs of cut elimination for linear logic relied on the associativity and commutativity of the underlying conjunction-like operation. The display-logic formulation demonstrates that this reliance is entirely unnecessary and beside the point: In

[2] In order to appreciate the property, one must be able to identify constituents (occurrences) as 'the same' in the course of transforming one consecution into another. Mere shape-alikeness is not to the point. For an exact statement of the property, see the account of the Display theorem in [4].

spite of what one might gather from standard proofs, associativity and commutativity have nothing whatever to do with the fact that linear logic has a cut-free formulation. In this and other cases, formulating a logic as a display logic, often in more than one way, has shed light in dark areas. Present space limitations require that for detailed illustrations we refer the reader to the following: [4, 5, 19, 12, 13, 14]. On the other hand, display logic has not assisted in tasks such as finding decision procedures. For a helpful appraisal, see [17].

The display property is not merely technical. It has a conceptual point. This point depends on an idea of Gentzen that we call 'the sufficiency idea':

THE SUFFICIENCY IDEA

> One can explain the meaning of a connective, say $\rightarrow$, by means of two sets of rules. (Let us stick to $\rightarrow$, just as an example.) One set gives sufficient conditions for the role of $A \rightarrow B$ as conclusion or consequent, telling us by what $A \rightarrow B$ is implied. The other set gives sufficient conditions for the role of $A \rightarrow B$ as premiss or antecedent, telling us what $A \rightarrow B$ implies. Since these are the only inferential roles that $A \rightarrow B$ can have, the explanation is complete.

The difficulty is that this account of the sufficiency idea is not an exact statement of what Gentzen himself does. In the first place, in Gentzen's consecution calculus for intuitionism, there is a familiar asymmetry between the grammar of where proof-theoretical premisses go–the antecedent X–and the grammar of where proof-theoretical conclusions go–the consequent Y. Briefly put, in $X \vdash Y, X$ can be a sequence of premisses, but Y must be a single conclusion. Gentzen's rules for $A \rightarrow B$ exhibit a companion asymmetry. The rules for $A \rightarrow B$ as conclusion are in fact in accord with the sufficiency idea as stated above, but the rules for $A \rightarrow B$ as premiss are not. It is necessary for Gentzen to state not merely how $A \rightarrow B$ behaves as premiss standing alone. He also must and does give us rules for $A \rightarrow B$ as premiss *in the context of other (conjunctively related) premisses.* He must tell us not only what $A \rightarrow B$ implies, but what $(A \rightarrow B)$-in-context implies. His rules are 'contextual'.

One might have supposed that when we pass to the *symmetrical* consecution calculus that Gentzen introduces for classical logic, the difficulty would be removed. Quite to the contrary. In Gentzen's system for classical logic, Y is no longer a single conclusion. The consequent, too, is a sequence. In this circumstance Gentzen must and does give us rules for $A \rightarrow B$ as conclusion *in the context of other (disjunctively related) conclusions.* Now *all* his rules are contextual.

The nub is this. If a rule for $\rightarrow$ only shows how $A \rightarrow B$ behaves *in context,* then that rule is not *merely* explaining the meaning of $\rightarrow$. It is also and inextricably explaining the meaning of the context. Suppose we give sufficient conditions for

$$A \rightarrow B, X \vdash Y$$

in part by the rule

$$\frac{X \vdash A \qquad B \vdash Y}{A \to B, X \vdash Y}$$

Then we are not explaining $A \to B$ alone. We are simultaneously involving the comma not just in our explicans (that would surely be all right), but in our explicandum. We are explaining two things at once. There is no way around this. You do not have to take it as defect, but it is a fact. It is as if you explained subtraction, '$b - a$', only in the context of addition, '$(b - a) + x$', so that you explained two operators at once. If you are a 'holist', probably you will not care; but then there is not much about which holists much care.

Proof-theorists tend not to be holists beyond necessity; proof-theorists prefer to articulate their proof theories so that they keep the elements as separate as possible. That's where the display property comes in. If a proof-theoretical system has the display property, one is not forced to explain the properties of the connectives simultaneously with the properties of the context. You can first explain how the contextual elements work (structural rules). Then, in a separate explanatory act, you can explain the role of e.g. $A \to B$ standing alone as antecedent, and standing alone as consequent. Finally, you can observe that the display property now guarantees that what you have done suffices to explain how $A \to B$ works in every context. In other words, it suffices to state with plain words what $A \to B$ implies, and by what it is implied, exactly in accordance with the sufficiency idea. This procedure is certainly not 'essential; but it is good.

2 THE DISPLAY SCHEMES **A** AND **P**

Two plans for securing the display property have been suggested. One comes from Wansing [19]; thinking of Amsterdam, we shall call it 'the **A** scheme'. The other comes from [4]; thinking of Pittsburgh, we shall call it 'the **P** scheme'. It appears that neither of these schemes is exactly a generalization of the other. We offer a perspective from which one can see how one might arrive at each of these two schemes.

2.1 *The grammar of the two schemes* **A** *and* **P**

The grammar of the two schemes is largely the same. Here are the key ideas.

- One key idea, due to [18], is that Gentzen's polyvalent comma shall be replaced by a strictly binary structure-connective, which we write $X \circ Y$. This replacement, while not essential, enormously simplifies all of our thinking about structural contexts, and in particular our thinking about the display property.

  Furthermore, just as Gentzen used his comma as having one significance on the left and another on the right (systematic ambiguity), we shall do the same.

Just as in Gentzen, one thinks of $\circ$ as signifying (something like) conjunction on the left, but (something like) disjunction on the right. These two structure connectives will be important for the matter of this paper.

- Another key idea is that there shall be a structure connective permitting one to flip from one side of the turnstile to the other–and back again. We use $*X$.[3] One may think of $*$ as representing (something like) negation. Accordingly if $*X$ is an antecedent [consequent] constituent, then X is a consequent [antecedent] constituent. The star structure connective will figure heavily in this paper.

  So far work on display logic has assumed that $*$ is *not* systematically ambiguous, but that assumption is not necessary. That is, freedom from context sensitivity is not essential for the fundamental properties of negative structuring that lead to the display property. That point is not, however, relevant to our present work.

- A further idea is that there shall be zero-place structure connectives, **I**, that do the work that Gentzen did with the empty symbol. As in Gentzen, they will be context sensitive (sysematically ambiguous). They represent propositional constants that are either truth-like or falsity-like, just as in Gentzen. The structure-connectives, **I**, are not part of the topic of this paper.

- There is no good reason for proof theory to restrict structural ideas to those on someone else's list. Wansing [19] adds a powerful one-place structural idea $\bullet X$ that helps enormously in perspicuously representing various modal logics. Like **I**, this idea is not part of our immediate concerns.

- Display logic in general permits the co-existence of many 'families' of structure-connectives of the kinds just described. This key idea, first exploited–but without the display property–independently by Dunn and by Minc (see [3, §61.1]), is also not a concern here.

In short, this paper is restricted to considering properties of the structural elements $X \circ Y$ and $*X$, although one may imagine the presence of other structure-connectives such as **I** or $\bullet$ or even structure-connectives from other families. We use X, Y, and Z to range over structures built from formulas by structure-connectives, and we reserve A and B for formulas.

2.2 *The* **A** *and* **P** *schemes defined*

We call the principles that secure the display property the 'display equivalences'. There is no reason to suppose that only one set will do. In fact the **A** scheme of [19] employs different display equivalences than those of the **P** scheme of [4]. Neither

[3] We follow [17] in using '$*X$' in place of the '$X*$' used in earlier papers.

scheme appears strictly more general than the other, nor is it plain exactly how they are related. What is offered here is a perspective for viewing the two schemes along side of each other. We begin by exhibiting the display equivalences postulated by each of the two schemes.

THE **P** SCHEME

The **P** scheme of [4] defines display equivalence as the smallest equivalence relation that makes equivalent all consecutions listed on the same line below.

$X \circ Y \vdash Z$	$X \vdash *Y \circ Z$	
$X \vdash Y \circ Z$	$X \circ *Y \vdash Z$	$X \vdash Z \circ Y$
$X \vdash Y$	$*Y \vdash *X$	$**X \vdash Y$

THE **A** SCHEME

The **A** scheme of [19] defines display equivalence, in so far as it concerns $\circ$ and $*$, as the smallest equivalence relation that makes equivalent all consecutions listed on the same line below.

$X \circ Y \vdash Z$	$X \vdash Z \circ *Y$	$Y \vdash *X \circ Z$
$X \vdash Y \circ Z$	$X \circ *Z \vdash Y$	$*Y \circ X \vdash Z$
$X \vdash Y$	$*Y \vdash *X$	$X \vdash **Y$

2.3 *Star in the two schemes*

The two schemes treat the negative structuring $*X$ in the same way. Each yields what amounts to full contraposition and double negation as structural principles. That is, consecutions on the same line below are display-equivalent each to each:

STAR DISPLAY EQUIVALENCES

$X \vdash Y$	$*Y \vdash *X$	$**X \vdash Y$	$X \vdash **Y$
$X \vdash *Y$	$Y \vdash *X$		
$*X \vdash Y$	$*Y \vdash X$		
$...X...$	$...**X...$		

The last line is supposed to convey a general intersubstitutability property. This general property follows from the first line of equivalences only in the presence of other display equivalences–different in the two schemes–sufficient to guarantee the display property.

There is no claim that these star display equivalences are *required* to be present in any sane scheme; it is just a fact that the **A** and **P** schemes share these principles. We hold them fixed in our further comparison of the two schemes and in our wondering about additional schemes.

3 HOW THE A AND P SCHEMES DIFFER

Where the **A** and **P** schemes differ is with respect to the *binary* structure connectives. This is not quite easy to see from the way in which each of the schemes was originally defined. We shall offer one good way of seeing the underlying principles. The idea is to envision each scheme as a different way of realizing the display property for the binary structure connectives. The remainder of the paper follows out this idea.

Consider the display-equivalent consecutions

$$X \circ Y \vdash *Z \qquad Z \vdash *(X \circ Y)$$

Obviously these two consecutions are equivalent solely by an agreed star principle (a form of contraposition), without consideration of binary structure connectives. Furthermore, because of double-star (double-negation) principles that we have agreed to hold fixed, there is no loss of generality in affixing a star to Z. Doing so has the virtue that all of X, Y, and Z are *antecedent* constituents of the given consecutions, so that we don't have to worry so much about redundant double entry bookkeeping in terms of antecedent and consequent constituents. Throughout the rest of the paper, we shall consistently exhibit consecutions in the fashion above, i.e. in 'contraposition-pairs' and with all of X, Y, and Z as antecedent constituents. It seems to help the eye.

Evidently in the above consecutions it is Z that is in effect displayed (as antecedent), while X and Y are grouped together in a certain way. This is critical relative to the display property. There are three other pairs of consecutions that display Z in this sense, while simultaneously grouping X and Y in one way or another. Repeating the above pair, this gives us altogether four contraposition-pairs, which we think of as a 'family'. We name it '$Z(XY)$' in order to remind us that Z is in effect set off by itself (displayed), whereas X and Y are grouped together.

$Z(XY)$ FAMILY

(1) $X \circ Y \vdash *Z$	$Z \vdash *(X \circ Y)$
(2) $Y \circ X \vdash *Z$	$Z \vdash *(Y \circ X)$
(3) $*(*X \circ *Y) \vdash *Z$	$Z \vdash *X \circ *Y$
(4) $*(*Y \circ *X) \vdash *Z$	$Z \vdash *Y \circ *X$

Then there are four more pairs in which X is displayed:

$X(YZ)$ FAMILY

(5) $X \vdash *Y \circ *Z$	$*\,(*Y \circ *Z) \vdash *X$
(6) $X \vdash *Z \circ *Y$	$*\,(*Z \circ *Y) \vdash *X$
(7) $X \vdash *(Y \circ Z)$	$Y \circ Z \vdash *X$
(8) $X \vdash *(Z \circ Y)$	$Z \circ Y \vdash *X$

And lastly, there are four pairs in which Y is displayed:
$Y(XZ)$ FAMILY

(9) $Y \vdash *X \circ *Z$	$*(*X \circ *Z) \vdash *Y$
(10) $Y \vdash *Z \circ *X$	$*(*Z \circ *X) \vdash *Y$
(11) $Y \vdash *(X \circ Z)$	$X \circ Z \vdash *Y$
(12) $Y \vdash *(Z \circ X)$	$Z \circ X \vdash *Y$

So there are twelve pairs in all, four pairs in each of three families. Given that X, Y, and Z are all taken to be *antecedent* consitituents, these possibilities are exhaustive. We summarize for easy reference.

THE FAMILIES SUMMARIZED

$Z(XY)$ family: (1), (2), (3), (4).
$X(YZ)$ family: (5), (6), (7), (8).
$Y(XZ)$ family: (9), (10), (11), (12).

THE DISPLAY PROBLEM

> The display property now sets as a problem that for each entry in each family, there must be at least one member of each of the other two families to which it is display equivalent.

Were it not so, the display property would fail. One may say that this problem–the *'display problem'*–is solved in different ways by the **A** and **P** schemes. In order better to appreciate their differences, we note in passing that there is an 'easy' way to solve the display problem: just declare all of (1)-(12) as display-equivalent. We call this the 'Easy scheme':

EASY SCHEME

> (1)-(12) are all taken as equivalent.

The Easy scheme is strong. By making (1)-(4) equivalent, it renders the operation $X \circ Y$ commutative wherever it occurs, and also identifies $X \circ Y$ with $*(*X \circ *Y)$ wherever it occurs. The Easy scheme nevertheless does not trivialize display logic. For example, unlike all standard Gentzen consecution calculuses–even those for e.g. linear logic that rely on multisets or on a polyvalent comma–the Easy scheme does *not* imply any associativity-type principle such as an equivalence between $W \circ (X \circ Y) \vdash Z$ and $(W \circ X) \circ Y \vdash Z$. But aside from that, we really know very little about the Easy scheme. We shall be using it just as a foil.

Let us turn now to the **P** and **A** schemes. Adjusting for relettering and double stars, the **P** scheme takes as fundamental the equivalence between (1), (5), and (6). It then 'notices' that this gives us an equivalence to (11), thus solving the display problem. Following out this line, we can see that on the **P** scheme, the twelve consecutions (or pairs of consecutions–we won't worry about this nicety) break down into three equivalence classes, where for each class, all the consecutions in the class are display-equivalent by the rules of the **P** scheme:

THE **P** SCHEME AS SOLVING THE DISPLAY PROBLEM

Equivalence class P_1. (1), (5), (6), (11).

Equivalence class P_2. (2), (7), (9), (10).

Equivalence class P_3. (3), (4), (8), (12).

Observe the following.

1. You can see by inspection that each class P_i contains a member from each of the three families. The **P** scheme therefore solves the display problem. In fact that is what we *mean* by 'solving the display problem'.

2. It is also easy to see that the equivalences in each class are just reletterings of the equivalences in each other class. For example, for the **P** scheme it suffices to postulate the equivalence of the members of Class P_1.

3. You can see that the **P** scheme is maximal relative to the Easy scheme: Adding any equivalence not already present generates the Easy scheme, i.e., the universal equivalence on (1)-(12). For example, making any consecution from class P_1 equivalent to any consecution from class P_2 would by transitivity make (1) equivalent to (2); which would then by relettering force the equivalence of (11) from P_1 with (12) from P_3, and thereby make all of (1)-(12) equivalent.

4. You can also see that Class P_1 contains not one but *two* members of the $X(YZ)$ family; and analogously for the other two P_i classes. In this respect the **P** scheme goes beyond the requirements set by the display problem. That's interesting.

The **A** scheme from [19] is different. Adjusting for relettering and double stars, the **A** scheme takes as fundamental the equivalence of (1), (6), and (9). By relettering and double-star principles, we can see that on that scheme, there are *two* equivalence classes rather than three:

THE **A** SCHEME AS SOLVING THE DISPLAY PROBLEM

Equivalence class A_1. (1), (4), (6), (7), (9), (12).

Equivalence class A_2. (2), (3), (5), (8), (10), (11).

Observe the following.

1. Evidently the **A** scheme contains at least one member from each family. The **A** scheme therefore solves the display problem.

2. Also the equivalences in each class are just reletterings of the equivalences in the other class. For the **A** scheme, it suffices to take e.g. the equivalences of the class A_1 as postulates.

3. You can see that the **A** scheme is maximal relative to the Easy scheme: Adding any equivalence not already present obviously generates the universal equivalence on (1)-(12).

4. You can also see that each class A_j contains exactly *two* members from each of the three families. In this respect the **A** scheme goes beyond the requirements set by the display problem–and in an elegantly symmetric fashion.

It is easy to see that the **A** and **P** schemes are incomparable, neither being a refinement of the other. For example, the **P** scheme identifies (1) and (5), whereas the **A** scheme does not; and the **A** scheme identifies (1) and (4), whereas the **P** scheme does not. Neither is, it would seem, more general than the other. Combining the schemes would force the equivalence of all twelve pairs–that is, would force the Easy scheme.[4]

In one respect, however, the schemes are different. The **P** scheme *relies* on the identification of (5) and (6), (9) and (10), and (3) and (4). The **A** scheme, in contrast, could, if it wished, treat its extra identifications, such as that between (1) and (4), as 'accidental'. You can see the truth of this two-part observation in the following way.

Start with Equivalence class P_1, and choose one of its entries from each family, say (1), (5), and (11). You need at least this much of P_1 in order to solve the display problem (though you could have solved it by picking (6) instead of (5).) Now since (11) also has the form of (1) (relettered), you can easily calculate that we are *forced* by transitivity to add (6) to the equivalence class generated by $\{(1), (5), (11)\}$; no choice. There is no way to avoid the identification of (5) and (6), given that (1), (5), and (11) are put in the same class.

The upshot is that there is no (proper) refinement of the **P** *scheme that solves the display problem.* Hence if, like the writer, one starts with the **P** scheme, one can wrongly be led to suppose that there is something 'unique' about the solution to the display problem that one has found.

In contrast, however, it is easy to find a refinement of the **A** *scheme, that is, a scheme that is more general (forces fewer provable equivalences) while still solving the display problem.* This is what we shall do in the final section of this paper.

---

[4] It is good to bear this in mind when examining how each scheme enriches itself by the addition of further structural postulates; otherwise, confusion is all too easy by which to come.

4 OTHER SOLUTIONS TO THE DISPLAY PROBLEM

The observation that leads immediately to a refinement of the **A** scheme is this. If we start with (1), (7), and (12) as equivalent, no transitivities ever force us to add to the class. In the same way, the entire following 'generalized **A** scheme'–we call it 'the **GA** scheme'–can be justified:

THE **GA** SCHEME

Equivalence class GA_{1a}. (1), (7), (12).

Equivalence class GA_{1b}. (4), (6), (9).

Equivalence class GA_{2a}. (2), (8), (11).

Equivalence class GA_{2b}. (3), (5), (10).

This scheme is minimal with respect to the display property: Each equivalence class contains exactly one member from each of the families. In contrast with the **P** and the **A** schemes, no 'unnecessary' identifications are made. One should not, however, infer too much: it remains true that the **GA** scheme is *not* a generalization of the **P** scheme; the **GA** scheme remains incomparable with the **P** scheme. As we indicated, the **P** scheme has *no* generalization (refinement) that solves the display problem. In other words, the **GA** scheme is minimal among those solving the display problem, and so is the **P** scheme.

Also striking is the following. Class GA_{2a} is a relettering of class GA_{1a}, and Class GA_{2b} is a relettering of Class GA_{1b}. But the equivalences of GA_{1a} leave those of GA_{1b} entirely undetermined. Having postulated GA_{1a}, you still need independently to postulate GA_{1b}, if you want the display property.

There is, however, one maximally elegant way in which one can have the **A** scheme without moving beyond the postulation of (only) the equivalences GA_{1a}. The idea is this.[5] In display logic $X \circ Y$ is context sensitive, having one meaning or the other depending on whether it occurs in antecedent position or in consequent position. In the **P** scheme this feature is taken as an excuse to take the two ideas ($X \circ Y$ in antecedent position, and $X \circ Y$ in consequent position) as each primitive. But given the **GA** scheme, it also makes perfectly good sense to take only one of them, say $X \circ Y$ in antecedent position, as primitive, while *defining* $X \circ Y$ in consequent position. In particular, let us define as follows:

DEFINITION 1 (**A**) *$X \circ Y$ in consequent position shall by definition be* $*(*Y \circ *X)$.

[5] Conversations with Dunn on his work in algebraic logic directly prompted this observation. Much more to the point, it seems inevitable that the separate and joint algebraic work of Dunn and Allwein and Hartonas listed below in the references should turn out to be deeply relevant to the development of logics with the display property–and perhaps vice versa. Perhaps, for instance, the Dunn-Allwein-Hartonas research can supply an enlightening algebraic or relational semantics for display logic. Certainly the ideas of residuation and Galois connection, so fundamental to the algebraic investigations, appear 'very like' the display property. Nothing, however, is definitely known.

Definition **A** appears circular. It is, however, not circular, because the operation $\circ$ in its definiendum occurs in *antecedent* position. Thus, the definition instructs us to trade in every occurrence of $\circ$ in consequent position for an occurrence of $\circ$ in antecedent position.

If we do this, we shall find that we obtain *for free*–without further postulation– the **A**-scheme equivalences that are between members of the same family, e.g., the **A**-scheme equivalence of (1) and (4), which are both in the $Z(XY)$ family. We obtain this equivalence, for example, just by Definition 1. This seems an especially beautiful feature of the **A** scheme. It follows that the **A** scheme can reduce the number of connectives that it needs to think of as primitive.

The **GA** scheme is a minimal solution to the display problem (but do not say '*the* minimal solution'!). It is therefore good to ask for its other coarsenings. Evidently we obtain another if we replace Definition **A** by Definition **A**$'$:

DEFINITION 2 (**A**$'$) *$X \circ Y$ in consequent position shall by definition be* $*(*X \circ *Y)$.

This gives us Scheme **A**$'$:

SCHEME **A**$'$

Equivalence Class A'_1. (1), (3), (5), (7), (10), (12).

Equivalence Class A'_2. (2), (4), (6), (8), (9), (11).

The **A**$'$ scheme is properly named since it is evidently but a notational variant of the **A** scheme. In fact one could obtain the **A**$'$ scheme by starting with the **A** scheme, and then simply defining a new operation $\circ'$ in consequent position by $X \circ' Y = Y \circ X$. Big deal.

There is, however, another (technically) nontrivial coarsening of the **GA** scheme. The **A** scheme resulted from combining GA_{1a} with GA_{1b}. The **A**$'$ scheme arose from combining GA_{1a} with GA_{2b}. One could instead decide to combine GA_{1a} and GA_{2a}. Then *independently* one could decide to combine GA_{1b} and GA_{2b}. We cannot guess, however, if these possibilities might lead to usefully different ways of constructing display logics. In a similar vein, the relations between the **P** scheme and the **A** scheme remain unexplored, partly because of not being clear on the best questions to ask. Perhaps it would help to have a complete survey of all ways of solving the display problem.

5 SUMMARY

Because it is easy to lose track of where we are, we conclude with an epistemic summary. So as not to finish in professed ignorance, we put what we don't know before what we do.

5.1 *Summary of what we don't know*

1. Metaquestion: What are good questions to ask about the relations between the **P** and the **A** schemes? Surely the two schemes don't just sit there. Are they good for different things, or can either do the work of the other, or ...?

2. We do not have a survey of the lattice of all ways of solving the display problem.

3. Is the extra generality provided by the **GA** scheme of use in constructing some particular display logic?

4. Is there an application for the logic defined by the Easy scheme?

5. Nearly every logic combines a variety of *formula*-connectives. Display logic analogously invites combining different families of *structure*-connectives into a single logic. For example, the display version of relevance logic in [4] involves two families of structure-connectives, one for the relevance connectives and one for the boolean connectives. As a second example, somewhat against the conventional wisdom, display logic permits intelligible combination of intuitionist logic and classical logic by allotting each a different family of structure-connectives. Heretofore, however, all of the actually-constructed examples using multiple families have postulated the same display-equivalences for each family, and have been differentiated only via structural rules that are *not* display equivalences. But mathematically and conceptually there is no necessity to maintain the same display equivalences for all the combined families. Thus we may wonder if anything of interest emerges when one combines two families that are based on different schemes. For example, is there an interesting logic that combines a family based on the **A** scheme with a family based on the **P** scheme, all in the same display logic?

6. One hopes that there is an algebraic meaning to all of this, but if so its nature remains to be determined. It may well be available through the work of Dunn and Allwein and Hartonas as listed in the references. And vice versa.

5.2 *Summary of what we know*

1. The **A** and **P** schemes each solve the display problem.

2. The **A** and **P** schemes are incomparable in the sense that each identifies consecutions kept separate by the other.

3. The **A** and **P** schemes are each maximal in the sense that adding any new equivalence to either yields the Easy scheme (the universal equivalence among (1)-(12)). Consequently they are unjoinable if one wishes not to make proof theory Too Easy.

4. The **P** scheme admits of no proper refinement that solves the display problem.

5. The **A** scheme admits of a proper refinement, the **GA** scheme, that (also) solves the display problem.

6. One may obtain the **A** scheme from the **GA** scheme by definition.

University of Pittsburgh, USA

REFERENCES

[1] G. Allwein, C. Hartonas. Duality for bounded lattices. Unpublished manuscript, Indiana University Logic Group Preprint Series, 1993.

[2] A. R. Anderson, N. D. Belnap. *Entailment: the logic of relevance and necessity*. Vol. I, Princeton University Press, Princeton, 1975.

[3] A. R. Anderson, N. D. Belnap, J. M. Dunn. *Entailment: the logic of relevance and necessity*. Vol. II, Princeton University Press, Princeton, 1992.

[4] N. Belnap. Display logic. *Journal of Philosophical Logic*, **11**, 375–417, 1982. Republished with minor changes as §62 of [3].

[5] N. Belnap. Linear logic displayed. *Notre Dame Journal of Formal Logic*, **31**, 14–25, 1990.

[6] H. B. Curry. *A theory of formal deducibility*. Notre Dame mathematical lectures, vol. no. 6, Notre Dame University Press, Notre Dame, 1950.

[7] J. M. Dunn. Gaggle theory: an abstraction of Galois connections and residuation with applications to negation and various logical operations. In: J. van Eijck (ed.), *Logics in AI: Proceedings European Workshop JELIA 1990*, pp. 31–51, Lecture Notes in Computer Sci. 478, Springer-Verlag, Berlin, 1990.

[8] J. M. Dunn. Partial-gaggles applied to sub-structural logics. In: P. Schroeder-Heister, K. Dosen (eds.), *Substructural logics*, pp. 63–108, Oxford University Press, Oxford, 1993.

[9] J. M. Dunn. A representation of relation algebras using Routley-Meyer frames. Unpublished manuscript, Indiana University Logic Group Preprint Series, 1993.

[10] J. M. Dunn. Perp and Star: two treatments of negation. In: J. Tomberlin (ed.), *Philosophical Perspectives*, vol. 7, *Language and Logic*, pp. 331–157, 1993.

[11] G. Gentzen. Untersuchungen über das logische Schliessen. *Mathematische Zeitschrift*, **39**, 176–210, 405–431, 1934. English translation (Investigations into logical deduction) by M. E. Szabo in: *The Collected Papers of Gerhard Gentzen*, M. E. Szabo (ed.), pp. 69–131, North-Holland, Amsterdam, 1969.

[12] R. Goré. Displayed intuitionistic modal logics. Technical report in preparation, Australian National University, April, 1994.

[13] R. Goré. Displaying sub-intuitionistic logics. Technical report in preparation, Australian National University, April 1994b.

[14] R. Goré. Intuitionistic logic redisplayed. Technical report, TR-ARP-1-1995, Australian National University, January 1995.

[15] C. Hartonas. Lattices with operators: a unified semantics for substructural logics. Unpublished manuscript, Indiana University Logic Group Preprint Series, 1993.

[16] C. Hartonas, J. M. Dunn. Duality theorems for partial orders, semilattices, and Galois connections and lattices. Unpublished manuscript, Indiana University Logic Group Preprint Series, 1993.

[17] M. Kracht. *Power and weakness of the modal display calculus*. Technical report, Free University of Berlin, Mathematics Institute II, August, 1993.

[18] R. K. Meyer. A general Gentzen system for implicational calculi. *Relevance Logic Newsletter*, **1**, 198–201, 1976.

[19] H. Wansing. *Sequent calculi for normal modal propositional logics*. Technical report, n. LP-92-12, University of Amsterdam, 1992, *Journal of Logic and Computation*, **4**, 124–142, 1994.

MARCUS KRACHT

POWER AND WEAKNESS OF THE MODAL DISPLAY CALCULUS

1 INTRODUCTION

The present paper explores applications of *Display Logic* as defined in [1] to modal logic. Acquaintance with that paper is presupposed, although we will give all necessary definitions. Display Logic is a rather elegant proof-theoretic system that was developed to explore in depth the possibility of total Gentzenization of various propositional logics. By *Gentzenization* I understand the strategy to replace connectives by structures. Gentzenization is something of an ingenious optical trick because it uses a single symbol to mean different things depending on the place it occupies in the sequent. In the original Gentzen system it was the *comma* that had to be interpreted as *and* when to the left of the turnstile and as *or* when to the right. The interpretation of the structures oscillates between two logical symbols depending on whether it is in the antecedent or in the consequent. This is why we call symbols like comma *Gentzen toggles*. These two symbols between which this toggle switches are the *Gentzen duals* of each other. So, *and* and *or* are *Gentzen duals*. The strength of Display logic lies in a rather general cut-elimination theorem. In [10] and [9], Heinrich Wansing has refined these methods for modal logics; he showed that contrary to Belnap's own Gentzenization of modal operators as binary structure operators, a unary one is more appropriate (not only from an esthetical point of view) and makes perfect sense semantically as well. The Gentzen dual of the modal operator $\Box$ is actually not – as one might expect – the possibility operator $\Diamond$, but the backward looking possibility operator, denoted here by $\diamondminus$. (To be consistent with that we write $\boxminus$ instead of $\Box$ and $\diamondminus$ instead of $\Diamond$.) The corresponding toggle is denoted by $\bullet$. The reason why this is so natural lies in the fact that it is the exact Display or Gentzen dual, for we have that the sequent $\bullet B \vdash A$ and the sequent $B \vdash \bullet A$ are equivalent if $\bullet$ is read as $\diamondminus$ if in the antecedent and if it is read as $\boxminus$ if in the consequent. Wansing uses this fact to display various modal and tense logics à la Belnap by providing some formula introduction rules and basic structural rules for **K** and **Kt** and then Gentzenizing the additional axioms. The benefit lies not only in the homogeneity with which all these systems are now handled and the rather clear intuitive background. The benefit lies in the possibility to use the general cut-elimination theorem of [1].

H. Wansing (ed.), Proof Theory of Modal Logic, 93–121.

During the summer of 1993 Rajeev Goré, Frank Wolter and myself have been intrigued by the possibility that Display Logic could be the key to rather simple decidability proofs via cut-elimination and some refined tricks of pushing around decidability. After successful proofs of the fact that all displayable logics are decidable – which we knew was wrong – and a subsequent investigation into the possibility of having proved the inconsistency of arithmetic we found that one crucial lemma was flawed. The headache was soon to follow. Not only were the theorems on decidability false, it is actually undecidable whether a display calculus for modal logics is decidable.

The negative results are now assembled in [5]. The present paper contains pretty much those parts of the original paper that have remained untouched by the disaster. I wish to thank Rajeev Goré for bringing my attention to Display Logic and his insistance that it is worth its while. Thanks for endlessly discussing this topic with me. Thanks also to Frank Wolter, Heinrich Wansing and Greg Restall for their criticism and to an anonymous referee for pointing out a number of mistakes in the first version.

2 THE BASIC CALCULUS

Below we outline the whole calculus for the basic tense logic **Kt**. This outline is a blend of [1], [10] and some own ideas. **Kt** is a special bimodal logic in which we have two pairs of modal operators, namely ⍈ and ⟢, as well as ⍇ together with ⟣. The pairs look in opposite directions of the basic relation of the Kripke frame. We assume ⍈ and ⟢ to look *with* the relation denoted here by $\lhd$ and ⍇ and ⟣ to look in direction of its converse $\rhd = \lhd^{\smile}$. The display calculus works with a special set of extra operators used to Gentzenize logical symbols into structures. These operators are **I**, $\circ$, $*$ and $\bullet$. Given a set $\mathcal{L}$ of formulas we write $\mathfrak{Struc}(\mathcal{L})$ for the algebra of *structures* over $\mathcal{L}$. $\mathfrak{Struc}(\mathcal{L})$ is actually nothing but the term algebra over $\mathcal{L}$ with the operators **I**, $*$, $\bullet$ and $\circ$. We distinguish *formulas* from *structures* from *sequents*. A sequent is of the form $X \vdash Y$ where both X and Y are structures. X is called the **antecedent** and Y the **succedent**. Succedents are called *consequents* in [1] and instances of rules are called *consecutions*. Formulas are denoted by upper case letters such as P, Q, R and structures by X, Y, Z. Sequents are pairs $X \vdash Y$ where X and Y are structures, and sequents are denoted by lower case letters such as s, t. We begin with the following fundamental *logical axioms* and *rules*:

$$(Id)\ \ p \vdash p \qquad (Cut)\ \frac{X \vdash P \quad P \vdash Y}{X \vdash Y}$$

and the following basic structural rules:

$$\frac{X \circ Y \vdash Z}{\overline{X \vdash Z \circ *Y}} \qquad \frac{X \circ Y \vdash Z}{\overline{Y \vdash *X \circ Z}}$$

$$\frac{X \vdash Y \circ Z}{\overline{X \circ *Z \vdash Y}} \qquad \frac{X \vdash Y \circ Z}{\overline{*Y \circ X \vdash Z}}$$

$$\frac{*X \vdash Y}{\overline{*Y \vdash X}} \qquad \frac{X \vdash *Y}{\overline{Y \vdash *X}}$$

$$\frac{**X \vdash Y}{\overline{X \vdash Y}} \qquad \frac{X \vdash **Y}{\overline{X \vdash Y}}$$

$$\frac{X \vdash \bullet Y}{\overline{\bullet X \vdash Y}}$$

Notice that contrary to the established notation we write $*$ as a prefix, which makes formulae much easier to read. The following rules are now derivable:

$$\frac{X \vdash Y}{\overline{*Y \vdash *X}}$$

These rules alone suffice to prove the following theorem, whose content gave rise to the name *display logic*.

THEOREM 1 (Display Theorem) *For every sequent s and every antecedent (succedent) part X of s there is a sequent s' structurally equivalent to s such that X is the antecedent (succedent) of s'.*

The correctness of this theorem depends on a proper definition of the terms *antecedent part* and *succedent part*. This is defined here via *positive occurrence in*, where an occurrence is **positive** if it is nested by an even number of $*$. Namely, if P, Q are formulae then $*P \vdash *Q$ is a sequent but neither P nor Q can be displayed in their original position. However, we can easily derive $Q \vdash P$, displaying both P and Q on the other side. So if $s = V \vdash W$ is a sequent and X occurred negatively in V, it will then occur positively in $*V$ and so it can be displayed in $*W \vdash *V$. The theorem is correct with the following definition as given in [1].

DEFINITION 2 *In a sequent $V \vdash W$ an occurrence of X is an* **antecedent part** *if it occurs positively in the antecedent or negatively in the succedent. An* *occurrence that is not an antecedent part is a* **succedent part**. *Equivalently, it is a succedent part if either it occurs positively in the succedent or negatively in the antecedent.*

It is actually possible to compute the sequent s' from s and the occurrence of X. Namely, we can consider s to be of the form $V \vdash W$, where V, say, contains an occurrence of X. This occurrence may be either positive or negative. Then we can think of V as obtained from this occurrence of X by applying only *unary* functions, for example prefixing with $\bullet$ or 'multiplying' from the left with Y, that is, applying the function $X \mapsto Y \circ X$, or multiplying with Y from the right. Each of these functions has a dual function; the function $\bullet : X \mapsto \bullet X$ is it's own dual, $X \mapsto Y \circ X$ has the dual $X \mapsto *Y \circ X$, $X \mapsto X \circ Y$ has the dual $X \mapsto X \circ *Y$. $*$ has no dual in this sense, which makes this story a bit complicated. Without $*$ we do the following. Let $V = f(X)$, where f is a unary polynomial in the termalgebra of structures, and let f^δ be the dual of f. Then we transform $f(X) \vdash W$ into $X \vdash f^\delta(W)$. This transformation is reversible, so from $X \vdash f^\delta(W)$ we can also derive $f(X) \vdash W$. As noted, $*$ creates a problem. To undo $*$ we have to make antecedent and consequent swap sides, so we move from $*X \vdash W$ to $*W \vdash X$. With this proviso, it is justifiable to say that $*$ is self-dual. It is now clear why the Display Theorem has this peculiar restriction concerning the part-of relation, because for X to be displayed on the same side as it occurred in the original sequent we need to perform this swapping an even number of times. The Display Theorem is brought here into the following form, writing $\mathbf{Pol}_1(\mathfrak{Struc})$ for the set of unary structure polynomials.

THEOREM 3 *Let $f \in \mathbf{Pol}_1(\mathfrak{Struc})$ which embeds an occurrence of its argument an even number of times with $*$. Then there exists a $f^\delta \in \mathbf{Pol}_1(\mathfrak{Struc})$ such that:*

$$\frac{f(X) \vdash W}{X \vdash f^\delta(W)}$$

displaying that particular occurrence of X. Moreover, for $f = g^\delta$, $f^\delta = g$ is an appropriate choice.

Let us note that the calculus has an inbuilt symmetry or self duality in the following sense. Define the **dual** $(X \vdash Y)^\Delta$ of a sequent by $(X \vdash Y)^\Delta = Y \vdash X$. The dual Π^Δ of a proof Π consists in the dualization of all sequents.

THEOREM 4 (Symmetry) *For every proof Π of $X \vdash Y$, Π^Δ is a proof of $Y \vdash X$.*

Now we define the following equivalence between structures. X and Y are called **similar**, in symbols $X \approx Y$, if for every Z the following two conditions hold:

$$\frac{X \vdash Z}{\overline{\overline{Y \vdash Z}}} \qquad \frac{Z \vdash X}{\overline{\overline{Z \vdash Y}}}$$

Similarity thus means that X and Y are interchangeable in a proof both as antecedent and succedent modulo some reversible rules. Replacement of Y for X in a given proof does not necessarily yield another proof but it can be made into one by adding some extra steps. The following theorem is a direct consequence of the Display Theorem.

THEOREM 5 *$\approx$ is a congruence on* $\mathfrak{Struc}(\mathcal{L})$.

Proof. Clearly, $\approx$ is an equivalence relation. Moreover, as in our calculi we have full substitutivity, the only thing to be checked is whether $\approx$ satisfies the genuine congruence property that if $\sharp$ is an n-ary function symbol and $X_1 \approx Y_1, \ldots, X_n \approx Y_n$ then also $\sharp(X_1, \ldots, X_n) \approx \sharp(Y_1, \ldots, Y_n)$. It can easily be seen that it is enough to show this for unary polynomials. Let f be such a polynomial, and consider an occurrence of its argument X. Let g be its dual with respect to this occurrence. Two cases have to be distinguished, namely whether or not $f(X)$ embeds X an even number of times. Let the number be odd. Then we deduce from $f(X) \vdash Z$ that $g(Z) \vdash X$ and then $g(Z) \vdash Y$, by assumption that $X \approx Y$, and then $f(Y) \vdash Z$. If f embeds X an even number of times, we deduce $X \vdash g(Z)$ instead, and then $Y \vdash g(Z)$ and finally $f(Y) \vdash Z$. Similarly for X and Y in the succedent. ■

3 THE MODAL DISPLAY CALCULUS

Let us now define the full calculus for the basic modal logic **K** and the basic tense logic **Kt**. It will be called **DLM**. In addition to the rules of the base calculus it has rules to introduce connectives and some more structural rules. The operational rules are the following:

$$(\vdash \top)\ \mathbf{I} \vdash \top \qquad\qquad (\top \vdash)\ \frac{\mathbf{I} \vdash X}{\top \vdash X}$$

$$(\vdash \bot)\ \frac{X \vdash \mathbf{I}}{X \vdash \bot} \qquad\qquad (\bot \vdash)\ \bot \vdash \mathbf{I}$$

$$(\vdash \neg)\ \frac{X \vdash *P}{X \vdash \neg P} \qquad\qquad (\neg \vdash)\ \frac{*P \vdash X}{\neg P \vdash X}$$

$$(\vdash\wedge)\ \frac{X \vdash P \quad Y \vdash Q}{X \circ Y \vdash P \wedge Q} \qquad (\wedge\vdash)\ \frac{P \circ Q \vdash X}{P \wedge Q \vdash X}$$

$$(\vdash\vee)\ \frac{X \vdash P \circ Q}{X \vdash P \vee Q} \qquad (\vee\vdash)\ \frac{P \vdash X \quad Q \vdash Y}{P \vee Q \vdash X \circ Y}$$

$$(\vdash\rightarrow)\ \frac{X \circ P \vdash Q}{X \vdash P \rightarrow Q} \qquad (\rightarrow\vdash)\ \frac{X \vdash P \quad Q \vdash Y}{P \rightarrow Q \vdash *X \circ Y}$$

$$(\vdash \boxright)\ \frac{\bullet X \vdash P}{X \vdash \boxright P} \qquad (\boxright \vdash)\ \frac{P \vdash X}{\boxright P \vdash \bullet X}$$

$$(\vdash \diamondright)\ \frac{X \vdash P}{*\bullet *X \vdash \diamondright P} \qquad (\diamondright \vdash)\ \frac{*\bullet *P \vdash X}{\diamondright P \vdash X}$$

$$(\diamondleft \vdash)\ \frac{P \vdash \bullet X}{\diamondleft P \vdash X} \qquad (\vdash \diamondleft)\ \frac{X \vdash P}{\bullet X \vdash \diamondleft P}$$

$$(\boxleft \vdash)\ \frac{P \vdash X}{\boxleft P \vdash *\bullet *X} \qquad (\vdash \boxleft)\ \frac{X \vdash *\bullet *P}{X \vdash \boxleft P}$$

The calculus for **K** will be obtained by deleting the introduction rules for the connectives $\boxleft$ and $\diamondleft$. Notice the complete duality between $\boxleft$ and $\diamondright$ as well as between $\diamondleft$ and $\boxright$. We can formalize it by extending the duality map as follows:

$$\begin{array}{rcl@{\qquad}rcl}
p^\Delta & = & p & (\neg P)^\Delta & = & \neg(P^\Delta) \\
(P \wedge Q)^\Delta & = & P^\Delta \vee Q^\Delta & (P \vee Q)^\Delta & = & P^\Delta \wedge Q^\Delta \\
(\boxright P)^\Delta & = & \diamondleft(P^\Delta) & (\diamondleft P)^\Delta & = & \boxright(P^\Delta) \\
(\boxleft P)^\Delta & = & \diamondright(P^\Delta) & (\diamondright P)^\Delta & = & \boxleft(P^\Delta) \\
\mathbf{I}^\Delta & = & \mathbf{I} & (*X)^\Delta & = & *(X^\Delta) \\
(X \circ Y)^\Delta & = & X^\Delta \circ Y^\Delta & (\bullet X)^\Delta & = & \bullet(X^\Delta)
\end{array}$$

THEOREM 6 (Duality) *For every proof* Π *of* $X \vdash Y$, Π^Δ *is a proof of* $Y^\Delta \vdash X^\Delta$.

Of course, this time we can only speak of *duality*, not of symmetry. Finally, the following structural rules are added. (If only a part of these rules is added, we have a substructural calculus. Much of what will be proved here applies to substructural logics as well; in a sense, the full calculus is the most difficult case, and this is the reason why we concentrate on this calculus.)

$$(\mathbf{I}l)\ \frac{X \vdash Z}{\mathbf{I} \circ X \vdash Z} \qquad (\mathbf{I}r)\ \frac{X \vdash Z}{X \vdash \mathbf{I} \circ Z}$$

$$(Ql)\ \frac{\mathbf{I} \vdash Y}{*\mathbf{I} \vdash Y} \qquad (Qr)\ \frac{X \vdash \mathbf{I}}{X \vdash *\mathbf{I}}$$

$$(Wl)\ \frac{X \vdash Z}{Y \circ X \vdash Z} \qquad (Wr)\ \frac{X \vdash Z}{X \circ Y \vdash Z}$$

$$(Al)\ \frac{X_1 \circ (X_2 \circ X_3) \vdash Z}{(X_1 \circ X_2) \circ X_3 \vdash Z} \qquad (Ar)\ \frac{Z \vdash X_1 \circ (X_2 \circ X_3)}{Z \vdash (X_1 \circ X_2) \circ X_3}$$

$$(Pl)\ \frac{X \circ Y \vdash Z}{Y \circ X \vdash Z} \qquad (Pr)\ \frac{Z \vdash X \circ Y}{Z \vdash Y \circ X}$$

$$(Cl)\ \frac{X \circ X \vdash Z}{X \vdash Z} \qquad (Cr)\ \frac{Z \vdash X \circ X}{Z \vdash X}$$

$$(Ml)\ \frac{\mathbf{I} \vdash Y}{\bullet\mathbf{I} \vdash Y} \qquad (Mr)\ \frac{X \vdash \mathbf{I}}{X \vdash \bullet\mathbf{I}}$$

This concludes the definition of **DLM**. There are a number of things which have to be explained. First of all, some rules originally proposed in [10] have been left out because they are derivable. On the other hand, a lot of structural rules have been added to the calculus, mainly the duals of existing rules. The one-sided rules are sufficient, but this has the effect of disturbing the duality on the level of proofs (not for provability, since the two-sided rules are derivable from the one-sided rules). The necessitation rules have been changed; they are now fully structural and do not require the use of formula variables. Furthermore, there are now two more necessitation rules derivable with the help of the newly introduced rules (Q), namely

$$(Dl)\ \frac{\mathbf{I} \vdash Y}{* \bullet *\mathbf{I} \vdash Y} \qquad (Dr)\ \frac{X \vdash \mathbf{I}}{X \vdash * \bullet *\mathbf{I}}$$

For example, the first is derived as follows:

$$\cfrac{\cfrac{\cfrac{\cfrac{\cfrac{\cfrac{\cfrac{\cfrac{X \vdash \mathbf{I}}{X \vdash * * \mathbf{I}}}{*\mathbf{I} \vdash *X}}{\mathbf{I} \vdash *X}}{\bullet\mathbf{I} \vdash *X}}{\mathbf{I} \vdash \bullet * X}}{*\mathbf{I} \vdash \bullet * X}}{\bullet * \mathbf{I} \vdash *X}}{X \vdash * \bullet *\mathbf{I}}$$

Define the following translation for sequents:

$$\tau(X \vdash Y) = \tau_1(X) \to \tau_2(Y)$$

$$\begin{array}{lcl@{\qquad}lcl}
\tau_1(P) & = & P & \tau_2(P) & = & P \\
\tau_1(\mathbf{I}) & = & \top & \tau_2(\mathbf{I}) & = & \bot \\
\tau_1(*X) & = & \neg\tau_2(X) & \tau_2(*X) & = & \neg\tau_1(X) \\
\tau_1(X \circ Y) & = & \tau_1(X) \wedge \tau_1(Y) & \tau_2(X \circ Y) & = & \tau_2(X) \vee \tau_2(Y) \\
\tau_1(\bullet X) & = & \Diamond\tau_1(X) & \tau_2(\bullet X) & = & \boxminus\tau_2(X)
\end{array}$$

With respect to this translation we have the following theorem.

LEMMA 7 *Let $X \vdash Y$ be a sequent. From $X \vdash Y$ the sequent $\boldsymbol{I} \vdash \tau_1(X) \to \tau_2(Y)$ is derivable in $\boldsymbol{DLM}$.*

Proof. We prove that if $X \vdash Y$ is derivable, then so is $\tau_1(X) \vdash \tau_2(Y)$. It is not difficult to put the proof to a conclusion from there. The remaining proof is an induction on the complexity of the sequent. In fact, we will again prove something stronger, namely that we can derive any $V \vdash W$ which arises from $X \vdash Y$ by replacing any given substructure by its proper translation (i. e. by its τ_1-translation or by its τ_2-translation depending on whether it occurs as antecedent part or as succedent part). We will not go through all the cases. But take the case of a sequent $X \circ Y \vdash Z$ and assume that the claim has been verified for X and Y. We wish to show it for $X \circ Y$. Here is a proof.

$$\begin{array}{c}
X \circ Y \vdash Z \\ \hline
X \vdash Z \circ *Y \\ \hline
\tau_1(X) \vdash Z \circ *Y \\ \hline
Y \vdash *\tau_1(X) \circ Z \\ \hline
\tau_1(Y) \vdash *\tau_1(X) \circ Z \\ \hline
\tau_1(X) \circ \tau_1(Y) \vdash Z \\ \hline
\tau_1(X) \wedge \tau_2(Y) \vdash Z \\ \hline
\tau_1(X \circ Y) \vdash Z
\end{array}$$

Another interesting case is $\bullet X \vdash Z$.

$$\begin{array}{c}
\bullet X \vdash Z \\ \hline
X \vdash \bullet Z \\ \hline
\tau_1(X) \vdash \bullet Z \\ \hline
\Diamond \tau_1(X) \vdash Z \\ \hline
\tau_1(\bullet X) \vdash Z
\end{array}$$

It is easy to supply the remaining cases. ■

THEOREM 8 *A sequent* $X \vdash Y$ *is* $\boldsymbol{DLM}$*-provable iff* $\tau(X \vdash Y)$ *is a theorem of the tense logic* **Kt**.

Proof. The correctness of the display calculus is a matter of straightforward verification. Notice that the display calculus contains modulo translation only rules which are derived rules of **Kt**, such as $(MN^+)\ P/\boxminus P$, and $(MN^-)\ P/\boxminus P$, which are consequences of the rules (Ml) and (Mr) (modulo some other rules). The completeness is somewhat tricky. Consider the set Θ of all formulae P such that $\mathbf{I} \vdash P$ is derivable in **DLM**. Moreover, consider the set $\Theta_{\vdash}$ of all $\tau(X \vdash Y)$ which are derivable in **DMLE**, defined below. By Lemma 9 below we know that a sequent is derivable in **DLME** iff it is derivable in **DLM**, and that sequents with identical τ-translation are interderivable. We conclude that $\Theta = \Theta_{\vdash}$. Now it remains to be seen that Θ is a normal modal logic, which will establish the completeness. If $P \in \Theta$, then also $\boxminus P \in \Theta$ and $\boxminus P \in \Theta$, as can easily be shown. Next, if $P \in \Theta$ and $P \rightarrow Q \in \Theta$, then not only $\mathbf{I} \vdash P \rightarrow Q$ is derivable, but also $P \vdash Q$. By an application of (Cut) we get $\mathbf{I} \vdash Q$, showing $Q \in \Theta$. It still needs to be shown that all boolean tautologies can be proved, but we refer here to [10]. ■

Here are the additional rules for the calculus **DLME**.

$$
(\circ \vdash)\ \frac{P \wedge Q \vdash Y}{P \circ Q \vdash Y} \qquad (\vdash \circ)\ \frac{X \vdash P \vee Q}{X \vdash P \circ Q}
$$

$$
(* \vdash)\ \frac{\neg P \vdash Y}{* P \vdash Y} \qquad (\vdash *)\ \frac{Y \vdash \neg P}{Y \vdash * P}
$$

$$
(\mathbf{I} \vdash)\ \frac{\top \vdash Y}{\mathbf{I} \vdash Y} \qquad (\vdash \mathbf{I})\ \frac{X \vdash \bot}{X \vdash \mathbf{I}}
$$

$$
(\bullet \vdash)\ \frac{\Diamondleft P \vdash Y}{\bullet P \vdash Y} \qquad (\vdash \bullet)\ \frac{X \vdash \boxright P}{X \vdash \bullet P}
$$

$$
(* \bullet * \vdash)\ \frac{\Diamondright P \vdash Y}{* \bullet * P \vdash Y} \qquad (\vdash * \bullet *)\ \frac{X \vdash \boxleft Q}{X \vdash * \bullet * Q}
$$

LEMMA 9 *All rules of $\boldsymbol{DLME}$ are admissible in $\boldsymbol{DLM}$. Moreover, if $X_1 \vdash Y_1$ and $X_2 \vdash Y_2$ have identical τ-translation, then they are interderivable in $\boldsymbol{DLME}$.*

Proof. The proof is by showing that the highest application in a proof tree of such a rule can be eliminated without adding new instances of such rules. To make life simple, we will assume that **DLM** admits cut-elimination. (This will be proved in the next section. This assumption is not strictly speaking necessary, but it simplifies the argument.) Thus we can assume our proofs to contain no cuts. Now take a highest instance of an elimination rule, say of $(\circ \vdash)$. It's premiss is of the form $P \wedge Q \vdash Y$. Now trace the occurrences of $P \wedge Q$ backwards. Each occurrence below the line has one (in the case of contraction two) counterparts above the line, unless, of course, $P \wedge Q$ is principal (see next section for a definition). This can only be in an application of $(\wedge \vdash)$. Replacing the traced occurrences of $P \wedge Q$ by $P \circ Q$ will transform valid instances of rules in valid ones, with the exception when $P \wedge Q$ is principal. Then replacing $P \wedge Q$ below by $P \circ Q$ will result in a trivial deduction, which can be omitted. ■

4 CUT-ELIMINATION AND THE SUBFORMULA PROPERTY

[1] lists eight conditions on a proper display logic called $(C1)$ – $(C8)$ and proves the following:

THEOREM 10 *Any display calculus satisfying $(C1)$ has the subformula property, that is, any cut-free proof of the sequent $X \vdash Y$ contains only structures over sub-*

formulas of formulas in X and Y.

THEOREM 11 *In any display calculus satisfying* $(C2)$ – $(C8)$ *(Cut) is eliminable.*

Now what are these conditions $(C1)$ – $(C8)$? We will give the original conditions in some slightly less general form. The difference is that we have stated a rule of simultaneous substitution $(C6/7)$, which is the appropriate choice to make in this context. Belnap assumes that in each rule we first stipulate a set of **constituents** and an equivalence relation on parameters called **congruence**. Here, **parameter** is an occurrence of a structure in a rule.

(C1) Each formula which is a constituent of some premiss of a rule ρ is a subformula of some formula in the conclusion of ρ.

(C2) Congruent parameters are occurrences of the same structure.

(C3) Each parameter is congruent to at most one constituent in the conclusion. Equivalently, no two constituents of the conclusion are congruent to each other.

(C4) Congruent parameters are either all antecedent or all succedent parts of their respective sequent.

(C5) If a formula is nonparametric in the conclusion of a rule ρ it is either the entire antecedent or the entire succedent. Such a formula is called **principal formula** of ρ.

(C6/7) Each rule is closed under simultaneous substitution of arbitrary structures for congruent parameters.

(C8) If there are inference rules ρ and σ with respective conclusions $X \vdash P$ and $P \vdash Y$ with P principal in both inferences (in the sense of $C5$) and if (Cut) is applied to yield $X \vdash Y$ then either $X \vdash Y$ is identical to either $X \vdash P$ or $P \vdash Y$; or it is possible to pass from the premisses of ρ and σ to $X \vdash Y$ by means of inferences falling under (Cut) where the cut-formula always is a proper subformula of P.

These conditions actually need some exegesis. First of all, the present formulation assumes that the so-called analysis is performed at the actual proofs, not at the rules as presented above. To see the difference, let us call a **rule** a consecution, and a **rule skeleton** a consecution formulated with the help of structure variables. A rule skeleton can be instantiated to a rule by substituting structures for structure variables. Notice that rules still contain variables, but only for formulae. We assume that any substitution instance of structures for structure variables and of formulas for formula variables is an instance of the rule. Rather than applying the analysis to the instances of the rules (which have no structure variables!), we perform the anaysis on the rule skeleta of the calculus **DLM** as presented above. Hence, for our purposes, a **parameter** may be thought of as an occurrence of a structure variable in a rule skeleton. In the actual Display Calculus, however, they are merely *metavariables* standing in for arbitary occurrences of structures. This is the way we will handle the Display Logic later as well. But let us suppose for the moment that they are genuine variables of the system. Then formula variables can never be parameters. Furthermore, if a structure X occurs parametric once, then all occurrences of X are parametric. Under our interpretation, then, $(C2)$ is trivially satisfied. Moreover, $(C6/7)$ is satisfied as well by force of our definition. Only the remaining conditions are ever to be checked in **DLM**. In the associativity law (A), for example, all variables are parameters, and all occurrences of the same variable are congruent. In a formula introduction rules such as $(\vdash \wedge)$ the new formula in the conclusion as well as the old ones in the premiss are typically not parameters.

With the exception of $(C8)$ the conditions are verified by direct inspection. We leave it to the reader to verify $(C1)$, $(C3)$, $(C4)$ and $(C5)$. For $(C8)$, [10] gives a proof that $(C8)$ holds for $\Diamond Q$ and $\boxminus Q$ as cut-formulas. The case of $\boxminus Q$ and $\Diamond Q$ is completely dual, i. e., obtained by swapping antecedent with succedent, so we might actually skip the proof here, but for the sake of completeness we give the corresponding proof of $(C8)$ in these cases. So let us first suppose that the cut-formula is $\Diamond Q$. Then we have

$$\frac{\dfrac{X \vdash Q}{\bullet X \vdash \Diamond Q} \qquad \dfrac{Q \vdash \bullet Y}{\Diamond Q \vdash Y}}{\bullet X \vdash Y}$$

The following, however, is a proof involving a cut on Q rather than $\Diamond Q$.

$$\dfrac{\dfrac{X \vdash Q \qquad Q \vdash \bullet Y}{X \vdash \bullet Y}}{\bullet X \vdash Y}$$

Suppose next that we have a proof involving $\boxminus Q$ as a cut-formula.

$$\frac{\dfrac{X \vdash *\bullet* Q}{X \vdash \boxminus Q} \qquad \dfrac{Q \vdash Y}{\boxminus Q \vdash *\bullet* Y}}{X \vdash *\bullet* Y}$$

The following is a proof with a cut on Q.

$$\cfrac{\cfrac{\cfrac{\cfrac{\cfrac{\cfrac{\cfrac{X \vdash *\bullet*Q}{\bullet*Q \vdash *X}}{*Q \vdash \bullet*X}}{*\bullet*X \vdash Q} \qquad Q \vdash Y}{*\bullet*X \vdash Y}}{*Y \vdash \bullet*X}}{\bullet*Y \vdash *X}}{X \vdash *\bullet*Y}$$

5 PROPERLY DISPLAYING EXTENSIONS OF KT

The usefulness of Display Logic shall be demonstrated with a theorem which shows that a large class of logics have a canonical proof system in Display Logic. We will analyse exactly which rules $\delta = P_1 \ldots P_m/Q$ can be incorporated into **DLM** by just adding another *structural rule* that does not destroy the properties $(C1)$ – $(C8)$. Such a rule captures a rule via its τ-translation in modal logic as follows. Let us be given a logic $\Lambda = \mathbf{Kt} + \Delta$ that is, an extension of (Hilbert-style) tense logic by a set of rules Δ, and assume that we add to **DLM** a set R of structural rules. Then we say that $\mathbf{DLM} + R$ **properly displays** $\mathbf{Kt} + \Delta$ if $(C1)$ – $(C8)$ are satisfied and every derived rule of $\mathbf{Kt} + \Delta$ is the τ-translation of a derived rule of $\mathbf{DLM} + R$. The latter condition can be rephrased as requiring that a rule transition is derivable in $\mathbf{DLME} + R$ iff its τ-translation is a derived rule of $\mathbf{Kt} + \Delta$. For, by the fact that the added structural rules preserve the properties $(C1)$ – $(C8)$ we have cut-elimination, and an analogue of Lemma 9 holds. Then in $\mathbf{DLME} + R$, τ-equivalent sequents are interderivable, while in $\mathbf{DLM} + R$ we can only go from sequents to sequents if we do not eliminate connectives. Nevertheless, it is easy to see that if a rule is $\mathbf{DLME} + R$-derivable, then there is another rule with identical τ-translation which is derivable in $\mathbf{DLM} + R$. In this section we will give a complete characterization of properly displayable **Kt**-calculi. We show first that the contribution of a rule $\rho = s_1 s_2 \ldots s_m/t$ can be directly computed as $\breve{\rho} = \tau(s_1) \ldots \tau(s_m)/\tau(t)$.

LEMMA 12 *Suppose that ρ is a structural rule, and let $\boldsymbol{DLM} + \rho$ satisfy* $(C1)$ – $(C8)$. *Then $\boldsymbol{DLM} + \rho$ properly displays $\boldsymbol{Kt} + \breve{\rho}$.*

Proof. Two things need to be seen. First, any extension by structural rules satisfying the display conditions axiomatizes a normal logic. And second, that it axiomatizes the logic as given. Let Θ be the set of all $\tau(X \vdash Y)$ which are derivable in $\mathbf{DLME} + \rho$. Since the rules of the calculus turn into axioms of $\mathbf{Kt} + \breve{\rho}$, we have proved now that we have displayed a logic at most as strong as $\mathbf{Kt}+\breve{\rho}$. On the other hand, suppose that for some substitution σ we have proved $P_1^\sigma, \ldots, P_m^\sigma$. Then there is a substitution ω of formulae to structure variables such that $P_i^\sigma = \tau(s_i{}^\omega)$. Then if we can derive $\tau(s_i{}^\omega)$ we can also derive $\tau(t^\omega) = Q^\sigma$. Thus $\breve{\rho}$ is a derived rule of the calculus $\mathbf{DLM} + \rho$ translated under τ. ■

For example, adding the rule

$$\frac{\mathbf{I} \vdash \bullet X}{\mathbf{I} \vdash X}$$

This rule translates into the rule $\boxminus P/P$, a rule which is actually admissible in **Kt**, but not derivable. In the calculus obtained by τ-translation it will however be a derived rule.

Now, modal logics are generally studied as axiomatic strengthenings of a basic system rather than a strengthening by proper rules. Therefore, let us concentrate on the question of displayability of axiomatic extensions. First we will consider which axioms admit a resolution into a structural rule. Denote by $\mathbf{Kt} \oplus \delta$ the least normal logic extending **Kt** which also contains δ. δ can always be written in the form $A \rightarrow B$, where A and B are free of $\rightarrow$. We can then pass from the axiom $A \rightarrow B$ to a sequent rule

$$\frac{B \vdash Y}{A \vdash Y}$$

This sequent rule is as powerful as the axiom. For putting $Y = p$, a variable not contained in A or B the τ-translation is the rule $\tau_1(B) \rightarrow p/\tau_1(A) \rightarrow p$. Now let $p = \tau_1(B)$; then the premiss of $\breve{\rho}$ becomes a theorem and we see that $\tau_1(A) \rightarrow \tau_1(B)$ is an axiom of the calculus axiomatized by this rule. The axiom allows to derive the rule, however, and so the two are equal in power. Assume now that both A and B are composed from $\top$ and propositional variables using only $\wedge$, $\vee$, $\diamondminus$ and $\diamondplus$. Then, by standard equivalences,

$$A. \leftrightarrow . \bigvee_{i \leq m} C_i \qquad B. \leftrightarrow . \bigvee_{j \leq n} D_j$$

where all C_i, D_j are composed from variables and $\top$ with the help of $\wedge$, $\diamondminus$ and $\diamondplus$ only. (Thus, disjunction has been eliminated.) Instead of the rule above we then equip our calculus with the rules $\rho_1, \ldots, \rho_m$ where each ρ_i is of the form

$$\frac{D_1 \vdash Y, \ldots, D_n \vdash Y}{C_i \vdash Y}$$

Again it is checked that this new calculus is equivalent in power with our logic. Finally, define the translation σ from formulae into structures via

$$\begin{array}{lcl} \sigma(\top) & = & \mathbf{I} \\ \sigma(p) & = & p \\ \sigma(P \wedge Q) & = & \sigma(P) \circ \sigma(Q) \\ \sigma(\overleftarrow{\Diamond} P) & = & \bullet\sigma(P) \\ \sigma(\overrightarrow{\Diamond} P) & = & *\bullet*\sigma(P) \end{array}$$

It is checked by induction that $\tau_1(\sigma(P)) = P$ for every formula P made from variables, $\top$, conjunction and possibility operators. Thus we have eliminated $\vee$. Finally, then, let us replace the above rules ρ_i by their Gentzenized sisters.

$$\frac{\sigma(D_1) \vdash Y, \ldots, \sigma(D_n) \vdash Y}{\sigma(C_i) \vdash Y}$$

So, for axioms of the form $A \to B$, where A and B are positive and free of $\overrightarrow{\Box}$ and $\overleftarrow{\Box}$ we have managed to write a display system that completely axiomatizes it. For the rules above $(C1)$ is obviously satisfied since we are strengthening the system for **Kt** by structural rules. $(C4)$ and $(C5)$ are verified by the eye. The condition $(C8)$ has to be checked only with respect to the rules which are introducing formulae, and this has been done already. We are left with $(C3)$. $(C3)$ is actually not automatically valid. In fact, we must place the restriction on the formula A that it may contain each variable *only once*. (Maybe only those which already occur in B, but the others can be eliminated.) Let us now agree to call a formula **primitive** if it is of the form $A \to B$ where both A and B contain only variables, $\top$, $\wedge$, $\vee$, $\overrightarrow{\Diamond}$ and $\overleftarrow{\Diamond}$ and that A contains each propositional variable at most once.

LEMMA 13 *Suppose Λ is a tense logic axiomatizable by primitive axioms. Then Λ can be properly displayed.*

Proof. Λ can be displayed in the way described above; the display calculus meets $(C1)$ – $(C8)$ and therefore enjoys cut-elimination and the subformula property. The system is complete for Λ in the sense that it derives $X \vdash Y$ iff $\tau_1(X) \vdash_\Lambda \tau_2(Y)$. ∎

Now let us tackle the question of what logics are defined by a display calculus. We will show that it is actually the same class of logics if we insist that additional rules are completely structural. To see this consider a rule

$$\rho \qquad \frac{X_1 \vdash Y_1, \ldots, X_n \vdash Y_n}{V \vdash W}$$

where all X_i, Y_j, W contain only structure variables. By $(C4)$, congruent formulae do not swap sides, and so they are either in the antecedent throughout or in the conclusion throughout. Now rewrite this rule into the following form.

$$\rho^* \qquad \frac{\mathbf{I} \vdash *X_1 \circ Y_1, \ldots, \mathbf{I} \vdash *X_n \circ Y_n}{\mathbf{I} \vdash *V \circ W}$$

ρ and ρ^* are equivalent and ρ^* satisfies $(C4)$ as well as $(C3)$ if ρ does. The other conditions are harmless. This rule translates into the axiom

$$\tau_2(*X_1 \circ Y_1), \ldots, \tau_2(*X_n \circ Y_n)/\tau_2(*V \circ W)$$

so that we now have to worry about which formula can occur as a translation of a structure under τ_2. Dually, we can solve the question of which formulae are translations under τ_1.

LEMMA 14 $P = \tau_1(X)$ *for some* X *containing no formula connectives exactly if* $P \dashv\vdash Q$ *for some* Q *built from variables and negated variables,* $\top$, $\wedge$ *and the diamonds* $\diamondminus$ *and* $\diamondminus$.

Proof. We know that if $X \approx Y$ then $\tau_1(X) \dashv\vdash \tau_1(Y)$ so that without loss of generality we can assume that X is in normal form (see Section 9)

$$X = p_{i_1} \circ \ldots \circ p_{i_k} \circ *p_{j_i} \circ \ldots \circ *p_{j_l} \circ \bullet X_i \circ \ldots \circ \bullet X_m \circ *\bullet Y_1 \circ \ldots \circ *\bullet Y_n$$

Then

$$\tau_1(X) = \bigwedge_{r \leq k} p_{i_r} \wedge \bigwedge_{s \leq l} \neg p_{j_s} \wedge \bigwedge_{t \leq m} \diamondminus\tau_1(X_t) \wedge \bigwedge_{u \leq n} \diamondminus\tau_1(*Y_u)$$

Now do induction on the number of nested occurrences of $\bullet$. ■

There remains now only the problem of negated occurrences of variables to prove the characterization. Therefore consider a variable p that occurs negated in $\tau_1(V \circ *W)$. Then replace p in the axiom throughout by $\neg p$. As it occurs negated throughout (by $(C4)$ and the definition of *part of*), after substitution it occurs doubly negated throughout and so we can eliminate the negation altogether for p. Similarly for variables occurring only in the premisses. Thus we have proved the following characterization.

THEOREM 15 *An extension of* **Kt** *by rules can be properly displayed iff each rule* $P_1 \ldots P_m \vdash Q$ *is such that all occurring formulae are built from* $\bot$ *and variables with the help of* $\wedge$, $\vee$ *and* ⧄, ⊟.

We can obtain as a special corollary a characterization of properly displayable axiomatic extensions. By the condition of completeness of **DLME** $+ \breve{\rho}$ with respect to the logic **Kt** $\oplus \delta$ it is possible to display **Kt** $\oplus \delta$ by rules of the form

$$\rho \qquad \frac{X_1 \vdash Y, \ldots, X_n \vdash Y}{V \vdash Y}$$

Call these rules **special**. They are characterized by the fact that one variable is shared as the common succedent or antecedent. For on the one hand, putting $Y = p$, a fresh variable, the rule corresponding to ρ is

$$\breve{\rho} \quad \tau_1(X_1) \to p, \ldots, \tau_1(X_n) \to p / \tau_1(V) \to p$$

Putting $p = \bigvee_i \tau_1(X_i)$ we derive the axiom $\tau_1(V) \to \bigvee_i \tau_1(X_i)$. This is a primitive formula. This formula if taken as an axiom is as least as strong as the rule $\breve{\rho}$. So, by the completeness of **DLM** $+ \rho$ for **Kt** $+ \breve{\rho}$ we get the completeness for **Kt** $\oplus \tau_1(V) \to \bigvee_i \tau_1(X_i)$. On the other hand, any axiom can be brought into the form $A \to B$, and hence is characterized by a special rule. This special rule must be derivable in **DLME**. By the condition that A and B are translations of structures, we ultimately arrive as before at the requirement that $A \to B$ is primitive. But then it can be properly displayed, as we have seen.

THEOREM 16 (Proper Display I) *An axiomatic extension of Hilbert-style* **tense** *logic can be properly displayed (by structural rules over* **DLM***) iff it is axiomatizable by a set of primitive axioms.*

6 SEMANTIC CHARACTERIZATION

It is possible to characterize exactly the semantic conditions that can be associated with primitive axioms. It is known from correspondence theory ([7] and [8]) that primitive formulae are canonical and therefore complete; in addition the condition that the formula places on the canonical and Kripke frames is elementary. Yet it is also important to know what condition a particular axiom expresses. [8] developed the method of substitutions to find the elementary equivalent of a Sahlqvist formula but we find this method not so user-friendly. Instead we use the technique of *decisive sets* as proposed in [4]. Some results have also been established in the somewhat simpler [3]. This method rests on the fact that the condition expressed by an axiom can be squeezed out by very special valuations. These valuations decide what elementary condition this axiom determines in the sense that to know what the condition is we just have to check these valuations. Equivalently, to test a Kripke frame for whether it accepts a given Sahlqvist formula we only have to use these decisive sets as values of the propositions (or, even better, this can be fine tuned to *decisive valuations* rather than decisive *sets*). On Kripke frames the question is then settled. On generalized frames we have two choices: either these valuations exist – and then the elementary condition is forced on the underlying Kripke frame by these general arguments – or the sets are the infinite intersection of admissible valuations. In this case if the frame is *compact* as a topological space it guarantees that a family of valuations converging to a decisive valuation has the same effect as the decisive valuation itself. In this case we can simply pretend that the decisive sets are internal in the generalized frame. (Or, equivalently, that they can be added without disturbing the sets of Sahlqvist formulae accepted by the frame.)

It turns out that for the axioms in question the singleton sets $\{w\}, w \in f$, are decisive. This is most welcome in practical computation, because then we can do what every student of modal logic (including the author) is at one stage always tempted to do; namely, to pretend that a variable stands for *being at a certain world*. For example, in the axiom $\Diamond p \wedge \Diamond q. \rightarrow .\Diamond(p \wedge q)$ let us pretend p means *being at* x_p and q means *being at* x_q. Then this axiom tells us that if we can see x_p and x_q then x_p and x_q must be equal. Hence, it tells us that any point has at most one successor. According to [4], the elementary properties defined by Sahlqvist tense formulae can be described as follows. They are of the form $(\forall x)\Phi$ where Φ is composed from positive formulae $x \lhd y$, $x = y$ with the help of $\wedge, \vee$ and the so-called *two way restricted quantifiers* $(\exists y \rhd x), (\exists y \lhd x), (\forall y \rhd x), (\forall y \lhd x)$ in such a way that in a subformula $x \lhd y$, $x = y$ at least one of x and y is *hereditarily universal*, which means that it is not inside an existential quantifier. Modal Sahlqvist formulae differ only with respect to the quantifiers $(\exists y \lhd x)$ and $(\forall y \lhd x)$, which may not be used. Let us call $\phi = (\forall x)\chi$ a **primitive** formula (modal or tense) if it is of the described form and no universal quantifier is in the scope of an existential quantifier; hence it has the form $(\forall)(\exists)\Psi$ (with the appropriate restricted quantifiers) where Ψ is positive and in an atomic subformula $x \lhd y$ or $x = y$ at least one of x and y is hereditarily universal.

THEOREM 17 *Suppose that $\mathfrak{F}$ is a class of modal or tense Kripke-frames described by some finite set of primitive sentences. Then the modal logic of $\mathfrak{F}$ can be properly displayed.*

Proof. It suffices to derive in the calculus of first-order equivalents as described in [4] that a negative, bounded $\exists\forall$-sentence is equivalent to a sequence $\mathfrak{N} \otimes \mathfrak{P}$ where

$\mathfrak{N}$ is negative and free of $\boxminus$, $\boxminus$ and $\mathfrak{P}$ is positive and free of $\diamondminus$, $\diamondminus$. ■

This seems surprising, because well-known systems such as **S5**, **Alt**$_1$ etc. are often axiomatized using $\Box$ ($= \boxminus$). But this need not be so. The axiom of transitivity can be rendered as $\Diamond\Diamond p \to \Diamond p$, the axiom of reflexivity as $p \to \Diamond p$, and the axiom of symmetry as $p \wedge \Diamond q \to \Diamond(q \wedge \Diamond p)$. The quasi-functionality can be axiomatized by $\Diamond p \wedge \Diamond q \to \Diamond(p \wedge q)$ rather than $\Diamond p \to \Box p$.

COROLLARY 18 *All elementary subframe logics can be properly displayed.*

Proof. If Λ is elementary, say, described by ϕ, then ϕ is a restricted universal sentence. Moreover, by a result in [4] if ϕ is universal and it's class is closed under p-morphisms, then ϕ is equivalent to a universal and positive sentence. [1] Thus elementary subframe logics fall under the class just mentioned in the theorem above. ■

COROLLARY 19 *All r-persistent subframe logics can be properly displayed.*

Proof. R-persistent logics are elementary, due to a result by [2]. ■

7 NICE RULES AND MODAL DISPLAYABILITY

This section discusses the possibility of writing nice rules for modal axioms. Two things will be shown; first, that the extra strength added by the fact that we have the tense dual of the modal operator allows to state rules quite concisely in some cases. And second, that the extra strength does not allow to axiomatize more logics. We will demonstrate the first point with a particular example. Suppose, we are interested in a display rule for **.3**.

$$\mathbf{.3}: \quad \diamondminus p \wedge \diamondminus q. \to .\diamondminus(p \wedge \diamondminus q) \vee \diamondminus(q \wedge \diamondminus p) \vee \diamondminus(p \wedge q)$$

This axiom is already in rather perspicuous form, it is primitive and we can translate it directly into a display rule.

$$\frac{*\bullet*(X \circ *\bullet*Y) \vdash Z \quad *\bullet*(X \circ Y) \vdash Z \quad *\bullet*(Y \circ *\bullet*X) \vdash Z}{*\bullet*X \circ *\bullet*Y \vdash Z}$$

[1] The proof there is actually highly incomplete, as I recently found out. A full proof can be found in [6].

Semantically, it has the following content.

$$(\forall w)(\forall x \rhd w)(\forall y \rhd w)(x = y \vee y \lhd x \vee x \lhd y)$$

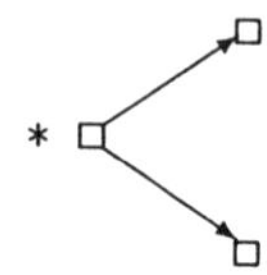

If we are sitting at the root of the tree (marked by a star), thus seeing two points, then these points are either identical or one of them sees the other.

In standard modal logic, one can do no better. However, although it is modal logic we are doing, the display calculus uses a Gentzen toggle and so we have implicitly the power of tense logic in the display calculus. The Gentzen toggle is there anyway, so we can use the extra power it gives us by rewriting the axiom **.3** according to the possibilities of tense logic. Notice, namely, that in tense logic we can generate the tree in the picture from any point we wish because we can look both ways. Consequently, we can shift the reference point of the axiom from the root of the tree to one of the branches.

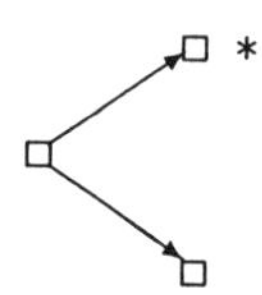

Read from there, the semantic characterization is as follows

$$(\forall w)(\forall x \lhd w)(\forall y \rhd x)(y = w \vee y \lhd w \vee y \rhd w)$$

Put into a tense formula it looks like this

$$\overleftarrow{\Diamond}\overrightarrow{\Diamond}p. \to .p \vee \overleftarrow{\Diamond}p \vee \overrightarrow{\Diamond}p$$

This axiom uses far less symbols and only one variable. In modal logic, there is no way to axiomatize **S4.3** over **S4** with the help of just one variable, even though there are different ways of writing **.3** (but effectively only one primitive way). Hence we can axiomatize **.3** in **DLM** with the following rule

$$\frac{X \vdash Y \quad \bullet X \vdash Y \quad * \bullet * X \vdash Y}{\bullet * \bullet * X \vdash Y}$$

Now for the second claim that we cannot axiomatize more logics. The Proper Display Theorem I states that a tense logic is properly displayable iff it can be axiomatized by a set of primitive formulae. This can be strengthened to modal logics using some model theoretic techniques. Let us begin by introducing *generalized restricted quantifiers*. We put

$$\begin{aligned}
(\forall y \rhd_0 x)\phi(x,y) &= \phi(x,x) \\
(\exists y \rhd_0 x)\phi(x,y) &= \phi(x,x) \\
(\forall y \rhd_{n+1} x)\phi(x,y) &= \phi(x,x) \wedge (\forall z \rhd x)(\forall y \rhd_n z)\phi(x,y) \\
(\exists y \rhd_{n+1} x)\phi(x,y) &= \phi(x,x) \vee (\exists z \rhd x)(\exists y \rhd_n z)\phi(x,y)
\end{aligned}$$

These quantifiers quantify over sets of points which can be reached within a fixed number of steps. Recall that a modal formula is called **restricted** if it is built from atomic formulae with the help of the quantifiers $(\exists y \rhd x)$, and $(\forall y \rhd x)$. It is equivalent to require that the generalized restricted quantifiers be used. Our aim is to show that if α is primitive and characterizes a modal class of frames then α is equivalent to a primitive modal formula. The problem here is to get rid of the quantifiers $(\forall y \lhd x)$ and $(\exists y \lhd x)$. In order to do this, we take a detour. Clearly, α can be written in the form $(\forall \overline{x})(\exists \overline{y})\phi(\overline{x},\overline{y})$, with $\phi(\overline{x},\overline{y})$ quantifier free, using unrestricted quantifiers. To simplify the notation, we use a single variable x instead of $\overline{x}$ and likewise a single variable y for $\overline{y}$. Define now the formulae $\delta_k(x) = (\exists y \rhd_k x)\phi(x,y)$. Then

$$\delta_0(x) \vdash \delta_1(x) \vdash \delta_2(x) \vdash \ldots \vdash (\exists y)\phi(x,y)$$

It is therefore enough if we show that there is a k such that $(\exists y)\phi(x,y) \vdash \delta_k(x)$. Suppose that this is not so. Then the following set is consistent.

$$\begin{aligned}
X \quad = \quad & \{(\forall x)(\exists y)\phi(x,y), (\exists y)\phi(u,y), \\
& (\forall y \rhd_0 u)\neg\phi(u,y), \\
& (\forall y \rhd_1 u)\neg\phi(u,y), \\
& (\forall y \rhd_2 u)\neg\phi(u,y), \ldots\}
\end{aligned}$$

Thus there is a model $\mathfrak{M}$ for X. Let $\mathfrak{N}$ be the submodel generated by u. This model consists of all points which can be reached from u in a finite number of steps. Since α is a modal formula, it is preserved under taking generated submodels and we find that $\mathfrak{N} \models \alpha$, by which also $\mathfrak{N} \models (\exists y)\phi(u,y)$. Hence for some $w \in \mathfrak{N}$, $\mathfrak{N} \models \phi(u,w)$. Since ϕ is constant, it is reflected under generated subframes. So $\mathfrak{M} \models \phi(u,w)$ as well and thus $\mathfrak{M} \models (\exists y \rhd_k u)\phi(u,y)$ for some k, since w can be reached from u in a finite number of steps, according to the definition of the submodel. This, however, contradicts our assumption on $\mathfrak{M}$. We conclude that X

is inconsistent. Hence, we can write α in the form $(\forall x)(\exists y \rhd_k x)\phi(x,y)$ with $\phi(x,y)$ quantifier-free.

The next problem are the universal quantifiers. Assume that

$$\alpha = (\forall x)(\forall y)(\exists z)\phi(x,y,z).$$

We have seen that we can strengthen the existential quantifier to a restricted one, keeping the matrix ϕ constant. Now we consider $\neg\alpha$. This sentence is preserved under generated subframes as well. It is of the form $(\exists x)(\exists y)\psi(x,y)$, where $\psi(x,y)$ is restricted. We can perform a similar argument as above, using $\psi(x,y)$ instead of $\phi(x,y,z)$. Notice, namely, that in the proof we have needed only that it is reflected under generated submodels. Thus we can effectively strengthen our formula to $(\forall x)(\forall y \rhd_k x)(\exists z \rhd_\ell x)\phi(x,y,z)$. The last task is to remove the negative atomic subformulae. This can be done. According to Theorem 31 of [4] since α is closed under p-morphic images it is equivalent to a formula $(\forall x)(\exists y \rhd_k x)(\forall z \rhd_\ell x)\psi(x,y,z)$ where $\psi(x,y,z)$ results from $\phi(x,y,z)$ by replacing all negative subformulae by either *false* or *true*.

THEOREM 20 (Proper Display II) *An axiomatic extension of Hilbert-style modal logic can be properly displayed (by structural rules over* $\mathbf{DLM}$*) iff it is axiomatizable by a set of primitive modal axioms.*

8 POLYMODAL LOGICS AND THE DECIDABILITY QUESTION

The most remarkable aspect of the display calculus is that it generalizes easily to logics with several modal operators. Just imagine we have instead of one modal operator (and its tense dual) a finite list $\boxminus_i$ of modal operators (possibly together with their tense duals $\diamondminus_i$). Then we proceed by redefining $\mathfrak{Struc}$, positing a Gentzen toggle $\bullet_i$ for each pair $\boxminus_i$, $\diamondminus_i$ and writing down the rules for introducing the operators for each operator independently. The conditions $(C1)-(C8)$ are immediately satisfied. Notice that the condition $(C8)$ is modular in the sense that only the calculi restricted to the individual operators have to be checked for $(C8)$. If they satisfy it, the overall calculus does so, too. We then have the subformula property and cut-elimination. Define *primitive* of an elementary condition as follows. A first-order n-modal sentence is **primitive** if it is if the form $(\forall x)\chi$ where χ is produced from atomic formulae $x = y$ with the help of $\wedge, \vee$ and two way restricted quantifiers $(\forall x \rhd_i y)$ $(\forall x \lhd_i y)$, $(\exists x \rhd_i y)$ and $(\exists x \lhd_i y)$ with the extra conditions that no universal quantifier is in the scope of an existential quantifier and in each atomic subformula $x = y$ at least one of x and y is inherently universal.

THEOREM 21 *Let $\mathfrak{F}$ be a class of Kripke polyframes defined by a finite set of primitive sentences. Then the logic of $\mathfrak{F}$ can be properly displayed.*

If such a logic can be properly displayed we can use axioms of the form $A \to B$ where both A and B are made from variables with the help of $\top$, $\wedge$, $\vee$, $\diamondminus_i$, ($\diamondplus_i$) such that A contains a variable only once. Again, we call these axioms **primitive**. All theorems of the preceding section hold in their canonical extension to polymodal logics.

We will use this fact to derive a rather negative result concerning the decidability of display logics. One might think that cut-elimination and the subformula property are enough to have decidability – but the subformula property is too weak to guarantee that we can give a priori bounds on the lengths of sequents occurring in a minimal proof of a given sequent. One might think, then, that it is just a question of being clever enough. However, the opposite is true. We will simply prove here the following theorem.

THEOREM 22 *It is undecidable whether or not a display calculus is decidable.*

The proof involves the simulation of a Thue-process. We produce here a counterexample based on a result found together with C. Grefe. Proofs and exact details can be found in [5].[2] Consider a bimodal logic with operators $\boxplus$ and $\boxtimes$. We assume that both operators satisfy the axiom $\mathbf{Alt_1}$, so that the relation on the Kripke frames satisfies the condition that each point has at most one successor in each relation. Suppose further that both satisfy the axiom **D**, so that in fact each point has *exactly* one successor. We will show how to code Thue-processes into such frames as extensions of the logic $\mathbf{Alt_1.D} \otimes \mathbf{Alt_1.D}$. It will turn out that the logics are undecidable if the corresponding Thue-process is and so we have plenty of undecidable finitely axiomatizable logics. So, let $\mathfrak{T} = \{u_i \sim v_i \mid i \in n\}$ be a set of equations over strings in the alphabet $\{a, b\}$. Recall that $\mathfrak{T}$ specifies a relation $\sim$ between words as follows. $w \sim_0 z$ if $w = z$; $w \sim_{n+1} z$ iff there is a y such that $w \sim_n y$ and $y = \overline{y}u_i\overline{\overline{y}}$, $z = \overline{y}v_i\overline{\overline{y}}$ for some $i \leq n$ or $y = \overline{y}v_i\overline{\overline{y}}$, $z = \overline{y}u_i\overline{\overline{y}}$. Then $\sim = \bigcup_k \sim_k$. Such equations can be mimicked by modal axioms. Define first a translation of strings into formulae.

$$a^t = \boxplus, b^t = \boxtimes, (vw)^t = v^t w^t$$

$$(u_i \sim v_i)^t = u_i^t p \leftrightarrow v_i^t p$$

For example, $aab \sim ba$ gets translated into $\boxplus \boxplus \boxtimes p \leftrightarrow \boxtimes \boxplus p$. Notice that the axioms are all equivalent to primitive formulae. It can now be shown that $w \sim z$ iff

[2] The relevant results have actually been known to Alexander Muchnik in 1974, as I have been told by Valentin Shehtman. He has not published his findings, though.

$w^t p \leftrightarrow z^t p$ is derivable in the logic $\mathbf{K.Alt_1.D} \otimes \mathbf{K.Alt_1.D}(\mathfrak{T})$. If $\mathfrak{T}$ is undecidable the corresponding logic is undecidable as well. Moreover, if $\mathfrak{T}$ is decidable, so is the corresponding logic. Hence, for the logics simulating Thue-processes the question of decidability is undecidable. Since they are all properly displayable logics, Theorem 22 is proved.

REMARK 23 *The proof of the theorem requires at least two modal operators. It is not known so far whether for monomodal logics the theorem holds as well.*

9 SPEEDING UP PROOFS

Despite its theoretic advantages, display logic is a rather clumsy tool in actual computation. Anyone having done or tried a few proofs in display logic will see this. We will therefore develop a calculus of compressed proofs that allows to speed up a display proof considerably. The way this problem is attacked is by noting that the rules of the calculus allow to compress the structures that appear during a proof into some smaller structures whose size is bounded a priori from the size of X and Y. Recall therefore the definition of $\approx$. The following holds in the classical calculus.

$$\begin{array}{lcl} X \circ \mathbf{I} & \approx & X \\ *(X \circ Y) & \approx & *X \circ *Y \\ **X & \approx & X \\ X \circ X & \approx & X \\ X \circ Y & \approx & Y \circ X \\ X \circ (Y \circ Z) & \approx & (X \circ Y) \circ Z \end{array}$$

These equivalences are not difficult to show. There might be more valid equations; however, it is enough if the ones given hold and so we fix $\approx$ to be the congruence defined by these particular equations. Especially useful is the normal form theorem.

DEFINITION 24 *A structure term is in normal form of rank 0 if it is of the form*

$$P_1 \circ P_2 \circ \ldots \circ P_k \circ *Q_1 \circ *Q_2 \circ \ldots \circ *Q_l$$

where the P_i, Q_j are formulae; it is called reduced if all P_i are different and all Q_j are different. A structure term is called in normal form of rank n+1 if it is of the form

$$V \circ \bullet X_1 \circ \bullet X_2 \circ \ldots \bullet \circ X_m \circ *\bullet Y_1 \circ *\bullet Y_2 \circ \ldots \circ *\bullet Y_n$$

where all V, X_i, Y_j are in normal form of rank $\leq n$. It is called reduced if all X_i are different and all Y_j are different as well, and if V is reduced.

THEOREM 25 (Normal Forms) *Every term $X \in \mathfrak{Struc}(\mathcal{L})$ can be brought effectively into normal form $Y \in \mathfrak{Struc}(\mathcal{L})$. Moreover, it can be brought effectively into reduced normal form.*

Let us denote by $\mathfrak{Comp}(\mathcal{L})$ the algebra $\mathfrak{Struc}(\mathcal{L})/\!\approx$. We will call it the algebra of **compressed $\mathcal{L}$-structures**. However, instead of working with compressed structures one can also work with reduced normal forms also called **reduced structures**. Notice that normal forms are not unique, but if X, Y are equivalent normal forms then they contain the same number of symbols, that is, they are of equal length. A reduced structure contains no stacked $*$ unless separated by a $\bullet$, and $*$ is obligatorily distributed over $\circ$. No double occurrences of a structure X are within the same nesting of $\bullet$.

Define now the *layer* of a structure as follows.

$$
\begin{aligned}
\lambda(P) &= 0\\
\lambda(\mathbf{I}) &= 0\\
\lambda(*X) &= \lambda(X)\\
\lambda(X \circ Y) &= \mathit{max}\{\lambda(X), \lambda(Y)\}\\
\lambda(\bullet X) &= 1 + \lambda(X)
\end{aligned}
$$

We can effectively count the number of reduced structures up to a given layer. The bounding number is computed recursively as follows.

$$
\begin{aligned}
n(0,\gamma) &= 2^{2\gamma}\\
n(\lambda+1,\gamma) &= n(0,\gamma)\cdot 2^{2n(\lambda,\gamma)}
\end{aligned}
$$

PROPOSITION 26 $\mathfrak{Comp}(\mathfrak{G})$ *contains exactly $n(\lambda, \sharp G)$ elements of layer $\leq \lambda$.*

Proof. It is enough to count the number of nonequivalent reduced normal forms. At layer 0, we have the form

$$P_1 \circ P_2 \circ \ldots \circ P_k \circ *Q_1 \circ *Q_2 \circ \ldots \circ *Q_l$$

Two such forms are equivalent iff the sets of unstarred formulae coincide and the sets of starred formulae coincide. There are 2^γ sets of formulae, where $\gamma = \sharp G$ since the generators are assumed different. Thus at layer 0 we have exactly $2^\gamma \times 2^\gamma = 2^{2\sharp G}$ elements. Suppose then that the elements of layer λ are counted by $n(\lambda, \sharp G)$. At layer $\lambda + 1$, every element is of the form

$$X \circ \bullet Y_1 \circ \bullet Y_2 \ldots \circ \bullet Y_m \circ * \bullet\, Z_1 \circ * \bullet\, Z_2 \circ \ldots \circ * \bullet\, Z_n$$

where X, the Y_i and the Z_j are of layer λ. The Y_i are all distinct as well as the Z_j, though the X need not be distinct from the Y_i and the Y_i need not be distinct from the Z_j. Counting the number of such elements gives $n(0, \sharp G) \cdot 2^{2n(\lambda, \sharp G)}$. ■

Let us define the **compressed calculus** to be the display-calculus read as a calculus over $\mathfrak{Comp}$ rather than $\mathfrak{Struc}$. A proof Π in the display calculus is translated into a **compressed proof** Π^c line-by-line, by translating $X \vdash Y$ to $X/\!\approx\, \vdash Y/\!\approx$. A compressed proof can be uncompressed by translating a sequent $X/\!\approx\, \vdash Y/\!\approx$ into a sequent $X' \vdash Y'$ where $X' \approx X$ and $Y' \approx Y$ and then adding some more proof steps. Although the representative can be chosen at random, there is always a way to complete this proof – by definition of $\approx$. We can, however, discern among these choices of uncompressions some good ones that make use only of reduced formulae. So let $u : \mathfrak{Comp}(\mathcal{L}) \to \mathfrak{Struc}(\mathcal{L})$ a map with $u(X/\approx) \approx X$, so u is picking representatives from each class. u is called *special* if $u(X)$ is always reduced. From now on u is assumed to be always special. Let $(\Pi^c)^u$ be the line-by-line translation of Π^c by u. As observed several times, $(\Pi^c)^u$ is not necessarily a *proof* in the strict sense, but there is a way to complete the uncompressed version. Namely if

$$\rho \; \frac{X_1 \vdash Y_1 \quad \ldots \quad X_n \vdash Y_n}{V \vdash W}$$

is a line in the uncompressed proof, then it is of the form $\rho = \sigma^{cu}$ where σ is an instance of a display-rule. So σ is of the form

$$\sigma \; \frac{X_1' \vdash Y_1' \quad \ldots \quad X_n' \vdash Y_n'}{V' \vdash W'}$$

where $X_i \approx X_i'$, $Y_j \approx Y_j'$, $V \approx V'$ and $W \approx W'$. We can then first move from $X_i \vdash Y_i$ to $X_i' \vdash Y_i'$ then conclude $V' \vdash W'$ and the go back to $V \vdash W$.

$$\dfrac{\dfrac{\dfrac{X_1' \vdash Y_1'}{\ldots}}{X_1 \vdash Y_1} \quad \ldots \quad \dfrac{\dfrac{X_n' \vdash Y_n'}{\ldots}}{X_n \vdash Y_n}}{\dfrac{\dfrac{V \vdash W}{\ldots}}{V' \vdash W'}}$$

But this proof is at least as complicated as Π itself, so this does not amount to a reduction in any sense. The most obvious waste that is produced this way is by completing steps that have been trivialized by this forth-and-back translation. Namely, the rules (A), (P), (C) are now completely empty. For example, $(X \circ X)^{cu} = X^{cu}$, so if ρ is an instance of (C) then $\rho^{cu} = id$, the identity transition. Evidently, we can get away by just forgetting about this rule altogether.

But there is a more economical way to complete the proof which goes as follows. Take, for example, the rule $(\vdash \wedge)$.

$$\frac{X \vdash P \quad X \vdash Q}{X \circ X \vdash P \wedge Q}$$

Let there be an instance of this rule and let us compress and uncompress this rule. Then we get

$$\frac{X^{cu} \vdash P \quad X^{cu} \vdash Q}{X^{cu} \vdash P \wedge Q}$$

There is no need to return to $X \vdash P$ from $X^{cu} \vdash P$ or to $X \vdash Q$ from $X^{cu} \vdash Q$. The rule can be applied directly to the new premisses but it yields the conclusion $X^{cu} \circ X^{cu} \vdash P \wedge Q$ rather than $X^{cu} \vdash P \wedge Q$. However, a single application of (Cl) brings the sequent into reduced form. In general, the antecedents are of the form

$$f(X_1, \ldots, X_k) \vdash g(X_1, \ldots, X_k)$$

where the X_i need not be different, but occur at most once in f and g. Let us write $f(\vec{X})$ instead of $f(X_1, \ldots, X_k)$, and $f(\vec{X}^{cu})$ for $f(X_1^{cu}, \ldots, X_k^{cu})$. We know that

$$f(\vec{X})^{cu} \approx f(\vec{X}^{cu})$$

as well as

$$g(\vec{X})^{cu} \approx g(\vec{X}^{cu})$$

by the fact that $\approx$ is a congruence. Moreover, $f(\vec{X})$ is structurally similar to $f(\vec{X}^{cu})$ and $g(\vec{X})$ is similar to $g(\vec{X}^{cu})$. Thus the rule in question can be applied to the premisses

$$f(\vec{X}^{cu}) \vdash g(\vec{X}^{cu})$$

Thus if the following line occurred in Π

$$\frac{f_1(\vec{X}) \vdash g_1(\vec{X}) \quad \ldots \quad f_n(\vec{X}) \vdash g_n(\vec{X})}{r(\vec{X}) \vdash s(\vec{X})}$$

then we will replace it by the following proof:

$$\cfrac{\cfrac{\cfrac{f_1(\vec{X})^{cu} \vdash g_1(\vec{X})^{cu}}{\cdots}}{f_1(\vec{X}^{cu}) \vdash g_1(\vec{X}^{cu})} \quad \cdots \quad \cfrac{\cfrac{f_n(\vec{X})^{cu} \vdash g_n(\vec{X})^{cu}}{\cdots}}{f_n(\vec{X}^{cu}) \vdash g_n(\vec{X}^{cu})}}{\cfrac{\cfrac{r(\vec{X}^{cu}) \vdash s(\vec{X}^{cu})}{\cdots}}{r(\vec{X}) \vdash s(\vec{X})}}$$

The starting sequents contain reduced structures and immediately before applying the rule we have structures built from at most k occurrences of reduced structures. So it has no more symbols than a term of k occurrences of reduced structures. This k varies from rule to rule. But we can fix $\widehat{\kappa}$ as the maximum of all k in our calculus. Moreover, we can let $\widehat{\kappa}$ be a bit smaller than that, namely the maximum of occurences of variables in an antecedent or succedent of a rule. Furthermore, rules that are emptied by compression can be left out of consideration. At present, $\widehat{\kappa} = 2$. (For (A), where $\widehat{\kappa} = 3$, the forth-and-back translation makes the rule void, so we do not have to care about it.)

It should be clear that the number of steps filling the dots in the proof is bounded a priori; one can namely give exact bounds as to how many steps are needed to reduce a structure, and similarly how many are needed to mediate between the sequent $X \vdash Z$ and $Y \vdash Z$ when $X \approx Y$. In order to supply a rigorous argument here we give bounds on the *size* of intermediate sequents that need occur when passing from $X \vdash Z$ to $Y \vdash Z$. First, let W be reduced and $W \approx X$. Then it is enough to bound the size of sequents for a transformation $X \vdash Z \rightsquigarrow W \vdash Z$. The crux is that one cannot simply count the symbols in $X \vdash Z$ and hope that one never needs to use sequents that are longer. Namely, intermediate calculations may involve displaying – and displaying usually means an increase in the length of the structure. Define therefore first the *symbol count* $sc(X)$ ($sc(X \vdash Y)$) to be the function counting every occurring symbol except brackets (and $\vdash$). Then let $\sharp$ be an n-ary function and $\flat_j$ it's left or right dual for the jth position. Define the **display factor** ∇ as follows.

$$\nabla := max\{\frac{sc(\flat_j(X_1, \ldots, X_n))}{sc(\sharp(X_1, \ldots, X_n))} \mid \sharp \text{ an } n\text{-ary function symbol and } j \leq n\}$$

By induction it is proved that

$$\nabla = sup\{\frac{sc(X \vdash f^\delta(W))}{sc(f(X) \vdash W)} \mid f \text{ a unary structure polynomial}\}$$

Notice that $*$ is included in the definition of ∇ if $*$ counts as it's own dual. The factor ∇ then gives an upper bound for the cost of displaying material at the same side. It is, however, also a bound for displaying at the other side. Namely,

$$sup\{\frac{sc(*X \vdash Y)}{sc(*Y \vdash X)} \mid X, Y \in \mathfrak{Struc}\} = 1$$

as well as

$$sup\{\frac{sc(X \vdash Y)}{sc(**X \vdash Y)} \mid X, Y \in \mathfrak{Struc}\} = 1$$

Notice that we do not have to consider the quotient $sc(**X \vdash Y)/sc(X \vdash Y)$ because we are interested in removing $*$ from a symbol not adding it. This allows to have $\nabla = 3/2$ for the current calculus rather than $\nabla = 2$.

The algorithm for reducing a sequent consists in displaying structure parts for which a reduction can be applied, applying reduction rules and then undoing the display. Thus, while the reduction parts will actually not increase the length of the sequents it is the display strategy that can increase the length of the structure by a factor ∇. But when we undisplay, the resulting sequent will not be longer than the one we started off with. Thus, the only price to be payed is an increase in the length of occurring structures by the factor ∇.

Free University of Berlin, Germany.

REFERENCES

[1] N. D. Jr. Belnap. Display logic. *Journal of Philosophical Logic*, **11**, 375–417, 1982.
[2] K. Fine. Some connections between elementary and modal logic. In: S. Kanger, (ed.), *Proceedings of the Third Scandinavian Logic Symposium*, pp. 15–31. North-Holland, Amsterdam, 1975.
[3] M. Kracht. *Internal Definability and Completeness in Modal Logic*. PhD thesis, FU Berlin, 1991.
[4] M. Kracht. How completeness and correspondence theory got married. In: M. de Rijke, (ed.), *Diamonds and Defaults*, pp. 175–214. Kluwer, Dordrecht, 1993.
[5] M. Kracht. Highway to the danger zone. *Journal of Logic and Computation*, **5**, 93–109, 1995.
[6] M. Kracht. Tools and Techniques in Modal Logic. Freie Universität Berlin, 1995.
[7] H. Sahlqvist. First and second order semantics for modal logic. In: S. Kanger, (ed.), *Proceedings of the Third Scandinavian Logic Symposium*, pp. 15–31. North-Holland, Amsterdam, 1975.
[8] J. van Benthem. Correspondence theory. In: D. M. Gabbay, F. Guenthner, (eds), *Handbook of Philosophical Logic*, Vol. 2, pp. 167–247. Reidel, Dordrecht, 1984.
[9] H. Wansing. A full circle theorem for simple tense logic. In: M. de Rijke, (ed.), *Recent Advances in Modal and Intensional Logic*. 1994. To appear.
[10] H. Wansing. Sequent calculi for normal modal propositional logics. *Journal of Logic and Computation*, **4**, 125–142, 1994.
[11] H. Wansing. Strong cut-elimination in Display Logic. Institute of Logic and Philosophy of Science, University of Leipzig, 1994.

HEINRICH WANSING

A PROOF-THEORETIC PROOF OF FUNCTIONAL COMPLETENESS FOR MANY MODAL AND TENSE LOGICS

1 INTRODUCTION

Suppose $\mathcal{L}$ is a propositional logic with a finite set Σ of primitive finitary operations. Moreover, assume that $\mathcal{S}$ is a (strongly) sound and complete semantics for $\mathcal{L}$, and Ω is the class of all finitary operations explicitly definable in $\mathcal{S}$. It may be that Ω is defined by imposing some natural constraints on semantical models, if $\mathcal{S}$ is a model-theoretic semantics. Of course, these constraints must leave the connectives in Σ explicitly definable from finite combinations of connectives in Ω. The problem of functional completeness for $\mathcal{L}$ consists in finding a proper subclass Γ of Ω such that every connective in Ω is explicitly definable by a finite number of compositions from the elements of Γ. In a *proof-theoretic semantics* for $\mathcal{L}$, the class Ω of permissible connectives is given by general schemata for introducing finitary operations into premises and conclusions; see, for instance, [6], [7], [9], [10], [11] for proof-theoretic proofs of functional completeness for minimal propositional logic, intuitionistic propositional logic, Nelson's constructive propositional logics and various substructural subsystems of these logics.

It seems, however, that this approach cannot directly be applied to the standard Gentzen-style proof systems for tense or modal logic. The reason is that the operational rules of these proof systems fail to be *explicit* or even *weakly explicit.* Ohnishi's and Matsumoto's [8] right rule for $\Box$ in their sequent system for $S5$, for instance, not only exhibits $\Box$ on both sides of the sequent arrow in the conclusion sequent, but even exhibits $\Box$ in the premise sequent:

$$\Box\Delta \rightarrow \Box\Gamma, A \vdash \Box\Delta \rightarrow \Box\Gamma, \Box A$$

(here Δ, Γ range over finite sets of formulas and $\Box\Delta = \{\Box A \mid A \in \Delta\}$). It is not clear how such rules could be captured by a proof-theoretic semantics in terms of introduction schemata.

H. Wansing (ed.), Proof Theory of Modal Logic, 123–136.

In the present paper, it is shown that the proof-theoretic approach towards functional completeness *can* be applied to normal propositional modal and tense logics, if these systems are presentable as systems of (a suitable version of) Belnap's display logic (**DL**), see [1], [2], [12], [13]. Due to an enriched structural language in display logic, **G** ("always in the future", alias $\Box$) and **H** ("always in the past") can be given explicit introduction rules in a display calculus LK_t for the minimal normal tense logic K_t. Moreover, many important axiom schemata like D, T, 4, 5, and B can be expressed as purely structural rules, which results in a *modular* Gentzen-style proof theory for the most important systems of modal and basic tense logic. We shall use **DL** to define a natural proof-theoretic semantics in terms of general introduction schemata. The set of connectives $\Phi = \{\mathbf{G}, \mathbf{H}, \neg, \wedge\}$ will turn out to be functionally complete for K_t, and $\Psi = \{\Box, \neg, \wedge\}$ will be shown to be functionally complete for K, relative to straightforwardly restricted introduction schemata. The demonstration of functional completeness is such that it immediately reveals Φ and Ψ to be functionally complete for any displayable extension of K_t and K, respectively.[1] Since the systems to be dealt with are based on *classical* propositional logic, there is no need to consider higher-level sequents as, for example, in the case of intuitionistic logic, cf. [6], [9]. Moreover, in contrast to the proof-theoretic semantics for *substructural* logics in [10], [11] and [14], we need only *one* right and *one* left introduction schema.

2 DISPLAY LOGIC

Display logic is a Gentzen-style proof-theoretic framework using a generalized structural language. This language comprises four structural connectives, the nullary **I**, the unary operations $*$ and $\bullet$, and the binary $\circ$. Every (tense logical or modal) formula is considered to be a structure, and the structural connectives are used to build up more complex structures. We shall use X, Y, Z, X_1, X_2 etc. to denote arbitrary structures and A, B, C etc. to denote arbitrary formulas. The structural language can be motivated by general considerations on what would be desirable and natural, if one is dealing with a deducibility relation $\rightarrow$ between possibly complex data and goals (cf. [4]). In this context, the structural connectives can be understood as follows:

- **I** is the empty structure.

[1] This notion of functional completeness must, of course, not be confused with the notion of temporal completeness over a class of Kripke frames, cf. [3, p. 116 f.]. However, there is a connection. Whereas certain natural temporal operations like "until", which are first-order definable on Kripke frames, cannot explicitly be defined from **G**, **H**, and the Boolean connectives, one may ask for a semantics with respect to which **G**, **H**, and the Boolean connectives turn out to be functionally complete. Such a semantics is presented in the present paper.

– $\circ$ is structure addition.

– $*$ shifts structures from one side of $\rightarrow$ to the other.

– $\bullet$ is an intensionality mark switching from one side of $\rightarrow$ to the other.

The structural meaning of the structural connectives becomes clear from the basic structural rules of **DL**.

Basic structural rules:

$$\begin{array}{ll}(1) & X \circ Y \rightarrow Z \dashv\vdash X \rightarrow Z \circ Y^* \dashv\vdash Y \rightarrow X^* \circ Z \\ (2) & X \rightarrow Y \circ Z \dashv\vdash X \circ Z^* \rightarrow Y \dashv\vdash Y^* \circ X \rightarrow Z \\ (3) & X \rightarrow Y \dashv\vdash Y^* \rightarrow X^* \dashv\vdash X \rightarrow Y^{**} \\ (4) & X \rightarrow \bullet Y \dashv\vdash \bullet X \rightarrow Y,\end{array}$$

where $X_1 \rightarrow Y_1 \dashv\vdash X_2 \rightarrow Y_2$ abbreviates $X_1 \rightarrow Y_1 \vdash X_2 \rightarrow Y_2$ and $X_2 \rightarrow Y_2 \vdash X_1 \rightarrow Y_1$. If two sequents are interderivable by means of (1) - (4), then these sequents are said to be *structurally* or *display equivalent*. The following pairs of sequents are structurally equivalent on the strength of (1) - (3):

$$\begin{array}{ll} X \circ Y \rightarrow Z & Z^* \rightarrow Y^* \circ X^* \\ X \rightarrow Y \circ Z & Z^* \circ Y^* \rightarrow X^* \\ X \rightarrow Y & X^{**} \rightarrow Y \\ X^* \rightarrow Y & Y^* \rightarrow X \\ X \rightarrow Y^* & Y \rightarrow X^*. \end{array}$$

This 'geometry of structures' is suitable for formulating introduction rules for quite a few logical operations known from the field of substructural logic. Since in this paper we are interested in normal tense and modal logics extending classical propositional logic, we shall, however, assume further structural inference rules which make void distinctions between logical operations that otherwise would be at our disposal.

Additional structural rules:

$$
\begin{array}{ll}
(\mathbf{I}+) & X \to Z \vdash \mathbf{I} \circ X \to Z \\
 & X \to Z \vdash X \circ \mathbf{I} \to Z \\
(\mathbf{I}-) & \mathbf{I} \circ X \to Z \vdash X \to Z \\
 & X \circ \mathbf{I} \to Z \vdash X \to Z \\
(\mathbf{1}\ ex) & \mathbf{I} \to X \vdash Z \to X \\
(ex\ \mathbf{0}) & X \to \mathbf{I} \vdash X \to Z \\
(\mathbf{A}) & X_1 \circ (X_2 \circ X_3) \to Z \dashv\vdash (X_1 \circ X_2) \circ X_3 \to Z \\
(\mathbf{P}) & X_1 \circ X_2 \to Z \vdash X_2 \circ X_1 \to Z \\
(\mathbf{C}) & X \circ X \to Z \vdash X \to Z \\
(\mathbf{M}) & X_1 \to Z \vdash X_1 \circ X_2 \to Z \\
 & X_1 \to Z \vdash X_2 \circ X_1 \to Z \\
(MN) & \mathbf{I} \to X \vdash \bullet\mathbf{I} \to X \\
 & \mathbf{I} \to X \vdash \bullet(X^*) \to \mathbf{I}^*.
\end{array}
$$

This menu of structural rules allows one to dispense with rules like permutation on the right hand side of $\to$, which would make the calculus more symmetric. The following context-sensitive reading of the structural connectives gives rise to the introduction rules for the usual logical operations of propositional modal and simple tense logic:

structural connective	lhs of $\to$	rhs of $\to$
$\mathbf{I}$	$\mathbf{1}$	$\mathbf{0}$
$\circ$	$\wedge$	$\vee$
$*$	$\neg$	$\neg$
$\bullet$	**P** "sometimes in the past"	**G** ($\Box$)

Operational rules:

$$
\begin{array}{ll}
(\to \mathbf{0}) & X \to \mathbf{I} \vdash X \to \mathbf{0} \\
(\mathbf{0} \to) & \vdash \mathbf{0} \to \mathbf{I} \\
(\to \mathbf{1}) & \vdash \mathbf{I} \to \mathbf{1} \\
(\mathbf{1} \to) & \mathbf{I} \to X \vdash \mathbf{1} \to X \\
(\to \neg) & X \to A^* \vdash X \to \neg A \\
(\neg \to) & A^* \to X \vdash \neg A \to X \\
(\to \mathbf{G}) & \bullet X \to A \vdash X \to \mathbf{G}A \\
(\mathbf{G} \to) & A \to X \vdash \mathbf{G}A \to \bullet X \\
(\to \mathbf{F}) & X \to A \vdash (\bullet(X^*))^* \to \mathbf{F}A \\
(\mathbf{F} \to) & (\bullet(A^*))^* \to Y \vdash \mathbf{F}A \to Y
\end{array}
$$

$(\rightarrow \mathbf{H})$	$X \rightarrow (\bullet(A^*))^* \vdash X \rightarrow \mathbf{H}A$
$(\mathbf{H} \rightarrow)$	$A \rightarrow X \vdash \mathbf{H}A \rightarrow (\bullet(X^*))^*$
$(\rightarrow \mathbf{P})$	$X \rightarrow A \vdash \bullet X \rightarrow \mathbf{P}A$
$(\mathbf{P} \rightarrow)$	$A \rightarrow \bullet X \vdash \mathbf{P}A \rightarrow X$
$(\rightarrow \wedge)$	$X \rightarrow A \quad Y \rightarrow B \vdash X \circ Y \rightarrow A \wedge B$
$(\wedge \rightarrow)$	$A \circ B \rightarrow X \vdash A \wedge B \rightarrow X$
$(\rightarrow \vee)$	$X \rightarrow A \circ B \vdash X \rightarrow A \vee B$
$(\vee \rightarrow)$	$A \rightarrow X \quad B \rightarrow Y \vdash A \vee B \rightarrow X \circ Y$
$(\rightarrow \supset)$	$X \circ A \rightarrow B \vdash X \rightarrow A \supset B$
$(\supset \rightarrow)$	$X \rightarrow A \quad B \rightarrow Y \vdash A \supset B \rightarrow X^* \circ Y.$

Finally, there are two

Logical rules:[2]

$$(\text{Id}) \quad \vdash A \rightarrow A \qquad \text{and} \qquad (\text{Cut}) \quad X \rightarrow A \quad A \rightarrow Y \vdash X \rightarrow Y.$$

Let us refer to the above sequent system as LK_t, and let us refer to the result of dropping the rules for **H** and **P** from LK_t as LK.[3]

Display logic derives its name from the fact that any substructure of a given sequent s may be displayed as the entire antecedent or succedent, respectively, of a structurally equivalent sequent s'. In order to state this fact precisely, we need a few definitions. An occurrence of a substructure in a given structure is said to be positive (negative), if it is in the scope of an even (uneven) number of *'s. The structure X (Y) is called the antecedent (succedent) of $X \rightarrow Y$. An antecedent (succedent) part of a sequent $X \rightarrow Y$ is a positive occurrence of a substructure of X or a negative occurrence of a substructure of Y (a positive occurrence of a substructure of Y or a negative occurrence of a substructure of X). One can then prove (see [1, p. 381]):

THEOREM 1 (Display Theorem) *For every sequent s and every antecedent (succedent) part X of s there exists a sequent s' structurally equivalent with s, such that X is the antecedent (succedent) of s'.*

Next, consider the following translation τ of sequents into (tense logical) formulas:

$$\tau(X \rightarrow Y) = \tau_1(X) \supset \tau_2(Y),$$

[2] It would be enough to require $\vdash p \rightarrow p$, for propositional variables p. (Id) can then be proved by induction on A.

[3] Thus, LK here does *not* denote the standard sequent system for classical predicate logic.

where τ_i $(i = 1, 2)$ is defined as follows:

$$\begin{array}{lcl}
\tau_i(A) & = & A \\
\tau_1(\mathbf{I}) & = & \mathbf{1} \\
\tau_2(\mathbf{I}) & = & \mathbf{0} \\
\tau_1(X^*) & = & \neg\tau_2(X) \\
\tau_2(X^*) & = & \neg\tau_1(X) \\
\tau_1(X \circ Y) & = & \tau_1(X) \wedge \tau_1(Y) \\
\tau_2(X \circ Y) & = & \tau_2(X) \vee \tau_2(Y) \\
\tau_1(\bullet X) & = & \mathbf{P}\tau_1(X) \\
\tau_2(\bullet X) & = & \mathbf{G}\tau_2(X).
\end{array}$$

THEOREM 2 *(i) If* $\vdash A$ *in* K_t*, then* $\vdash \mathbf{I} \rightarrow A$ *in* LK_t*. (ii) If* $\vdash X \rightarrow Y$ *in* LK_t*, then* $\vdash \tau(X \rightarrow Y)$ *in* K_t*.*

Proof. See [12], [13]. ■

COROLLARY 3 *(i)* $\vdash A$ *in* K_t *iff* $\vdash \mathbf{I} \rightarrow A$ *in* LK_t*. (ii)* $\vdash A$ *in* K *iff* $\vdash \mathbf{I} \rightarrow A$ *in* LK*.*

Proof. (i) By the previous theorem. (ii) By the obvious fact that each normal propositional tense logic which is complete with respect to a class of Kripke frames is a conservative extension of its fragment in $\{\mathbf{0}, \mathbf{1}, \mathbf{G}, \mathbf{F}, \neg, \wedge, \vee, \supset\}$.[4] ■

A normal propositional tense (modal) logic S is said to be *properly displayable* if (i) S can be presented as a calculus of sequents which satisfies certain conditions ensuring the subformula-property and the eliminability of Cut (see [1], [15]), and (ii) S is the result of extending LK_t (LK) by purely structural inference rules. The class of all properly displayable normal propositional tense logics has recently been characterized by Marcus Kracht [5]. In fact, many important axiom schemata can be captured by purely structural rules of inference, thereby giving us a *modular* sequent-style proof theory for extensions of K_t and K. If R is any of these axiom schemata, we shall associate with R a structural inference rule R', for instance:

R		R'	
D	$\mathbf{G}A \supset \neg\mathbf{G}\neg A$	D'	$\bullet X \circ \bullet Y \to \mathrm{I}^* \vdash X \to Y^*$
T	$\mathbf{G}A \supset A$	T'	$X \to \bullet Y \vdash X \to Y$
4	$\mathbf{G}A \supset \mathbf{GG}A$	$4'$	$X \to \bullet Y \vdash X \to \bullet\bullet Y$
5	$\neg\mathbf{G}\neg A \supset \mathbf{G}\neg\mathbf{G}\neg A$	$5'$	$(\bullet(X^*))^* \to Y \vdash \bullet((\bullet(X^*))^*) \to Y$
B	$A \supset \mathbf{G}\neg\mathbf{G}\neg A$	B'	$(\bullet(X^*))^* \to Y \vdash \bullet X \to Y$
$A2$	$\mathbf{GG}A \supset \mathbf{G}A$	$A2'$	$\bullet\bullet X \to Y \vdash \bullet X \to Y$
$A3$	$A \supset \mathbf{FP}A$	$A3'$	$(\bullet((\bullet X)^*))^* \to Y \vdash X \to Y$
$A4$	$A \supset \mathbf{PF}A$	$A4'$	$\bullet((\bullet(X^*))^*) \to Y \vdash X \to Y$
$A5$	$\mathbf{F}A \supset \mathbf{G}(\mathbf{F}A \vee A \vee \mathbf{P}A)$	$A5'$	$(\bullet(X^*))^* \to Y \quad X \to Y \quad \bullet X \to Y \vdash$ $\vdash \bullet((\bullet(X^*))^*) \to Y$
$A6$	$\mathbf{P}A \supset \mathbf{G}(\mathbf{F}A \vee A \vee \mathbf{P}A)$	$A6'$	$(\bullet(X^*))^* \to Y \quad X \to Y \quad \bullet X \to Y \vdash$ $\vdash (\bullet((\bullet X)^*))^* \to Y$

Let Δ be the set of all of the above axiom schemata R and $\Gamma \subseteq \Delta$. Then $\Gamma' = \{R' \mid R \in \Gamma\}$.

THEOREM 4 *(i) In* $LK_t \cup \Gamma'$, $\vdash \mathbf{I} \to A$ *iff* $\vdash A$ *in* $K_t \cup \Gamma$. *(ii) Cut is admissible in* $LK_t \cup \Gamma'$.

Proof. (i): see [12], [13]; (ii): see [1]. ■

3 PROOF-THEORETIC SEMANTICS

We shall assume the structural language, the logical rules, and the (basic and additional) structural rules of the previous section. This is our 'structural framework'; and we will refer to it as **B**. The idea of the proof-theoretic semantics is to delimit the class of admissible connectives by specifying general schemata for introducing (finitary) connectives of a propositional language L into premises and conclusions, that is, on both sides of the sequent arrow. Of course, these schemata must not be arbitrary, cf. [6]. A rule schema for an n-ary connective F should exhibit no other connective than F, and the deductive role of formulas $F(A_1, \ldots, A_n)$ should depend only on the deductive relationships among the parameters $A_1, \ldots, A_n$. Moreover, the schemata should be conservative (or non-creative), that is, each proof of an F-free formula A should be convertible into a proof of A with no applications of rules characterizing F. This condition boils down to the requirement that applications of cut, in which both premise sequents are derived by introducing the cut-formula $F(A_1, \ldots, A_n)$, can be eliminated.

DEFINITION 5 *The general right introduction schema is this:*

$$(\to F) \quad X_1 \to Y_1 \; \ldots \; X_k \to Y_k \;\vdash\; X_1 \circ \ldots \circ X_k \to F(A_1, \ldots, A_n),$$

[4] **F** is the operation "sometimes in the future".

where every X_i $(i = 1, \ldots, k)$ is an unspecified structure, every Y_i contains only formulas from A_1, ..., A_n, and every A_j $(j = 0, \ldots, n)$ occurs in some Y_i.

We have to state a left introduction schema such that the eliminability constraint is satisfied. In order to be able to succinctly formulate such a schema, we define certain operations on sequents.

DEFINITION 6 *The operations $(\cdot)^l$, $(\cdot)^r$ on sequents are simultaneously defined as follows:*

$$(Y \to Z)^l = \begin{cases} \mathbf{I} \to Z & \text{if } Y = \mathbf{I} \\ A \to Z & \text{if } Y = A \\ (W \to Z)^l & \text{if } Y = \bullet W \\ (Z \to W)^r & \text{if } Y = W^* \\ (W_1 \to Z_1)^l \ (W_2 \to Z_2)^l & \text{if } Y = W_1 \circ W_2 \end{cases}$$

$$(Z \to Y)^r = \begin{cases} Z \to \mathbf{I} & \text{if } Y = \mathbf{I} \\ Z \to A & \text{if } Y = A \\ (Z \to W)^r & \text{if } Y = \bullet W \\ (W \to Z)^l & \text{if } Y = W^* \\ (Z_1 \to W_1)^r \ (Z_2 \to W_2)^r & \text{if } Y = W_1 \circ W_2, \end{cases}$$

where Z, Z_1, and Z_2 are unspecified structures.

Clearly $(Y \to Z)^l$ is a sequence of sequents each of which has either the shape $A \to X$, $X \to A$, $\mathbf{I} \to X$, or $X \to \mathbf{I}$. Moreover, for every occurrence of a formula A in Y, there is exactly one sequent $A \to X$ or $X \to A$ in $(Y \to Z)^l$. We shall call X the structure *corresponding to* this occurrence of A in $(Y \to Z)^l$.

DEFINITION 7 *The general left introduction schema is:*

$$\begin{array}{ll} (F \to) & (Y_1 \to W)^l \ \vdash \ F(A_1, \ldots, A_n) \to (Y_1)^\sharp \\ & \vdots \\ & (Y_k \to W)^l \ \vdash \ F(A_1, \ldots, A_n) \to (Y_k)^\sharp, \end{array}$$

where W is an unspecified structure and $(Y_i)^\sharp$ is the result of replacing every occurrence of a formula in Y_i by its corresponding structure in $(Y_i \to W)^l$.

EXAMPLE 8 Consider the following instantiation of $(\to F)$:

$$X \to (A^* \circ \bullet B) \circ \bullet(\mathbf{I}^*) \vdash X \to F(A, B).$$

We have

$$
\begin{array}{rl}
& ((A^* \circ \bullet B) \circ \bullet(\mathbf{I}^*) \to W)^l \\
= & (A^* \circ \bullet B \to W_1)^l \; (\bullet(\mathbf{I}^*) \to W_2)^l \\
= & (A^* \to W_{1_1})^l \; (\bullet B \to W_{1_2})^l \; (\mathbf{I}^* \to W_2)^l \\
= & (W_{1_1} \to A)^r \; (B \to W_{1_2})^l \; (W_2 \to \mathbf{I})^r \\
= & W_{1_1} \to A \;\; B \to W_{1_2} \;\; W_2 \to \mathbf{I}
\end{array}
$$

and obtain the following instantiation of $(F \to)$:

$$
W_{1_1} \to A \;\; B \to W_{1_2} \;\; W_2 \to \mathbf{I} \vdash F(A, B) \to (W_{1_1}^* \circ \bullet W_{1_2}) \circ \bullet(\mathbf{I}^*).
$$

We must verify that the eliminability constraint is indeed satisfied.

OBSERVATION 9 *For every proof*

$$
\dfrac{(1)\;\; \dfrac{X_1 \to Y_1 \;\ldots\; X_k \to Y_k}{X_1 \circ \ldots \circ X_k \to F(A_1, \ldots, A_n)} \qquad (2)\;\; \dfrac{(Y_i \to W)^l}{F(A_1, \ldots, A_n) \to (Y_i)^\sharp}}{(3)\;\; X_1 \circ \ldots \circ X_k \to (Y_i)^\sharp}
$$

in which the cut-formula $F(A_1, \ldots, A_n)$ is introduced in the inferences ending in (1) and (2) and in which (3) is not identical to one of (1) or (2), there is a proof of (3) in $\mathbf{B} + (\to F) + (F \to)$ from the premises of (1) and (2) in which every cut-fromula of any application of cut is a proper subformula of $F(A_1, \ldots, A_n)$.

Proof. Suppose $(Y_i \to W)^l = s_{i_1} \ldots s_{i_m}$. Consider the sequent s_{i_1}. If $s_{i_1} = A \to V$ or $V \to A$, then it is clear from the definition of $(\cdot)^l$ and $(\cdot)^r$ that a certain occurrence of A in Y_i is a succedent part of $X_i \to Y_i$ or an antecedent part of $X_i \to Y_i$, respectively. We transform $X_i \to Y_i$ into a structurally equivalent sequent $U \to A$ or $A \to U$, respectively, and then apply cut to obtain $U \to V$ or $V \to U$, respectively. Then we iterate this process for $s_{i_2}, \ldots, s_{i_m}$ (if possible) and finally obtain $X_i \to (Y_i)^\sharp$. Applying (**M**) gives $X_1 \circ \ldots \circ X_k \to (Y_i)^\sharp$. ■

EXAMPLE 4 (continued) We show in some detail how the procedure is applied. Our starting point is the following application of cut:

$$
\dfrac{\dfrac{X \to (A^* \circ \bullet B) \circ \bullet(\mathbf{I}^*)}{X \to F(A, B)} \qquad \dfrac{W_{1_1} \to A \quad B \to W_{1_2} \quad W_2 \to \mathbf{I}}{F(A, B) \to (W_{1_1}^* \circ \bullet W_{1_2}) \circ \bullet(\mathbf{I}^*)}}{X \to (W_{1_1} \circ \bullet W_{1_2}) \circ \bullet(\mathbf{I}^*)}
$$

We first display A in $X \rightarrow (A^* \circ \bullet B) \circ \bullet(\mathbf{I}^*)$ and use it as cut-formula:

$$\dfrac{W_{1_1} \rightarrow A \qquad \dfrac{\dfrac{\dfrac{X \rightarrow (A^* \circ \bullet B) \circ \bullet(\mathbf{I}^*)}{X \circ (\bullet(\mathbf{I}^*))^* \rightarrow A^* \circ \bullet B}}{(X \circ (\bullet(\mathbf{I}^*))^*) \circ (\bullet B)^* \rightarrow A^*}}{A \rightarrow ((X \circ (\bullet(\mathbf{I}^*))^*) \circ (\bullet B)^*)^*}}{W_{1_1} \rightarrow ((X \circ (\bullet(\mathbf{I}^*))^*) \circ (\bullet B)^*)^*}$$

Then we display B in $W_{1_1} \rightarrow ((X \circ (\bullet(\mathbf{I}^*))^*) \circ (\bullet B)^*)^*$ and use it as cut-formula:

$$\dfrac{\dfrac{\dfrac{\dfrac{\dfrac{W_{1_1} \rightarrow ((X \circ (\bullet(\mathbf{I}^*))^*) \circ (\bullet B)^*)^*}{(X \circ (\bullet(\mathbf{I}^*))^*) \circ (\bullet B)^* \rightarrow W_{1_1}^*}}{X \circ (\bullet(\mathbf{I}^*))^* \rightarrow W_{1_1}^* \circ \bullet B}}{W_{1_1} \circ (X \circ (\bullet(\mathbf{I}^*))^*) \rightarrow \bullet B}}{\bullet(W_{1_1} \circ (X \circ (\bullet(\mathbf{I}^*))^*)) \rightarrow B} \qquad B \rightarrow W_{1_2}}{\bullet(W_{1_1} \circ (X \circ (\bullet(\mathbf{I}^*))^*)) \rightarrow W_{1_2}}$$

The conclusion sequent we have reached is structurally equivalent with the sequent we wanted to derive, viz. $X \rightarrow (W_{1_1}^* \circ \bullet W_{1_2}) \circ \bullet(\mathbf{I}^*)$.

One might be inclined to object that the schema $(\rightarrow F)$ is not general enough, since it permits only connectives with exactly one right introduction rule. This is, however, no real restriction, since multiple right introduction rules can be interpreted disjunctively. We shall illustrate this by means of a simple example.[5]

EXAMPLE 10 Consider the four-place connective $\lhd$ with the following right introduction rules:

$$\begin{array}{ll}(\rightarrow\lhd) & X_{1_1} \rightarrow A \;\; X_{1_2} \rightarrow B \vdash X_{1_1} \circ X_{1_2} \rightarrow\lhd (A, B, C, D) \\ & X_{2_1} \rightarrow C \;\; X_{2_2} \rightarrow D \vdash X_{2_1} \circ X_{2_2} \rightarrow\lhd (A, B, C, D).\end{array}$$

$(\rightarrow\lhd)$ is interreplaceable with the single rule

$$\begin{array}{ll}(\rightarrow\lhd)' & X \rightarrow A \circ C \;\; X \rightarrow A \circ D \;\; X \rightarrow B \circ C \;\; X \rightarrow B \circ D \vdash \\ & \vdash X \rightarrow\lhd (A, B, C, D).\end{array}$$

[5] If there is more than one right introduction rule, it seems to be natural to require that every A_j $(j = 0, \ldots, n)$ occurs in a premise sequent of some right rule.

$(\rightarrow\triangleleft) \Rightarrow (\rightarrow\triangleleft)'$: Consider the following derivation:

$$\dfrac{\dfrac{\dfrac{\dfrac{\dfrac{X \rightarrow A \circ C}{X \circ C^* \rightarrow A} \quad \dfrac{X \rightarrow B \circ C}{X \circ C^* \rightarrow B}}{(X \circ C^*) \circ (X \circ C^*) \rightarrow F(A,B,C,D)}}{X \circ C^* \rightarrow F(A,B,C,D)}}{X \rightarrow F(A,B,C,D) \circ C}}{F(A,B,C,D)^* \circ X \rightarrow C}$$

Similarly we derive $F(A,B,C,D)^* \circ X \rightarrow D$ and continue as follows:

$$\dfrac{\dfrac{\dfrac{\dfrac{\dfrac{\dfrac{F(A,B,C,D)^* \circ X \rightarrow C \quad F(A,B,C,D)^* \circ X \rightarrow D}{(F(A,B,C,D)^* \circ X) \circ (F(A,B,C,D)^* \circ X) \rightarrow F(A,B,C,D)}}{F(A,B,C,D)^* \circ X \rightarrow F(A,B,C,D)}}{F(A,B,C,D)^* \rightarrow F(A,B,C,D) \circ X^*}}{F(A,B,C,D)^* \circ F(A,B,C,D)^* \rightarrow X^*}}{F(A,B,C,D)^* \rightarrow X^*}}{X \rightarrow F(A,B,C,D).}$$

$(\rightarrow\triangleleft)' \Rightarrow (\rightarrow\triangleleft)$: Consider the following derivations:

$$\dfrac{\dfrac{\dfrac{X_{1_1} \rightarrow A}{X_{1_1} \circ X_{1_2} \rightarrow A}}{(X_{1_1} \circ X_{1_2}) \circ C^* \rightarrow A}}{X_{1_1} \circ X_{1_2} \rightarrow A \circ C} \qquad \dfrac{\dfrac{\dfrac{X_{1_2} \rightarrow B}{X_{1_1} \circ X_{1_2} \rightarrow B}}{(X_{1_1} \circ X_{1_2}) \circ C^* \rightarrow B}}{X_{1_1} \circ X_{1_2} \rightarrow B \circ C.}$$

Similarly we derive $X_{1_1} \circ X_{1_2} \rightarrow A \circ D$ and $X_{1_1} \circ X_{1_2} \rightarrow B \circ D$. Applying $(\rightarrow\triangleleft)'$ gives $X_{1_1} \circ X_{1_2} \rightarrow F(A,B,C,D)$.

Note that $(\rightarrow\triangleleft)$ is also interreplaceable with

$$(\rightarrow\triangleleft)'' \quad X_1 \rightarrow A \circ C \quad X_2 \rightarrow A \circ D \quad X_3 \rightarrow B \circ C \quad X_4 \rightarrow B \circ D \vdash \\ \vdash X_1 \circ X_2 \circ X_3 \circ X_4 \rightarrow\triangleleft (A,B,C,D).$$

For the proof of functional completeness, we shall need a replacement theorem. Let C_A denote an L-formula, which contains a certain occurrence of A as a subformula, and let C_B denote the result of replacing this occurrence of A in C by B. We use $[B/A]X$ to denote the result of replacing every occurrence of A in the structure X by B. The degree of A ($d(A)$) is defined as the number of occurrences of propositional connectives in A. Let $\vdash X \leftrightarrow Y$ abbreviate $\vdash X \rightarrow Y$ and $\vdash Y \rightarrow X$.

THEOREM 11 *If* $\vdash A \leftrightarrow B$ *in* $\mathbf{B} + (\rightarrow F) + (F \rightarrow)$, *then* $\vdash C_A \leftrightarrow C_B$ *in* $\mathbf{B} + (\rightarrow F) + (F \rightarrow)$.

Proof. If $d(C_A) - d(A) = 0$, there is nothing to prove. Suppose therefore that $d(C_A) - d(A) > 0$. Then C_A has the form $F(A_1, \ldots, A_n)_A$. For every $i = 1, \ldots, k$, we consider a certain instantiation of $(Y_i \rightarrow W)^l = s_{i_1} \ldots s_{i_m}$. If s_{i_q} has the shape $A \rightarrow V$ or $V \rightarrow A$ ($q = 1, \ldots, m$), we instantiate V by B. If s_{i_q} has the shape $\mathbf{I} \rightarrow V$ or $V \rightarrow \mathbf{I}$, we instantiate V by $p \circ p^*$, for some propositional variable p. If s_{i_q} has the shape $D \rightarrow V$ or $V \rightarrow D$ and $A \neq D$, we instantiate V by D. Obviously, every instantiation of any s_{i_q} is provable. Therefore by $(F \rightarrow)$ we obtain $\vdash F(A_1, \ldots, A_n)_A \rightarrow [B/A]Y_i$, and by $(\rightarrow F)$ together with (**C**) it follows that $\vdash C_A \rightarrow C_B$. Analogously, we can show $\vdash C_B \rightarrow C_A$. ■

4 FUNCTIONAL COMPLETENESS

We shall first show that $\Phi' = \{\mathbf{0}, \mathbf{1}, \neg, \mathbf{G}, \mathbf{P}, \wedge, \vee\}$ is functionally complete for K_t and then prove functional completeness of $\Psi' = \{\mathbf{0}, \mathbf{1}, \neg, \mathbf{G}, \wedge, \vee\}$ for K. Functional completeness of Φ for K_t and Ψ for K follows from standard definitions. By induction on the complexity of X, we can prove

LEMMA 12 *In* LK_t*: (i)* $\vdash X \rightarrow \tau_1(X)$. *(ii)* $\vdash \tau_2(X) \rightarrow X$. *(iii)* $\vdash X \rightarrow Y$ *implies* $\vdash \tau_1(X) \rightarrow Y$. *(iv)* $\vdash Y \rightarrow X$ *implies* $\vdash Y \rightarrow \tau_2(X)$.

THEOREM 13 Φ' *is functionally complete for* K_t.

Proof. Suppose that the rules for F are instantiations of $(\rightarrow F)$ and $(F \rightarrow)$. Then

	$\vdash \tau_2(Y_i) \rightarrow Y_i$	by Lemma 12 (ii)
$\Rightarrow$	$\vdash \tau_2(Y_1) \circ \ldots \circ \tau_2(Y_k) \rightarrow F(A_1, \ldots, A_n)$	by $(\rightarrow F)$
$\Rightarrow$	$\vdash \tau_1(\tau_2(Y_1) \circ \ldots \circ \tau_2(Y_k)) \rightarrow F(A_1, \ldots, A_n)$	by Lemma 12 (iii)

Moreover,

	$\vdash F(A_1, \ldots, A_n) \rightarrow Y_i$	by (Id) and $(F \rightarrow)$
$\Rightarrow$	$\vdash F(A_1, \ldots, A_n) \rightarrow \tau_2(Y_i)$	by Lemma 12 (iv)
$\Rightarrow$	$\vdash F(A_1, \ldots, A_n) \rightarrow \tau_2(Y_1) \circ \ldots \circ \tau_2(Y_k)$	by (**M**) and display equivalence
$\Rightarrow$	$\vdash F(A_1, \ldots, A_n) \rightarrow \tau_1(\tau_2(Y_1) \circ \ldots \circ \tau_2(Y_k))$	by Lemma 12 (i)

By the previous theorem, it follows that $F(A_1, \ldots, A_n)$ and $\tau_1(\tau_2(Y_1) \circ \ldots \circ \tau_2(Y_k))$ are provably interreplaceable.

It still has to be shown that for every primitive operation $\sharp$ of K_t, either the rules for $\sharp$ are (interreplaceable with) instantiations of $(\rightarrow F)$ and $(F \rightarrow)$ or $\sharp$ is explicitly definable by a finite composition of operations from Φ'. This can be verified by eye and the following straightforward observations:(i) $(\mathbf{0} \rightarrow)$ is interreplaceable with $(\mathbf{0} \rightarrow)'$ $\quad \mathbf{I} \rightarrow X \vdash \mathbf{0} \rightarrow \mathbf{I}$. (ii) $(\neg \rightarrow)$ is interreplaceable with $(\neg \rightarrow)'$ $\quad X \rightarrow A \vdash \neg A \rightarrow X^*$. (iii) The 'multiplicative' rules for $\wedge$ are interreplaceable with their 'additive' formulations:

$$
\begin{array}{ll}
(\rightarrow \wedge)' & X \rightarrow A \quad X \rightarrow B \vdash X \rightarrow (A \wedge B) \\
(\wedge \rightarrow)' & A \rightarrow X \vdash (A \wedge B) \rightarrow X \\
 & B \rightarrow X \vdash (A \wedge B) \rightarrow X.
\end{array}
$$

(iv) The rules for $\supset$ are interreplaceable with:

$$
\begin{array}{ll}
(\rightarrow \supset)' & X \rightarrow A^* \circ B \vdash X \rightarrow (A \supset B) \\
(\supset \rightarrow)' & A^* \rightarrow X \quad B \rightarrow Y \vdash A \supset B \rightarrow X \circ Y.
\end{array}
$$

(v) $\mathbf{1}$ is definable as $(p \vee \neg p)$ for some propositional variable p. (vi) $\mathbf{F}A$ is definable as $\neg\mathbf{G}\neg A$; $\mathbf{H}A$ is definable as $\neg\mathbf{P}\neg A$. In the logical language of K_t, we thus have $\vdash X \rightarrow A$ in K_t iff $\vdash X \rightarrow A$ in $\mathbf{B} + (\rightarrow F) + (F \rightarrow)$. ■

EXAMPLE 4 (continued) We get $(\neg A \vee \mathbf{G}B) \vee \mathbf{G}\neg\mathbf{1}$ as definiens for $F(A, B)$.

THEOREM 14 *Ψ' is functionally complete for K.*

Proof. The proof is completely analogous to the proof for K_t except that now the schemata $(\rightarrow F)$ and $(F \rightarrow)$ have to be suitably restricted: in every Y_i the structural connective $\bullet$ may occur only within the scope of an even number of *'s. Then the definiens $\tau_1(\tau_2(Y_1) \circ \ldots \circ \tau_2(Y_k))$ remains in the language of modal logic. ■

The cut rule remains eliminable, if the conditions on inference rules defining properly displayable logics are slightly generalized, so as to allow to present, for example, classical linear propositional logic as a cut-free display calculus, see [2], [15]. If S (i) satisfies these more general conditions, and (ii) S is an extension of LK_t (LK) by purely structural rules, then S is said to be a *displayable* normal propositional tense (modal) logic.

COROLLARY 15 *(i) Φ is functionally complete for every displayable normal propositional tense logic.(ii) Ψ is functionally complete for every displayable normal propositional modal logic.*

Proof. Adding *purely structural* rules does not affect the above proofs of functional completeness. ■

ACKNOWLEDGEMENTS

I would like to thank Marcus Kracht and Grigori Mints for their useful comments on an earlier version of this paper and Nuel Belnap for giving me the opportunity to present it at the Department of Philosophy of Pittsburgh University.

University of Leipzig, Germany.

REFERENCES

[1] N. Belnap. Display Logic. *Journal of Philosophical Logic*, **11**, 375–417, 1982.
[2] N. Belnap. Linear Logic Displayed. *Notre Dame Journal of Formal Logic*, **31**, 14–25, 1990.
[3] J. Burgess. Basic Tense Logic. In: D. M. Gabbay, F. Guenthner (eds.), *Handbook of Philosophical Logic*, Vol. II, pp. 89–133, Reidel, Dordrecht, 1984.
[4] D. Gabbay. A general theory of structured consequence relations. In: K. Došen, P. Schroeder-Heister (eds.), *Substructural Logics*, pp. 109–151, Oxford University Press, Oxford, 1993.
[5] M. Kracht. Power and Weakness of the Modal Display Calculus, 1993. In: H. Wansing (ed.), *Proof Theory of Modal Logic*, this volume.
[6] F. von Kutschera. Die Vollständigkeit des Operatorensystems $\{\neg, \wedge, \vee, \supset\}$ für die intuitionistische Aussagenlogik im Rahmen der Gentzensemantik. *Archiv für Mathematische Logik und Grundlagenforschung*, **11**, 3–16, 1968.
[7] F. von Kutschera. Ein verallgemeinerter Widerlegungsbegriff für Gentzenkalküle. *Archiv für Mathematische Logik und Grundlagenforschung*, **12**, 104–118, 1969.
[8] M. Ohnishi and K. Matsumoto. Gentzen Method in Modal Calculi. *Osaka Mathematical Journal*, **9**, 113–130, 1957.
[9] P. Schroeder-Heister. A Natural Extension of Natural Deduction. *Journal of Symbolic Logic* **49**, 1284–1300, 1984.
[10] H. Wansing. Functional Completeness for Subsystems of Intuitionistic Propositional Logic. *Journal of Philosophical Logic*, **22**, 303–321, 1993.
[11] H. Wansing. *The Logic of Information Structures*. Lecture Notes in AI 681, Springer-Verlag, Berlin, 1993.
[12] H. Wansing. Sequent Calculi for Normal Modal Propositional Logics. *Journal of Logic and Computation*, **4**, 125–142, 1994.
[13] H. Wansing. A Full-Circle Theorem for Simple Tense Logic. To appear in: M. de Rijke (ed.), *Advances in Intensional Logic*, 1995.
[14] H. Wansing. Tarskian Structured Consequence Relations and Functional Completeness. *Mathematical Logic Quarterly*, **41**, 73–92, 1995.
[15] H. Wansing. Strong cut-elimination in Display Logic. Manuscript, Institute of Logic and Philosophy of Science, University of Leipzig, 1994.

RAJEEV GORÉ

ON THE COMPLETENESS OF CLASSICAL MODAL DISPLAY LOGIC

1 COMPLETENESS OF CMDL

We wish to prove that the CMDL system of Wansing [2] and Kracht [1] is sound and complete with respect to the tense logic **Kt**. Fortunately the direction corresponding to soundness of CMDL is easy, leaving us with the task of proving completeness.

The known proofs of completeness for CMDL proceed via the following translation τ of structures (denoted by X, Y and Z) into formulae (denoted by A):

$$\tau(X \vdash Y) = (\tau_1(X) \to \tau_2(Y))$$

$$\begin{array}{lcllcl}
\tau_1(A) & = & A & \tau_2(A) & = & A \\
\tau_1(I) & = & \top & \tau_2(I) & = & \bot \\
\tau_1(\bullet X) & = & \blacklozenge\tau_1(X) & \tau_2(\bullet X) & = & \Box\tau_2(X) \\
\tau_1(*X) & = & \neg\tau_2(X) & \tau_2(*X) & = & \neg\tau_1(X) \\
\tau_1(X \circ Y) & = & \tau_1(X) \wedge \tau_1(Y) & \tau_2(X \circ Y) & = & \tau_2(X) \vee \tau_2(Y)
\end{array}$$

With respect to this translation, the statement of completeness is: if $\tau(X \vdash Y)$ is a theorem of **Kt** then $X \vdash Y$ is provable in CMDL.

Note that completeness here is stated at the level of arbitrary *structures* X and Y, not arbitrary *formulae* as in the original proof of Wansing recorded below as Lemma 1.

Kracht [1] has already proved completeness at the level of structures, but his proof is quite complicated involving not only a reference to Hilbert characterisations of classical modal logics, but also the addition of extra rules which then have to be shown to be admissible. Here we give a greatly simplified and completely syntactic proof of the completeness of CMDL at the level of structures.

H. Wansing (ed.), Proof Theory of Modal Logic, 137–140.

LEMMA 1 *The sequent $I \vdash A$ is provable in CMDL iff the formula A is a theorem of the tense logic* **Kt**. *See Wansing [2].*

In particular, if we know that the formula $\tau(X \vdash Y)$ is a theorem of tense logic **Kt** then we know that $I \vdash \tau_1(X) \to \tau_2(Y)$ is provable in CMDL.

LEMMA 2 *If $I \vdash \tau_1(X) \to \tau_2(Y)$ is provable in CMDL then $\tau_1(X) \vdash \tau_2(Y)$ is also provable in CMDL.*

Proof. Below, (dp) stands for "display postulates".

$$\dfrac{I \vdash \tau_1(X) \to \tau_2(Y) \qquad \dfrac{\tau_1(X) \vdash \tau_1(X) \qquad \tau_2(Y) \vdash \tau_2(Y)}{\tau_1(X) \to \tau_2(Y) \vdash *\tau_1(X) \circ \tau_2(Y)}\,(\to\vdash)}{I \vdash *\tau_1(X) \circ \tau_2(Y)}\,(\text{cut})$$

$$\dfrac{\dfrac{I \vdash *\tau_1(X) \circ \tau_2(Y)}{I \circ \tau_1(X) \vdash \tau_2(Y)}\,(\text{dp})}{\tau_1(X) \vdash \tau_2(Y)}\,(I^- \vdash)$$

■

LEMMA 3 *For any structure Z both $Z \vdash \tau_1(Z)$ and $\tau_2(Z) \vdash Z$ are provable in CMDL.*

Proof. By simultaneous induction on the structure of Z. The lemma holds for all formulae A since $\tau_1(A) = \tau_2(A) = A$. Similarly, it holds for I since $\tau_1(I) = \top$ and $\tau_2(I) = \bot$, and both $I \vdash \top$ and $\bot \vdash I$ are provable in CMDL by the rules $(\vdash \top)$ and $(\bot \vdash)$ respectively.

Suppose the lemma holds for all structures of degree k and suppose Z is of degree $k + 1$. We consider each case in turn depending on the outermost connective of Z. The induction hypothesis is identified by I.H. below:

$$\dfrac{\dfrac{\dfrac{\text{I. H.}}{X \vdash \tau_1(X)} \qquad \dfrac{\text{I. H.}}{Y \vdash \tau_1(Y)}}{X \circ Y \vdash \tau_1(X) \wedge \tau_1(Y)}}{X \circ Y \vdash \tau_1(X \circ Y)}\,(\tau_1) \qquad \dfrac{\dfrac{\dfrac{\dfrac{\text{I. H.}}{\tau_2(X) \vdash X}}{*X \vdash *\tau_2(X)}}{*X \vdash \neg\tau_2(X)}}{*X \vdash \tau_1(*X)}\,(\tau_1) \qquad \dfrac{\dfrac{\dfrac{\text{I. H.}}{X \vdash \tau_1(X)}}{\bullet X \vdash \blacklozenge\tau_1(X)}}{\bullet X \vdash \tau_1(\bullet X)}\,(\tau_1)$$

$$
\dfrac{\dfrac{\dfrac{\text{I. H.}}{\tau_2(X) \vdash X} \quad \dfrac{\text{I. H.}}{\tau_2(Y) \vdash Y}}{\tau_2(X) \lor \tau_2(Y) \vdash X \circ Y}}{\tau_2(X \circ Y) \vdash X \circ Y}\,(\tau_2)
\qquad
\dfrac{\dfrac{\dfrac{\dfrac{\text{I. H.}}{X \vdash \tau_1(X)}}{*\tau_1(X) \vdash *X}}{\neg\tau_1(X) \vdash *X}}{\tau_2(*X) \vdash *X}\,(\tau_2)
\qquad
\dfrac{\dfrac{\dfrac{\text{I. H.}}{\tau_2(X) \vdash X}}{\Box\tau_2(X) \vdash \bullet X}}{\tau_2(\bullet X) \vdash \bullet X}\,(\tau_2)
$$

■

THEOREM 4 *If $\tau(X \vdash Y)$ is a theorem of tense logic* **Kt** *then any sequent $\widehat{X} \vdash \widehat{Y}$ with $\tau(X \vdash Y) = \tau(\widehat{X} \vdash \widehat{Y})$ is provable in CMDL. In particular, the sequent $X \vdash Y$ is provable in CMDL.*

Proof. First note that if $\tau(X \vdash Y) = \tau(\widehat{X} \vdash \widehat{Y})$ then $\tau_1(X) = \tau_1(\widehat{X})$ and $\tau_2(Y) = \tau_2(\widehat{Y})$ thus giving us one line proofs of $\tau_1(\widehat{X}) \vdash \tau_1(X)$ and $\tau_2(Y) \vdash \tau_2(\widehat{Y})$ in CMDL (depicted as note 1 below). Second, the fact that $\tau(X \vdash Y)$ is a theorem of tense logic **Kt**, together with Lemmas 1 and 2 gives us that $\tau_1(X) \vdash \tau_2(Y)$ is provable in CMDL (depicted as note 2 below). Now we can complete the proof by constructing a CMDL proof of $\widehat{X} \vdash \widehat{Y}$ as follows (using the cut rule which we know to be eliminable):

$$
\dfrac{\dfrac{\text{Lemma 3}}{\widehat{X} \vdash \tau_1(\widehat{X})} \qquad \dfrac{\dfrac{\text{note 1}}{\tau_1(\widehat{X}) \vdash \tau_1(X)} \qquad \dfrac{\dfrac{\dfrac{\text{note 2}}{\tau_1(X) \vdash \tau_2(Y)} \quad \dfrac{\text{note 1}}{\tau_2(Y) \vdash \tau_2(\widehat{Y})}}{\tau_1(X) \vdash \tau_2(\widehat{Y})} \quad \dfrac{\text{Lemma 3}}{\tau_2(\widehat{Y}) \vdash \widehat{Y}}}{\tau_1(X) \vdash \widehat{Y}}}{\tau_1(\widehat{X}) \vdash \widehat{Y}}}{\widehat{X} \vdash \widehat{Y}}
$$

Lemma 3 is also at the heart of Kracht's proof for it allows us to *derive* his extra (inverted) operational rules by using cut. For example:

$$\dfrac{\dfrac{\text{Lemma 3}}{A \circ B \vdash \tau_1(A \circ B)} \quad \dfrac{A \wedge B \vdash X}{\tau_1(A \circ B) \vdash X}\,(\tau)}{A \circ B \vdash X} \qquad \dfrac{\dfrac{X \vdash \Box A}{X \vdash \tau_2(\bullet A)}\,(\tau) \quad \dfrac{\text{Lemma 3}}{\tau_2(\bullet A) \vdash \bullet A}}{X \vdash \bullet A}$$

derivation of Krachts's $(\circ \vdash)$ derivation of Krachts's $(\vdash \bullet)$

The point is that each structural connective (in the appropriate position) is an *exact* proxy for the logical connective it replaces. ■

ACKNOWLEDGEMENTS

Thanks to Marcus Kracht for his timely and constructive comments on an earlier version of this paper.

Australian National University, Australia.

REFERENCES

[1] M. Kracht. Power and weakness of the modal display calculus. In: H. Wansing, (ed.), *Proof Theory of Modal Logic*. This volume.

[2] H. Wansing. Sequent calculi for normal modal propositional logics. *Journal of Logic and Computation*, **4**, 125–142, 1994.

CLAUDIO CERRATO

MODAL SEQUENTS

1 INTRODUCTION

We present some results about non-standard calculi of sequents for normal modal logics (NLs). While for standard calculi the attention is on possible extensions to more complex NLs (e.g., see Goré [9]), or on searching for a unified theory, like the subformula property (as explored in Takano [18]), non-standard calculi have been developed by changing the rules of the game, searching an unified view by modifying the structure of the calculus.

A first line is outlined (for different purposes) in Došen [6], where higher levels were introduced to control metalinguistic behaviours. In Cerrato [1, 2, 4], the attention is focussed on developing an uniform treatment of sequent calculi for the main 15 NLs between **K** and **S5**, namely for **K**, **KB**, **KD**, **KT** (=**T**), **K**4, **K**5, **KBD**, **KBT** (=**B**), **KB**4, **KD**4, **KD**5, **KD**45, **K**45, **KT**4 (=**S4**), **KT**5 (=**S5**). Techniques are different, but the trend is the same: to find a structure that work uniformly well for all those NLs. In sequent with modal signs (see [1]) metalinguistic signs were added to separate modal behaviours from propositional ones; the treatment is uniform, but it does not seem to work well for cut-elimination (as pointed out by Goré [10]). Semantic modal sequents directly introduce Kripke accessibility relation into the structure of the calculus, leading to an uniform treatment that is cut-free for all those NLs. The semantic proof exhibited (following the line of Girard [8]) bypasses the problem of syntactic cut-elimination, that is afforded by modal tree-sequents, in [4]: those sequents are a variant of semantic modal-sequents, where the tree-structure remains, but the accessibility is absorbed into peculiar modal rules; syntactic cut-elimination is proved by extending the usual proof (see Takeuti [17]) to modal sequents.

In this work, the language is $\mathbf{L} = \{\mathcal{P}, \wedge, \vee, \neg, \rightarrow, \Box\}$; we define the other operators as follows:

equivalence	$A \leftrightarrow B$ =def	$(A \rightarrow B) \wedge (B \rightarrow A)$
possibility	$\Diamond A$ =def	$\neg\Box\neg A$
strict implication	$A \Rightarrow B$ =def	$\Box(A \rightarrow B)$
strict equivalence	$A \Leftrightarrow B$ =def	$(A \Rightarrow B) \wedge (B \Rightarrow A)$

H. Wansing (ed.), Proof Theory of Modal Logic, 141–166.

Furthermore, $\top$ and $\bot$ denote a generic theorem and a generic contradiction, respectively. Actually, for sequents with modal signs we use possibility as primitive operator, since its dual role w.r.t. necessity is managed by specific rules.

As to notation, we use Λ to refer to a system among **K**, **KB**, **KD**, **KT** (=**T**), **K4**, **K5**, **KBD**, **KBT** (=**B**), **KB4**, **KD4**, **KD5**, **KD45**, **K45**, **KT4** (=**S4**), **KT5** (=**S5**) (see Chellas [5]). We exhibit a basic calculus for **K**, and specific rules corresponding to axioms $\top, 4, 5, B, D$,

$$\begin{array}{ll} \top: & \Box A \to A \\ 4: & \Box A \to \Box\Box A \\ 5: & \Diamond A \to \Box\Diamond A \\ B: & A \to \Box\Diamond A \\ D: & \Box A \to \Diamond A \end{array}$$

obtaining any Λ-calculus by adding the rules corresponding to the specific modal axioms of Λ to the **K**-calculus. When proving soundness of NLs-calculi, we really prove that for any NL Λ if there exists a sequent-style Λ-proof of a formula A (written, A is Λ-*provable*) then there exists an Hilbert-style Λ-proof of it (written, A is a Λ-*theorem* or $\vdash_\Lambda A$), so that (by soundness of usual Hilbert-style systems) A is true in any Λ-model (written, A is Λ-*valid*). As to completeness, we exhibit different proofs depending on the nature of sequents.

As a standard line, first we develop our calculus for the minimal normal modal logic **K**, exhibiting the main ideas, and then we extend them to the other systems. In the Hilbert style (see [5]), the system **K**:

1. is a modal system, i.e. a system closed under the rule of inference

$$RPL: \frac{A_1, A_2, \ldots, A_n}{A}$$

   where A is a tautological consequence of $A_1, A_2, \ldots, A_n$

2. is closed under the rule of necessitation

$$RN: \frac{A}{\Box A}.$$

3. contains the axiom schema

$$K: \Box(A \to B) \to (\Box A \to \Box B)$$

4. is closed under the rule

$$RK : \frac{A_1 \wedge \ldots \wedge A_n \to A}{\Box A_1 \wedge \ldots \wedge \Box A_n \to \Box A}.$$

5. contains the axiom schema

$$Df\Diamond : \Diamond A \leftrightarrow \neg\Box\neg A.$$

Condition (5) depends on which language we adopt, while the other conditions are not mutually independent: so, for sequents with modal signs, we consider the conditions (1), (4), (5), whereas both for semantic modal sequents and for modal tree-sequents, we consider the conditions (1)–(3).

2 SEQUENTS WITH MODAL SIGNS

Sequent calculi for normal modal logics with separated modal and propositional behaviours are presented in [1], proving a cut-elimination theorem for the basic system **K**.

In these sequents formulas can occur in a modal (possible or necessary) way; to denote this fact, we add to the usual modal language two metalinguistic symbols, namely $\langle\ \rangle$ and $[\ ]$, and we sign by such symbols formulas occurring in a modal way.

Sequents range on sets of formulas either unsigned or signed by $\langle\ \rangle$ or $[\ ]$, and our calculus transforms the modal way formulas occur into modal operators.

Signed formulas, e.g. $\langle A\rangle$, appear quite similar to the corresponding usual modal formulas, e.g. $\Diamond A$, but they are not really the same thing. In fact, $\langle A\rangle$ stresses the modal nature of the formula, and so modal rules may work on it, while $\Diamond A$ stresses the propositional nature, and so usual propositional rules may work on it. So, from a semantic point of view $\langle A\rangle$ and $\Diamond A$ ($[A]$ and $\Box A$) are the same thing, while, from a procedural point of view they are different. In this sense the metalinguistic signs $\langle\ \rangle$ and $[\ ]$ have nothing in common with the semantic metalinguistic signs used in semantic tableaux (see [7]).

Furthermore, we allow directly transforming $\langle A\rangle$ into $\Diamond A$, but we do not allow the vice versa; in such a way we introduce an ordering in applying rules (firstly the modal ones, then the propositional ones), without any lack in the expressive power (in fact, completeness also holds).

The rules of the calculi are the usual ones, plus general rules to treat modalities, plus peculiar rules, each one for each modal axiom instead of each one for each modal system, as usual (e.g. see [7, 13, 14]). Formally, let $\langle\ \rangle, [\ ]$ be two new metalinguistic symbols:

$\langle A\rangle, [A]$ are *signed formulas*, where A is a formula of **L**;
$A, \langle A\rangle, [A]$ are *expressions*, where A is a formula of **L**;
$\Gamma \vdash \Delta$ is a *sequent* where Γ, Δ are (finite) sets of expressions.

As to notation, we use capital latin letters for formulas of **L**, small greek letters for expressions and capital ones for sets of expressions. A signed formula is made of two parts:

1. a formula of the language **L**, possibly with explicit modal operators;
2. a metalinguistic sign whose role is to stress the way the formula occurs; this sign has no reference to explicit modal operators (if any) occurring in the formula.

Metalinguistic symbols cannot be joined together by connectives to form other signed formulas. The expressions are formulas together with the indication of the way they occur:

1. non-modal way
2. 'possibly' way
3. 'necessarily' way.

In the first case expressions are really formulas of **L**, and we call them unsigned formulas, otherwise they are signed formulas. Actually, the expressions are the objects of our calculus, so the sequents are made of expressions and rules apply to sequents of expressions. We divide the rules in:

Logical rules	regarding logical connectives
Structural rules	regarding sequents' structure
Duality rules	regarding transformation of modalities
Modal rules	characteristic of each modal system

Logical rules are the usual ones $\neg\vdash, \vdash\neg, \wedge\vdash, \vdash\wedge, \vee\vdash, \vdash\vee, \rightarrow\vdash, \vdash\rightarrow$ (see [16, 17]) and affect only unsigned formulas; so e.g. the rule $\vee\vdash$ ($\vee$: left in [17]) is:

$$\frac{C, \Gamma \vdash \Delta \quad D, \Gamma \vdash \Delta}{C \vee D, \Gamma \vdash \Delta}$$

where C, D are formulas and Γ, Δ are sets of expressions. The structural rules are:

$$\text{Thinning}\vdash \frac{\Gamma \vdash \Delta}{\alpha, \Gamma \vdash \Delta} \qquad \vdash\text{Thinning}\ \frac{\Gamma \vdash \Delta}{\Gamma \vdash \Delta, \alpha}$$

where α is an expression and Γ, Δ are sets of expressions, and:

$$\text{Cut}\ \frac{\Gamma \vdash \Delta, D \quad D, \Gamma' \vdash \Delta'}{\Gamma, \Gamma' \vdash \Delta, \Delta'}$$

where D is a formula and $\Gamma, \Gamma', \Delta, \Delta'$ are sets of expressions. Thinning is like weakening in [17]; furthermore, we need neither contraction nor exchange rules because we use sets (as in [16]) instead of sequences (as in [17]).

Duality rules convert metalinguistic signs into modal operators or transform 'possibly' into 'necessarily' and vice versa. Such rules apply only to signed formulas:

$$\Box\vdash\ \frac{[A], \Gamma \vdash \Delta}{\Box A, \Gamma \vdash \Delta} \qquad \vdash\Box\ \frac{\Gamma \vdash \Delta, [A]}{\Gamma \vdash \Delta, \Box A}$$

$$\Diamond\vdash\ \frac{\langle A\rangle, \Gamma \vdash \Delta}{\Diamond A, \Gamma \vdash \Delta} \qquad \vdash\Diamond\ \frac{\Gamma \vdash \Delta, \langle A\rangle}{\Gamma \vdash \Delta, \Diamond A}$$

$$\Box\Diamond\vdash\ \frac{\Gamma \vdash \Delta, [A]}{\langle\neg A\rangle, \Gamma \vdash \Delta} \qquad \vdash\Box\Diamond\ \frac{[A], \Gamma \vdash \Delta}{\Gamma \vdash \Delta, \langle\neg A\rangle}$$

$$\Diamond\Box\vdash\ \frac{\Gamma \vdash \Delta, \langle A\rangle}{[\neg A], \Gamma \vdash \Delta} \qquad \vdash\Diamond\Box\ \frac{\langle A\rangle, \Gamma \vdash \delta}{\Gamma \vdash \delta, [\neg A]}$$

where A is a formula and Γ, Δ are sets of expressions.

The notions of *proof* and of *end-sequent* are defined as in a standard calculus of sequents [17]; an *initial sequent* is one of the kind $A \vdash A$; a formula A is *provable* iff $\vdash A$ is an end-sequent.

To develop a calculus of modal sequents for **K** we introduce the first modal rules, namely K-rules:

$$K\vdash \frac{A, \Gamma \vdash \Delta}{\langle A\rangle, [\Gamma] \vdash \langle\Delta\rangle} \qquad \vdash K \frac{\Gamma \vdash \Delta, A}{[\Gamma] \vdash \langle\Delta\rangle, [A]}$$

where A is a formula, Γ and Δ are sets of formulas, and, given a set of formulas $\phi = \{F_0, \ldots, F_n\}, [\phi] = [F_0, \ldots, F_n] = [F_0], \ldots, [F_n], \langle\phi\rangle = \langle F_0, \ldots, F_n\rangle - \langle F_0\rangle, \ldots, \langle F_n\rangle, [\varnothing] = \langle\varnothing\rangle = \varnothing$. Note that in K-rules the premises contain only unsigned formulas, while the conclusions contain only signed ones. By K-rules, we can equivalently assume $\alpha \vdash \alpha$ as initial sequents, where α is an expression.

THEOREM 1 (Soundness of the **K**-calculus) *If A is* **K**-*provable then A is a* **K**-*theorem.*

Proof. We translate sequents to formulas suitably adapting to the modal sequents case the *Schütte translation* [15]: let $*$ be the translation from expressions to formulas of **L** defined as:

$$\begin{aligned} A^* &= A \\ \langle A\rangle^* &= \Diamond A \\ [A]^* &= \Box A \end{aligned}$$

where A is a formula; we extend that translation to sets of expressions, $\psi^* = \{\psi_0, \ldots, \psi_n\}^* = \{\psi_0^*, \ldots, \psi_n^*\}$, to sequents, $(\Gamma \vdash \Delta)^* = \wedge(\Gamma^*) \to \vee(\Delta^*)$, and to sequent-style inferences into Hilbert-style inferences,

$$\left(\frac{\text{premise}_0 \quad \text{premise}_n}{\text{conclusion}}\right)^* = \frac{(\text{premise}_0)^*, \ldots, (\text{premise}_n)^*}{(\text{conclusion})^*}$$

By $*$ initial sequents are clearly translated to tautologies, and sequent-style inferences are translated to Hilbert-style inferences, so that the translation of a **K**-provable sequent is also Hilbert-style provable in **K** (see [1, Theorem 1]); so if A is a **K**-provable formula, i.e. if $\varnothing \vdash A$ is an end-sequent, then $\vdash_{\mathbf{K}} (\varnothing \vdash A)^*$, i.e. (by the definition of $*$) $\vdash_{\mathbf{K}} A$, that is the thesis.

To develop calculi for other NLs we introduce the modal rules $T, 4, 5, B, D$. These rules are quite immediate translations of modal axioms into the formalism of the calculus of sequents (this correspondence is explicitly shown in the following Theorem 2). However, these rules respect the 'introduction' nature of the calculus, and in fact involved expressions do not decrease in complexity.

$$T\vdash \frac{A,\Gamma\vdash\Delta}{[A],\Gamma\vdash\Delta} \qquad \vdash T \frac{\Gamma\vdash\Delta,A}{\Gamma\vdash\Delta,\langle A\rangle}$$

$$4\vdash \frac{\langle A\rangle,\Gamma\vdash\Delta}{\langle\Diamond A\rangle,\Gamma\vdash\Delta} \qquad \vdash 4 \frac{\Gamma\vdash\Delta,[A]}{\Gamma\vdash\Delta,[\Box A]}$$

$$5\vdash \frac{[A],\Gamma\vdash\Delta}{\langle\Box A\rangle,\Gamma\vdash\Delta} \qquad \vdash 5 \frac{\Gamma\vdash\Delta,\langle A\rangle}{\Gamma\vdash\Delta,[\Diamond A]}$$

$$B\vdash \frac{A,\Gamma\vdash\Delta}{\langle\Box A\rangle,\Gamma\vdash\Delta} \qquad \vdash B \frac{\Gamma\vdash\Delta,A}{\Gamma\vdash\Delta,[\Diamond A]}$$

$$D\vdash \frac{\langle A\rangle,\Gamma\vdash\Delta}{[A],\Gamma\vdash\Delta} \qquad \vdash D \frac{\Gamma\vdash\Delta,[A]}{\Gamma\vdash\Delta,\langle A\rangle}$$

where A is a formula of **L** and Γ,Δ are sets of expressions. ■

THEOREM 2 (Soundness of Normal-Calculi) *If A is Λ-provable then it is a Λ-theorem, for any NL Λ.*

Proof. The proof is the same used for Theorem 1, with a slight modification; we must only prove that also the translations of rules $T, 4, 5, B, D$ are Hilbert-style inferences. That is an easy check; e.g., for $\vdash T$

1)	$(\Gamma\vdash\Delta,A)^*$	hypothesis
2)	$\bigwedge(\Gamma^*)\to\bigvee(\Delta^*)\vee A$	definition of $*$
3)	$\Box\neg A\to\neg A$	axiom T applied to $\neg A$
4)	$A\to\Diamond A$	3, **PC**, $Df\Diamond$
5)	$\bigwedge(\Gamma^*)\to\bigvee(\Delta^*)\vee\Diamond A$	2, 4, **PC**
6)	$(\Gamma\vdash\Delta,\langle A\rangle)^*$	definition of $*$.

■

THEOREM 3 (Completeness of NLs-calculi) *If A is a Λ-theorem then A is Λ-provable, for any NL Λ.*

Proof. We transform Hilbert-style Λ-proofs into sequent-style Λ-proofs. Fixed an

Hilbert style Λ-proof, for each formula proved at some step we define a *degree* ∂ of dependence on modal inferences as:

$\partial(A) = 0$	if either $\vdash_{\mathbf{PC}} A$ or A is a modal axiom
$\partial(A) = n$	if $B_0, \ldots, B_s \vdash_{\mathbf{PC}}$ where each B_i is proved at some previous step and $\max.\{\partial(B_i)\}_{0 \leq i \leq s} = n$
$\partial(A) = n + 1$	if $B \vdash_{\Lambda} A$ by a direct use of rule RK, with $\partial(B) = n$.

That is a good definition (fixed a proof), and the degree counts the maximal number of nested uses of rule RK in the proof. Moreover, we can always reduce the second case to the following alternatives: either each B_i is a modal axiom or is proved by a direct use of RK, or (only if $\partial(A) = 0$) we can also reduce this case to $\vdash_{\mathbf{PC}} A$.

Recalling that ($\bullet$) if $\Gamma \vdash_{\mathbf{PC}} A$ then the sequent $\Gamma \vdash A$ is **PC**-provable (see Kleene [11]), we transform each Hilbert-style Λ-proof into a sequent-style Λ-proof by an induction on the degree, by suitable uses of cut (for details see [1, Theorems 3, 4]). ■

THEOREM 4 (Cut-elimination for **K**) *If a sequent is* **K**-*provable then it is* **K**-*provable without a cut.*

Proof. We add to the usual cut-elimination proof for **PC** (see [17]) the cases regarding duality and modal rules. As in [1], we allow cut on expressions (that is sound), so to apply the inductive hypothesis to the duality rules $\vdash \Box, \Box \vdash, \vdash \Diamond, \Diamond \vdash$, too.

The rank of a proof P, $r(P)$, is defined as usual, while the grade of P, $g(P)$, is, as usual, the grade of the principal formula of its last inference, $g(\alpha)$, that indicates the complexity of α with respect to the non-structural inferences, but we compute it as usual with a slight modification: in fact, in [17] the only non-structural rules are the logical ones, each one introducing a new (logical) symbol in the principal formula; so the grade of α is immediately determined only by counting the number of logical connectives occurring in α. In our case, each non-structural rule introduces a new (logical or modal or metalinguistic) symbol, but $\Box \vdash, \vdash \Box, \Diamond \vdash$ and $\vdash \Diamond$ also delete a metalinguistic one: so the global number of symbols does not increase; however, only such rules introduce modal operators, so that we solve the problem assigning to each modal operator a grade 2 (1 for itself $+1$ for the deleted metalinguistic symbol). So:

$g(P) = g(\alpha) =$
- the number of logical connectives occurring in α, plus
- the number of metalinguistic symbols occurring in α, plus
- twice the number of modal operators occurring in α;

for example, we have $g(A) < g(\langle A \rangle) < g(\Diamond A)$.

The rest of the proof is a technical routine, whose details can be found in [1, Theorem 5]. ■

Takano, [19], noticed that for system **K** the duality rules $\Box\Diamond\vdash, \vdash\Box\Diamond$, $\Diamond\Box\vdash, \vdash\Diamond\Box$ 'are derived ones'; for example $\Box\Diamond\vdash$ is provable by:

$$\dfrac{\Gamma \vdash \Delta, [A] \quad \dfrac{\dfrac{A \vdash A}{\neg A, A, \vdash}}{\langle \neg A \rangle, [A] \vdash} K\vdash}{\langle \neg A \rangle, \Gamma \vdash \Delta} \text{cut}$$

and 'omission of these rules causes no difficulty in the cut-elimination proof for **K**'.

Goré [10] noticed that any attempt to prove the **S4**-theorem $\Diamond\Box(\Diamond P \to \Box\Diamond P)$ without cut fails; the reader can construct a cut-free proof using the semantic modal sequents presented in the next section, and can see that such a proof requires using non-modal rules (e.g. weakening) between nested modal rules. That is not allowed by sequents with modal signes, since their structure is too simple, so that we must use cut. For a simpler example, let us consider how a cut-free proof of the **K4**-theorem $\Box A \to \Box(B \to \Box A)$ should appear:

$$\dfrac{\dfrac{\dfrac{\dfrac{[A] \vdash [A]}{[A] \vdash [\Box A]} \vdash 4}{[A] \vdash [B \to \Box A]} \text{weakening} \vdash \text{and} \vdash\to \text{into } [\ldots]}{\Box A \vdash \Box(B \to \Box A)} \vdash K}{\vdash \Box A \to \Box(B \to \Box A)}$$

So calculi of sequents with modal signs work uniformly well for the main 15 NLs, but do not seem enjoy cut-elimination for systems other than **K**. In certain a sense, we have distinguished between two levels, an usual one, and a metalinguistic level controlled by modal signs, making that distinction on formulas. But rules corresponding to the axioms that affect more than one nested modality (e.g. B) must use both modal signs and modal operators together (e.g. $\langle \Box A \rangle$), since do not have at their disposal more than one metalevel; furthermore, we cannot completely manage formulas when contained into modal signs. So, as a natural evolution, we extend those non-standard sequents adopting many levels organized in a tree-hierarchy, and making that distinction on sequences of formulas (actually on usual sequents). The calculi we develop offer a wider uniformity, solving the problem of cut-elimination for the main NLs.

3 SEMANTIC MODAL SEQUENTS

Cut-free calculi of sequents for normal modal logics are developed in [2] by using semantic modal sequents, which are trees of usual sequents plus an accessibility relation, and by introducing modal operators when moving formulas along the branches of such trees.

We use only two general modal rules for all NLs $\Box\vdash$ and $\vdash\Box$ (with a technical exception for systems containing the axiom schema D); we vary the first rule when varying the system depending on the accessibility, while we fix the second rule for all. We prove the completeness of our calculi for NLs, giving a semantic proof of cut-elimination, and obtain a uniform treatment that work well in every case. A *semantic modal sequent* is a triple $\langle W, \rightarrow, R\rangle$ where W is a non-empty set of (occurrences of) usual sequents called worlds (i.e. $\Gamma\vdash\Delta$, where Γ,Δ are sequences of formulas of $\mathbf{L}$), $\rightarrow$ is a strict tree-ordering on W and R is a binary relation on W that extends $\rightarrow$, called accessibility. We often call 'modal sequent' a semantic modal sequent, and just only 'sequent' a usual sequent; furthermore, the locution 'occurrences of' indicates that several instances of the same sequent can occur as different worlds in a modal sequent.

The first two components of semantic modal sequents, W and $\rightarrow$, give rise to *trees of sequents*, that we can also inductively define by:

$\Gamma\vdash\Delta$	is a tree of sequents, where Γ,Δ are sequences of formulas of $\mathbf{L}$;
$\begin{array}{c}\Gamma\vdash\Delta\\ \swarrow \ldots \searrow\\ \lambda_0 \ldots \lambda_n\end{array}$	is a tree of sequents, where Γ,Δ are sequences of formulas of $\mathbf{L}$ and $\lambda_0,\ldots,\lambda_n (n\geq 0)$ are trees of sequents.

We use capital latin letters for formulas, capital greek letters for sequences of formulas, small latin letters for worlds and small greek letters for trees. The other component of semantic modal sequents, the relation R, puts the corresponding aspects of the semantic accessibility relation into sequents: namely, for a system having as modal axioms $Ax_1,\ldots,Ax_n$ the accessibility R is the minimal relation containing $\rightarrow, R(Ax_1),\ldots,R(Ax_n)$, where the correspondence between axioms and relations is given by the following table

Ax	$R(Ax)$
K	$\to$
T	the reflexive closure of $\to$
4	the transitive closure of $\to$
5	the euclidean closure of $\to$
B	the symmetric closure of $\to$

In the case of the axiom D, since we cannot uniquely determine R as the serial closure of $\to$, we consider a new rule, namely the empty rule; when proving completeness, that rule (read upward) really allows us only to add unessential accessible worlds that maintain seriality in the countermodel under construction.

The rules for propositional connectives are $\neg\vdash, \vdash\neg, \wedge\vdash, \vdash\wedge, \vee\vdash, \vdash\vee, \to\vdash, \vdash\to$ (see [17]) with suitable modifications for the use of trees. First, as to notation, two semantic modal sequents $\alpha = \langle W, \to, R\rangle$ and $\beta = \langle W', \to', R'\rangle$ are similar by (w, w'), where $w \in W$ and $w' \in W'$, (written as $\alpha \ {}_{w}\!\sim_{w'} \beta$) iff they become the same by changing only w with w', i.e. iff $W - \{w\} = W' - \{w'\}$, and, given the bijection f between W and W' that is the identity except that $f(w) = w'$, we have both $w_0 \to w_1$ iff $f(w_0) \to' f(w_1)$ and $w_0 R w_1$ iff $f(w_0) R' f(w_1)$, for every $w_0, w_1 \in W$. Then, given a usual **PC**-rule $\frac{a' a''}{b}$, where a', a'', b are usual sequents, the corresponding tree version of that rule is $\frac{\alpha' \alpha''}{\beta}$, where α', α'', β are semantic modal sequents and $\alpha' \ {}_{a'}\!\sim_{\alpha''} \alpha'' \ {}_{a''}\!\sim_{b} \beta \ {}_{b}\!\sim_{a'} \alpha'$ (we can apply a similar reasoning to rules with only one premise). So, the tree version of a **PC**-rule leaves unaltered both the structures of the tree and of all the unaffected usual sequents. For example, now the rule $\to\vdash$ appears as:

$$
\frac{\begin{array}{c}\gamma\\ \downarrow\\ \Gamma\vdash\Delta, C\\ \swarrow\cdots\searrow\\ \delta_1\ldots\delta_n\end{array}\qquad\qquad\begin{array}{c}\gamma\\ \downarrow\\ D,\Pi\vdash\Lambda\\ \swarrow\cdots\searrow\\ \delta_1\ldots\delta_n\end{array}}{\begin{array}{c}\gamma\\ \downarrow\\ C\to D,\Gamma,\Pi\vdash\Delta,\Lambda\\ \swarrow\cdots\searrow\\ \delta_1\ldots\delta_n\end{array}}
$$

where $\gamma, \delta_1, \ldots, \delta_n$ are trees of sequents (when $n = 0$ no δ_i appears), $\Gamma, \Delta, \Pi, \Lambda$ are sequences of formulas, and C, D are formulas of **L**. The structural rules are the tree adaptation of the usual weakening $\vdash$, $\vdash$weakening, exchange $\vdash$, $\vdash$ exchange, contraction $\vdash$, $\vdash$ contraction (see [17]).

The modal rules are $\Box\vdash$ and $\vdash\Box$, and move formulas along trees. Rule $\Box\vdash$ varies when varying modal system, depending on the accessibility relation R:

$\Box\vdash$ from a modal squent α having two worlds $\Gamma\vdash\Delta$ and $A,\Pi\vdash\Lambda$ with $(\Gamma\vdash\Delta)R(A,\Pi\vdash\Lambda)$ infer the modal sequent β obtained from α by substituting (both in the domain and in the relations $\rightarrow$ and R) those worlds with $\Box A,\Gamma\vdash\Delta$ and $\Pi\vdash\Lambda$, respectively.

Rule $\vdash\Box$ is the same for all modal system, since it depends on $\rightarrow$ instead of on R:

$\vdash\Box$ from a modal sequent α having two worlds $\Gamma\vdash\Delta$ and $\vdash A$, where $\vdash A$ is a terminal leaf and $(\Gamma\vdash\Delta)\rightarrow(\vdash A)$, infer the modal sequent β obtained from α by substituting (both in the domain and in the relations $\rightarrow$ and R) the former word with $\Gamma\vdash\Delta,\Box A$ and by completely deleting both the latter world from the domain and any occurrence of it from $\rightarrow$, and consequently modifying R.

For example, when R is the transitive closure of $\rightarrow$ those rules appear as :

$$\Box\vdash \qquad \begin{array}{c} \gamma \\ \downarrow \\ \Gamma\vdash\Delta \\ \swarrow\cdots\downarrow\cdots\searrow \\ \cdots\quad\vdots\quad\cdots \\ \downarrow \\ A,\Pi\vdash\Lambda \\ \swarrow\cdots\searrow \\ \delta_1\ldots\delta_n \\ \hline \gamma \\ \downarrow \\ \Box A,\gamma\vdash\Delta \\ \swarrow\cdots\downarrow\cdots\searrow \\ \cdots\quad\vdots\quad\cdots \\ \downarrow \\ \Pi\vdash\Lambda \\ \swarrow\cdots\searrow \\ \delta_1\ldots\delta_n \end{array} \qquad\qquad \begin{array}{c} \gamma \\ \downarrow \\ \Gamma\vdash\Delta \\ \swarrow\cdots\downarrow\cdots\searrow \\ \sigma_1\ldots\sigma_r \quad \vdash A \quad \sigma_{r+1}\ldots\sigma_m \\ \hline \gamma \\ \downarrow \\ \Gamma\vdash\Delta,\Box A \\ \swarrow\cdots\searrow \\ \sigma_1\ldots\sigma_m \end{array} \qquad \vdash\Box$$

when $n=0$ no δ_i appears. Note that the sequent $\vdash A$ is a terminal leaf to do not introduce a disconnection in the tree of sequents.

Finally, for 'serial' systems, i.e. for those systems containing the axiom D, the empty rule holds:

empty rule: from a modal sequent α having a world $\vdash$ as terminal leaf infer the modal sequent β obtained from α by deleting that world (when not occurring as top node of α) from W and from both the relations R and $\rightarrow$:

$$\frac{\begin{array}{c}\gamma\\ \downarrow\\ \vdash\end{array}}{\gamma}$$

The notions *proof* and *end-sequent* are used as in standard calculi of sequents (e.g. see [17]), an *initial sequent* is one of the form $\langle W, \rightarrow, R\rangle$ where $A \vdash A \in W$ for some formula A, an usual sequent $\Gamma \vdash \Delta$ is *provable* when $\langle W, \rightarrow, R\rangle$ is an end-sequent with $W = \{\Gamma \vdash \Delta\}$, and a formula A is *provable* when $\vdash A$ is provable.

THEOREM 5 (Soundness of the NLs-calculi) *For any NL Λ, if A is Λ-provable then A is a Λ-theorem, and so it is Λ-valid.*

Proof. As usual, we translate semantic modal sequents into formulas by suitably modifying the Schütte translation; namely, we define the translation $*$ by an induction on the complexity of trees (see [2, Theorem 1]):

$$\begin{array}{rcl}(\Gamma \vdash \Delta)^* & = & \bigwedge\Gamma \rightarrow \bigvee\Delta\\ \left(\begin{array}{c}\Gamma \vdash \Delta\\ \swarrow \cdots \searrow\\ \lambda_0 \ldots \lambda_n\end{array}\right)^* & = & \bigwedge\Gamma \rightarrow (\bigvee\Delta \vee \Box\lambda_0^* \vee \ldots \vee \Box\lambda_n^*)\end{array}$$

where A is a formula of $\mathbf{L}, \Gamma$ and Δ are sequences of formulas, $\lambda_0, \ldots, \lambda_n$ are modal sequents and, by definition, $\bigwedge\varnothing = \top$ and $\bigvee\varnothing = \bot$. Only the relation $\rightarrow$ influences the translation $*$, since there is no mention of R, which is indirectly reintroduced in this theorem by using modal axioms. We extend the translation to sequent-style inferences into Hilbert-style inferences as usual, proving that initial sequents are translated to theorems (see [2, theorem 4]), and that sequent-style inferences are translated to Hilbert-style inferences (see [2, Theorem 5]), so proving, as usual, that if a formula A is Λ-provable, i.e. if $\vdash A$ is an end-sequent, then A is a Λ-theorem. ∎

Now we prove completeness and semantic cut-elimination together. A sequent proof appears as an inverted, at most binary, finite tree, whose root is the end-sequent (at the bottom), and whose terminal leaves are initial sequents (at the top). The way of moving along such reversed trees allows us to prove completeness: in fact, when moving downward we consider that tree as a 'proof', otherwise when moving upward we consider it as an attempt of constructing a 'countermodel'; we really prove that when a usual sequent is not cut-free provable we can construct a countermodel of it (really of its Schütte translation), so that it is not valid.

THEOREM 6 (Completeness of the NLs-calculi) *For any NL Λ, if A is Λ-valid then A is Λ-provable (without any use of cut).*

Proof. As usual, we prove the statement by contraposition: if A is not Λ-provable (without any use of cut—anyway we can rewrite it in the modal-sequent formalism) then A is not Λ-valid. We proceed by several steps, using a simplified version of the proof used in [8] for **PC**, adapted to modal systems (note we have only propositional symbols, and transform Schütte valuations directly into Kripke valuations). Namely, for any NL Λ:

1. we fix a class of functions on Kripke frames (the Schütte valuations) that preserve the values true **t** and false **f** of formulas when going along a proof-tree as countermodel and when requiring every formula that appears on the left part of a sequent to be true, while any on the right to be false. Really, we define such functions by imposing conditions that must be respected when going along the tree as a proof, so that we can define them by an induction on the complexity of formulas (that increases downward);

2. we prove that for any usual sequent $\Gamma \vdash \Delta$ that is not cut-free provable there is a Schütte valuation on a Kripke frame such that for some world the value of $\bigwedge \Gamma$ is **t** while the value of $\bigvee \Delta$ is **f**;

3. we prove that any Schütte valuation can be reduced to a binary valuation (i.e. a valuation that can take only the values **t** and **f** not **u**) that preserves the values **t** and **f** (so, there is a Kripke model where $\Gamma \vdash \Delta$ is not valid, proving completeness);

4. we introduce three-valued models, proving both that any Schütte valuation can be reduced to a three-valued model valuation that preserves the values **t** and **f** (so, there is a three-valued model where $\Gamma \vdash \Delta$ is not valid) and that any three-valued model valuation can be reduced to a binary valuation that preserves the values **t** and **f** (so, proving completeness in another way).

Let us examine each step (for more details see [2, Theorem 6]):

1. Given the set of well-founded formulas of the language, wff($\mathbf{L}$), a set of worlds W, a strict tree-ordering on W, $\rightarrow$, a relation R extending $\rightarrow$ and according with the properties of the accessibility for the system Λ, and a set of values $\{\mathbf{t}, \mathbf{f}, \mathbf{u}\}$, we extend a Schütte valuation $S : W \times \text{wff}(\mathbf{L}) \longrightarrow \{\mathbf{t}, \mathbf{f}, \mathbf{u}\}$ (see [8, Definition 3.1.3]) to modal operators: for any $w, w' \in W$,

$$\begin{array}{ll} \text{if exists } w' \text{ s.t. } wRw' \text{ and } S(w', A) \neq \mathbf{t} & \text{then } S(w, A) \neq \mathbf{t} \\ \text{if for every } w' \text{ s.t. } wRw', S(w', A) \neq \mathbf{f} & \text{then } S(w, A) \neq \mathbf{f}. \end{array}$$

2. We define the domain W, the relations $\rightarrow$ and R, and the function S that falsifies $\Gamma \vdash \Delta$ on the basis of an infinite branch of an inductively constructed reversed tree of semantic modal sequents. Namely, we start from $\Gamma \vdash \Delta$, and at any level of depth, we examine every sequent of that level, giving rise to the sequents of the next level: since any sequent can construct at most two other sequents, at any level we have a finite number of sequents; for any of those sequents, we examine only one formula of only one world (if any) terminating the examination of that level in a finite number of steps. At the level n, for any sequent we need a list of its worlds, and for any world we need a list of its formulas: we examine the first formula of the first world in those lists, possibly we add new worlds or new formulas at the end of such lists (to not bypass the queue) and then we put the just examined world and formula at the end of the corresponding lists (cyclically rotating them, so that in a finite number of steps we can reach any formula of any world).

   For the sake of simplicity, an index n stresses only the level where a sequent γ_n occurs, without specifying the branch (in fact we are really interested only in one infinite branch of that tree); list_n and $\text{list}_n(v)$ denote the corresponding list of worlds and, for any world v, the list of formulas, respectively.

   Finally, since the worlds of a semantic modal sequent are really usual sequents, with a left and a right part, and since those parts can change at any step (really, can only increase), we need two other lists, namely $\text{left}_n(v)$ and $\text{right}_n(v)$, the left and the right sequence of formulas of the world v at the step n, respectively. So, let $\gamma_0 = \langle W_0, \rightarrow_0, R_0 \rangle$ with $W_0, = \{v_0\}$, $\text{left}(v_0) = \Gamma$, $\text{right}(v_0) = \Delta, \rightarrow_0 = \varnothing$ and let R_0 be the suitable closure of $\rightarrow_0$; let also $\text{list}_0 = v_0$ (the listing of the worlds of W_0) and $\text{list}_0(v_0) = \Gamma, \Delta$ (the listing of the formulas of v_0). Given $\gamma_n = \langle W_n, \rightarrow_n, R_n \rangle$ with $\text{list}_n = v_0, \ldots, v_s$ (never empty, since only increasing) and $\text{list}_n(v_0) = A_1, \ldots, A_p$ (possibly empty), let γ_{n+1} (possibly γ'_{n+1} and γ''_{n+1}), list_{n+1} and $\text{list}_{n+1}(v)$ (for $v \in W_{n+1}$) be defined in the way shown in [2, Theorem 6], that is an adaptation of the way indicated in [8, Theorem 3.1.9]; here we show only modal cases:

i) if A_1 is an atomic formula and the system Λ contains the axiom D, then let $W_{n+1} = W_n \cup \{v_{s+1}\}, \rightarrow_{n+1} = \rightarrow_n \cup \{v_0 \rightarrow v_{s+1}\}$, R_{n+1} be the suitable closure of $\rightarrow_{n+1}$ (it is easy to prove that it contains R_n), list$_{n+1}$ $= v_1, \ldots, v_s, v_{s+1}, v_0$, list$_{n+1}(v_0) = A_2, \ldots, A_p, A_1$, left$(v_{s+1}) = \varnothing$, right $(v_{s+1}) = \varnothing$ and list$_{n+1}(v_{s+1}) = \varnothing$ (we really add only a serial queue for v_0);

ii) if A_1 is $\Box A$, and it occurs in the left part of v_0, and $R(v_0) = \{v \in W : v_0 R_v\} \neq \varnothing$, then let left$_{n+1}(v_0)$=left$_n(v_0)$, list$_{n+1}(v_0) = A_2, \ldots, A_p, A_1$, left$_{n+1}(v) = A$, left$_n(v)$ and list$_{n+1}(v)$=list$_n(v)$, A for any $v \in R(v_0)$; let also $W_{n+1} = W_n, \rightarrow_{n+1}=\rightarrow_n, R_{n+1} = R_n$, list$_{n+1} = v_1, \ldots, v_s, v_0$;

iii) if A_1 is $\Box A$, and it occurs in the left part of v_0, $R(v_0) = \varnothing$, and the system Λ does not contain the axiom D, then let $\gamma_{n+1} = \gamma_n$, but list$_{n+1}(v_1) = A_2, \ldots, A_p, A_1$, and list$_{n+1} = v_1, \ldots, v_s, v_0$;

iv) if A_1 is $\Box A$, and it occurs in the left part of v_0, $R(v_0) = \varnothing$, and the system Λ contains the axiom D, then let $W_{n+1} = W_n \cup \{v_{s+1}\}, \rightarrow_{n+1}=\rightarrow_n \cup \{v_0 \rightarrow v_{s+1}\}, R_{n+1}$ be the suitable closure of $\rightarrow_{n+1}$ (it easy to prove that it contains R_n), list$_{n+1} = v_1, \ldots, v_s, v_{s+1}, v_0$, left$_{n+1}(v_0)$=left$_n(v_0)$, list$_{n+1}(v_0) = A_2, \ldots, A_p, A_1$, left$(v_{s+1}) = \varnothing$, right$(v_{s+1}) = \varnothing$ and list$_{n+1}(v_{s+1}) = \varnothing$ (we really add only a serial queue for v_0);

v) if A_1 is $\Box A$, and it occurs on the right part of v_0 , then let $W_{n+1} = W_n \cup \{v_{s+1}\}, \rightarrow_{n+1}=\rightarrow_n \cup\{v_0 \rightarrow v_{s+1}\}, R_{n+1}$ be the suitable closure of $\rightarrow_{n+1}$ (it contains R_n), list$_{n+1} = v_1, \ldots, v_s, v_{s+1}, v_0$, right$_{n+1}(v_0)$= right$_n(v_0)$, list$_{n+1}(v_0) = A_2, \ldots, A_p, A_1$, right$(v_{s+1}) = A$, left$(v_{s+1}) = \varnothing$ and list$_{n+1}(v_{s+1}) = A$.

If γ_{n+1} is (both γ'_{n+1} and γ''_{n+1} are) cut-free provable, then γ_n is cut-free provable too; so, since we stop the construction of the tree on initial sequents, if the tree is finite then it is a cut-free proof of γ_0, against the hypothesis; thus, the tree must be an infinite, denumerable, (at most binary) tree of sequents; by König lemma, an infinite branch exists: let $\gamma_0, \ldots, \gamma_i, \ldots, (i < \omega)$ be the list of the sequents on that branch; let $W = \bigcup\{W_i : W_i \in \gamma_i\}, \rightarrow = \bigcup\{\rightarrow_i : \rightarrow_i \in \gamma_i\}, R = \bigcup\{R_i : R_i \in \gamma_i\}$ (for non-serial systems) or let R be $\bigcup\{R_i : R_i \in \gamma_i\}$ plus the reflexive closure of non-serial worlds (for serial systems); let also left$(w) = \bigcup\{$left$_i(w) : i < \omega\}$ and right$(w) = \bigcup\{$right$_i(w) : i < \omega\}$, for any $w \in W$. R satisfies the properties required for the accessibility relation of the system Λ.

Now, let S be the valuation defined as $S(w, A) = \mathbf{t}$ iff A occurs in left(w), $S(w, A) = \mathbf{f}$ iff A occurs in right(w), $S(w, A) = \mathbf{u}$ otherwise. S is well-defined and is a Schütte valuation: for example, we prove by contraposition that S respects the last modal condition on Schütte valuations: let us assume $S(w, \Box A) = \mathbf{f}$; by definition of $S, \Box A \in$right(w); so, by definition of W, from an index r on, $\Box A$ must appear on the right of w; by construction of the tree, when $\Box A$ is to be examined after a finite number m of steps, A must appear on the right of some w' with $w \rightarrow_{r+m} w'$; so $S(w', A) = \mathbf{f}$ with wRw' (by the definitions of $W, \rightarrow$ and R).

Finally, by construction, $S(v_0, \bigwedge \Gamma) = \mathbf{t}$ and $S(v_0, \bigvee \Delta) = \mathbf{f}$. This completes the step 2.

3. Let $\lhd$ be an ordering on $\{\mathbf{t}, \mathbf{f}, \mathbf{u}\}$ (in [8] was used the symbol $\rhd$) defined as $\mathbf{u} \lhd \mathbf{f}, \mathbf{u} \lhd \mathbf{t}, \mathbf{u} \lhd \mathbf{u}, \mathbf{f} \lhd \mathbf{f}, \mathbf{t} \lhd \mathbf{t}$; $\lhd$ induces on ordering on the Schütte valuations:

$$S \lhd T \quad \text{iff} \quad S(A) \lhd T(A) \text{ for every } A \in \text{wff}\,(\mathbf{L});$$

it is easy to see that any $\lhd$-chain has an upper bound, so that for any Schütte valuation S there is a $\lhd$-maximal valuation V such that $S \lhd V$; but a maximal valuation must be a binary one, and, by the conditions on Schütte valuations, must be a Kripke valuation. Since $\lhd$ maintains the values **t** and **f**, $V(v_0, \bigwedge \Gamma) = \mathbf{t}$ and $V(v_0, \bigvee \Delta) = \mathbf{f}$, so that $V(v_0, \bigwedge \Gamma \rightarrow \bigvee \Delta) = \mathbf{f}$; as particular case, when the sequent is $\vdash A$ we have $V(v_0, A) = \mathbf{f}$, proving completeness.

4. A three-valued model is a triple $\langle W, R, m \rangle$ where W is a non-empty set, R is a binary relation on W, according with the properties of accessibility for any system $\Lambda, m : W \times \text{wff}(\mathbf{L}) \longrightarrow \{\mathbf{t}, \mathbf{f}, \mathbf{u}\}$ is a function defined by an induction on the complexity of formulas as usual (see [8, Definition 3.2.2]), plus the following modal cases: given $w \in W$ and $R(w) = \{w' : wRw'\}$,

$$\begin{array}{ll} \text{if for every } w' \in R(w)\ m(w', A) = \mathbf{t} & \text{then } m(w, \Box A) = \mathbf{t} \\ \text{if exists } w' \in R(w) \text{ with } m(w', A) = \mathbf{f} & \text{then } m(w, \Box A) = \mathbf{f} \\ \text{if for every } w' \in R(w)\ m(w', A) \neq \mathbf{f} & \\ \quad \text{and exists } w' \in R(w)\ m(w', A) = \mathbf{u} & \text{then } m(w, \Box A) = \mathbf{u}. \end{array}$$

A three-valued valuation is really a Schütte valuation where the values of formulas are strictly determined by the values of their subformulas [8]: in fact, when both the value of A and the value of B is **t** then the value of $A \wedge B$ could be either **t** or **u** for a Schütte valuation, while must be only **t** for a three-valued valuation.

To stress the link between three-valued valuations and Schütte valuations, we introduce a refinement of the ordering $\lhd$, namely we define the relation $\ll$ on the Schütte valuations as:

$$S \ll T \quad \text{iff} \quad S \lhd T \text{ and } S\,|_{\mathcal{P}} = T\,|_{\mathcal{P}}$$

it easy to see that any $\ll$-chain has an upper bound, so that for any Schütte valuation S there is a $\ll$-maximal valuation m such that $S \ll m$, and that such a maximal valuation must be a three-valued valuation. Since $\ll$ (as $\lhd$) maintains the values **t** and **f** $m(v_0, \bigwedge \Gamma) = \mathbf{t}$, $m(v_0, \bigvee \Delta) = \mathbf{f}$. Furthermore, we can restrict the relation $\lhd$ to the three-valued valuations: reasoning as above, for any three-valued valuation m there is a $\lhd$-maximal valuation V such that $m \lhd V$; such a V must be a Kripke valuation and $V(v_0, \bigwedge \Gamma \to \bigvee \Delta) = \mathbf{f}$; as a particular case, when the sequent is $\vdash A$ we have $V(v_0, A) = \mathbf{f}$, proving, in another way, completeness.

■

Accessible worlds for modal logics were introduced by Kripke [12], where a tableau system based on auxiliary tableaux was presented to prove completeness for **S5**. The construction of countermodels in our completeness theorem for modal-sequent calculi gives an immediate technique for transforming our semantic modal-sequents into Kripke's tableaux, once we have accepted to allow formulas to be reused in a tableau, renouncing to discharge formulas or worlds.

4 MODAL TREE-SEQUENTS

Sequents with modal signs adopt a two-levels structure and syntactical rules; semantic modal sequents use a tree structure and semantical rules based on accessibility. By modal tree-sequents (see [4]) we try to combine the rich tree-structure of latter sequents with the syntactical approach of the former. Our goal is to prove that only reinforcing the structure of sequents sufficies to obtain calculi for NLs that work uniformly well, but instead of confirming in our opinion, modal tree-sequents open new questions. In fact, individuating rules that work well for NLs is quite easy (and allow us to link in natural way tree-sequents both with semantic modal sequents and with systems of natural deduction based on strict implication, presented in [3]), while eliminating cut requires to mimic accessibility by syntactical rules, so requires extra rules both when considering single axioms, as for axiom 5, and when combining many axioms, as for axioms B with 4.

Modal tree-sequents are trees of sequences of formulas with specific modal rules in correspondence to different modal axioms, which characterize modal operators by the way formulas move along trees. Tree-sequents adopt the tree-structure of semantic modal sequents (using trees of sequences instead of trees of sequents), but rules only have syntactical behaviours, since do not use accessibility. Without accessibility, a direct proof of completeness (in [8] style, like in [2]) is not easily practicable, so that we must translate Hilbert-style systems into our calculi. To do that we need cut rule, that eliminate in a syntactical way, adapting to tree-sequents the usual proof in Takeuti [17].

We define *tree-sequents* by induction on complexity of trees:

Γ is a sequent, where Γ is a (possibly empty, $\varnothing$) sequence of formulas of $\mathbf{L}$;

$$\Gamma\left\{\begin{array}{l}\delta_0\\ \vdots\\ \delta_n\end{array}\right.$$

is a sequent, where Γ is a (possibly empty sequence of formulas of $\mathbf{L}$ and $\delta_0, \ldots, \delta_n (n \geq 0)$ are sequents.

We call *trivial* the former and *non-trivial* the latter. An old usual sequent $\Gamma \vdash \Delta$ is now represented by the tree-sequent $\Gamma\{\Delta$, while the tree-structure of a semantic modal sequent (see [4]) like

$$\Gamma_0 \vdash \Delta_0 \left\{\begin{array}{l}\Gamma_1 \vdash \Delta_1\\ \Gamma_2 \vdash \Delta_2\end{array}\right. \quad \text{is now represented by} \quad \Gamma_0 \left\{\begin{array}{l}\Delta_0\\ \Gamma_1\{\Delta_1\\ \Gamma_2\{\Delta_2\end{array}\right.$$

As to notation, we use capital latin letters for formulas, capital greek letters for sequences of formulas, small greek letters for tree-sequents. Furthermore, '$\bullet$' represents the 'empty tree' (i.e. a tree without nodes and branches), and, for sake of simplicity, we use a single tree-sequent to refer to the many branches departing from the same node: e.g.

$$\alpha\{\Gamma\{\beta \text{ denotes } \alpha\{\Gamma\left\{\begin{array}{l}\beta_0\\ \vdots\\ \beta_n\end{array}\right., \text{ where '}\bullet\{\beta\text{' actually represents } \bullet\left\{\begin{array}{l}\beta_0\\ \vdots\\ \beta_n\end{array}\right..$$

The structural and **PC**-connectives rules are $\neg\vdash, \vdash\neg, \wedge\vdash, \vdash\wedge, \vee\vdash, \vdash\vee$, $\rightarrow\vdash, \vdash\rightarrow$, weakening $\vdash$, $\vdash$ weakening, exchange $\vdash$, $\vdash$ exchange, contraction $\vdash$, $\vdash$ contraction, with suitable modifications for the uses of trees. Namely, in the rules involving two premises, the two upper sequents must be equal in the structure of trees and in all the non-affected nodes; rules work only on non-trivial sequents, and right rules work on the terminal leaves, while left ones work on the other sequences. As to **PC**-rules, for example $\rightarrow\vdash, \vdash\rightarrow$ are:

$$\rightarrow\vdash \quad \frac{\alpha\{\Gamma\left\{\begin{array}{l}\beta\\ C, \Delta\end{array}\right. \qquad \alpha\{D, \Pi\left\{\begin{array}{l}\beta\\ \Lambda\end{array}\right.}{\alpha\{C \rightarrow D, \Gamma, \Pi\left\{\begin{array}{l}\beta\\ \Delta, \Lambda\end{array}\right.} \qquad \vdash\rightarrow \quad \frac{\alpha\{C, \Gamma\left\{\begin{array}{l}\beta\\ \Delta, D\end{array}\right.}{\alpha\{\Gamma\left\{\begin{array}{l}\beta\\ \Delta, C \rightarrow D\end{array}\right.}$$

We only need an extra rule to manage degenerated trees; that presented in [4] is

merge

$$\frac{\alpha\left\{\begin{array}{l}\varnothing\\ \beta\end{array}\right.}{\alpha\{\beta}$$

$\alpha \neq \bullet$

Modal rules control moving formulas along trees; we have a right rule common to all NLs:

$\vdash \Box$

$$\frac{\alpha\{\varnothing\{C}{\alpha\{\Box C}$$

where $\alpha \neq \bullet$, and a left rule depending on the system, decomposed into many rules, one for each modal axiom: K:

$$\frac{\alpha\{\Pi\left\{\begin{array}{l}\sigma\\ C,\Gamma\{\beta\end{array}\right.}{\alpha\{\Box C,\Pi\left\{\begin{array}{l}\sigma\\ \Gamma\{\beta\end{array}\right.}$$

where $\beta \neq \bullet$.

T:

$$\frac{\alpha\{C,\Gamma\{\beta}{\alpha\{\Box C,\Gamma\{\beta}$$

where $\beta \neq \bullet$.

4:

$$\frac{\alpha\{\Pi\left\{\begin{array}{l}\sigma\\ \Box C,\Gamma\{\beta\end{array}\right.}{\alpha\{\Box C,\Pi\left\{\begin{array}{l}\sigma\\ \Gamma\{\beta\end{array}\right.}$$

where $\beta \neq \bullet$.

5:

$$\frac{\alpha\left\{\begin{array}{l}\Pi\{\delta\\ C,\Gamma\{\beta\end{array}\right.}{\alpha\left\{\begin{array}{l}\Box C,\Pi\{\delta\\ \Gamma\{\beta\end{array}\right.}$$

where $\beta, \delta \neq \bullet$.

B:

$$\frac{\alpha\{C,\Gamma\left\{\begin{array}{l}\sigma\\ \Pi\{\beta\end{array}\right.}{\alpha\{\Gamma\left\{\begin{array}{l}\sigma\\ \Box C,\Pi\{\beta\end{array}\right.}$$

where $\beta \neq \bullet$.

D:

$$\frac{\alpha\{\varnothing\{\varnothing}{\alpha\{\varnothing}$$

where $\alpha \neq \bullet$.

For modal rules, as for structural ones, we consider a formula as occurring in the right part of a sequent when it belongs to a terminal leaf, while we say it occurs in the left part of that sequent when it belongs to any other node. So, in the rule $\vdash \Box$ the affected formula remains on the right, while in the rules $K, T, 4, 5, B, D$ the affected formula remains on the left: all modal rules do not change the side where the affected formulas occur, but move them towards the top of the tree, like putting them on an upper metalinguistic level. The notions of *proof* and of *end-sequent* are defined as in standard calculi of sequents (see [17]); an *initial sequent* is one of the kind

$$\varepsilon\left\{\begin{array}{l}\varepsilon\\ A\{A\end{array}\right.$$

where εs denote generic trees made only of empty sequences $\varnothing$; a formula A of **L** is *provable* iff $\varnothing\{A$ is an end-sequent.

Weakening, exchange and contraction of branches are derived rules (see [4]):

tree-weakening tree-exchange tree- contraction

$$\frac{\alpha\{\beta}{\alpha\left\{\begin{array}{l}\delta\\ \beta\end{array}\right.} \qquad \frac{\alpha\left\{\begin{array}{l}\beta\\ \beta\end{array}\right.}{\alpha\{\beta} \qquad \frac{\alpha\left\{\begin{array}{l}\delta\\ \beta\end{array}\right.}{\alpha\left\{\begin{array}{l}\beta\\ \delta\end{array}\right.}$$

$$\alpha, \beta \neq \bullet \qquad \alpha \neq \bullet \qquad \alpha \neq \bullet.$$

THEOREM 7 (Soundness of NLs-calculus) *If A is Λ-provable then A is a Λ-theorem, for any NL Λ.*

Proof. We define the Schütte translation $*$ by an induction on the complexity of non-trivial tree-sequents (using also a companion function $^{\Box}$, that also works on trivial tree-sequents) as follows:

$$(\Delta)^{\Box} = \bigvee \Delta$$

$$\Gamma\left\{\begin{array}{l}\delta_0\\ \vdots\\ \delta_n\end{array}\right.^{*} = \bigwedge \Gamma \rightarrow (\delta_0^{\Box} \vee \ldots \vee \delta_n^{\Box})$$

$$\gamma\left\{\begin{array}{l}\delta_0\\ \vdots\\ \delta_n\end{array}\right.^{\Box} = \Box\left(\gamma\left\{\begin{array}{l}\delta_0\\ \vdots\\ \delta_n\end{array}\right.^{*}\right) = \bigwedge \Gamma \Rightarrow (\delta_0^{\Box} \vee \ldots \vee \delta_n^{\Box})$$

where Γ and Δ are sequences of formulas, $\delta_0, \ldots, \delta_n$ are tree-sequents.

So, when translating a sequent, the first implication is material, as usual, while the other ones are strict (resembling that done for natural deduction based on strict implication for NLs in [3]); in this sense any level of depth along branches can be viewed as a metalevel w.r.t. the deeper ones; in particular, the first level is like a metalevel w.r.t. all the others. As usual, we extend the translation $*$ to sequent-style inferences into Hilbert-style inferences, and prove, in a way similar to semantic modal sequents, that initial sequents are translated to tautologies, and sequent-style inferences are translated to Hilbert-style inferences, so to immediately obtain the thesis (for a detailed proof see [4]). ∎

THEOREM 8 (Completeness) *If A is a Λ-theorem then A is Λ- provable, for any NL Λ.*

Proof. As standard proof, we show that modal axioms are provable and that our calculus is closed under rules RPL and Necessitation.

1. If A is **PC**-deducible from $A_1, \ldots, A_n$ then the sequent $A_1, \ldots, A_n \vdash A$ is **PC**-provable (see [11]), i.e. since the **PC**-fragment of any Λ- calculus is really a **PC**-calculus, $A_1, \ldots, A_n\{A$ is **PC**-provable; so if $A_1, \ldots, A_n$ are Λ- provable (e.g. $\varnothing\{A_1, \ldots, \varnothing\{A_n$ are end-sequents) by n uses of cut A is Λ- provable, too; i.e. the rule RPL.

2. Any modal axiom among $K, T, 5, 4, B, D$ is provable in a calculus containing rules $K, T, 5, 4, B, D$, respectively; for example:

   K:

   hypothesis

$$\dfrac{\dfrac{\dfrac{\dfrac{\varnothing\{A\{A \quad \varnothing\{B\{B}{\varnothing\{A \to B, A\{B}\;{\to\vdash}}{\Box A, \Box(A \to B)\{\varnothing\{B}\;{\Box\vdash\text{ twice}}}{\Box A, \Box(A \to B)\{\Box B}\;{\vdash\Box}}{\varnothing\{\Box(A \to B) \to (\Box A \to \Box B)}\;{\vdash\to\text{ twice}}$$

   4:

   hypothesis

$$\dfrac{\dfrac{\dfrac{\varnothing\{\Box A\{\Box A}{\Box A\{\varnothing\{\Box A}\;4}{\Box A\{\Box\Box A}\;{\vdash\Box}}{\varnothing\{\Box A \to \Box\Box A}\;{\vdash\to}$$

3. Our calculus is closed under Necessitation, i.e. if A is Λ-provable then $\Box A$ is Λ-provable, too. In fact, we can easily modify all of the proof of A (i.e. of the sequent $\varnothing\{A)$ by substituting any sequent α (especially the initial ones) with $\varnothing\{\alpha$, so obtaining a proof of $\varnothing\{\varnothing\{A$; by rule $\vdash\Box$ we immediately have $\varnothing\{\Box A$, i.e. a proof of $\Box A$.

■

For cut-elimination the question is unexpectedly complicated: in fact we can easily adapt the proof in [17], once we have proved that a left formula can be moved along a tree according with the corresponding accessibility relation (see [4]). But e.g. rule 5 allows us to transport a formula from a sequence to its brother, but not upward or downward the tree, and combining rules B with 4 (and with usual $\Box\vdash$) we can move a formula as upwards as we like, but only one step downward.

For example, we cannot prove the **K5**-theorem $\Diamond A \to \Box\Box\Diamond A$, which requires a formula A moving to the son of the brother when becoming $\Box A$ (or equivalently requires $\Box A$ moving downward); a cut-free proof should appear as:

$$\dfrac{\dfrac{\varnothing\left\{\begin{array}{l}\neg A\{\neg A\\ \varnothing\{\varnothing\{\varnothing\end{array}\right.}{\varnothing\left\{\begin{array}{l}\varnothing\{\neg A\\ \varnothing\{\Box\neg A\{\varnothing\end{array}\right.}\text{ not allowed by the only 5 rule}}{\varnothing\left\{\begin{array}{l}\varnothing\{\neg A\\ \varnothing\{\varnothing\{\neg\Box\neg A\end{array}\right.}\vdash\neg$$

$$\vdots$$

So we must allow a formula C to move according with euclidean accessibility (when becoming $\Box C$): in [4] that movement is splitted in four distinct rules: $\Box\vdash$, which transforms C into $\Box C$ moving of one step upward, 5_1, which moves $\Box C$ to the father (when is not a top- node), 5_2, which moves $\Box C$ to the brother (when is not a top-node), 5_3, which moves $\Box C$ to the son:

5_1

$$\dfrac{\alpha\{\Pi\left\{\begin{array}{l}\lambda\\ \Box C,\Gamma\{\delta\end{array}\right.}{\alpha\{\Box C,\Pi\left\{\begin{array}{l}\lambda\\ \Gamma\{\delta\end{array}\right.}\qquad \alpha,\delta,\lambda\neq\bullet$$

5_2

$$\dfrac{\alpha\left\{\begin{array}{l}\Pi\{\lambda\\ \Box C,\Gamma\{\beta\end{array}\right.}{\alpha\left\{\begin{array}{l}\Box C,\Pi\{\lambda\\ \Gamma\{\beta\end{array}\right.}\qquad \alpha,\delta,\lambda\neq\bullet$$

5_3

$$\dfrac{\alpha\{\Box C,\Pi\left\{\begin{array}{l}\lambda\\ \Gamma\{\delta\end{array}\right.}{\alpha\{\Pi\left\{\begin{array}{l}\lambda\\ \Box C,\Gamma\{\delta\end{array}\right.}\qquad \delta,\lambda\neq\bullet$$

Rule 5_3 just sufficies to prove completeness (in fact, rules $\Box\vdash$ and 5_3 immediately give rule 5), but all of them needs to eliminate cut (e.g. proving $\Box(\Box A \to \Box\Box A)$ requires rule 5_1).

That opens a question: are not trees (of sequences or of sequents) the right structure for syntactical sequent calculi for NLs, or can only usual modalities reach a portion of the expressive power offered by trees (so requiring the use e.g. of tense operators to completely managing trees)? To be continued.

5 CONCLUSIONS AND ACKNOWLEDGEMENTS

Non-standard calculi for NLs show hidden aspects of deductive systems: sequents with modal signs stress the different role of modal and propositional behaviours in NLs proofs; that difference assumes an algorithmic value.

Semantic modal sequents combine sequents with Kripke models, directly introducing the semantic feature of accessibility into the calculus; the consequence is a uniform treatment for NLs which works well, is cut-free, and offers a natural completeness theorem.

Modal tree-sequents combine a tree-structure of levels with syntactical rules, so representing a link between the other two formulations; those calculi work well and are cut-free, but the rules necessary for eliminating cut are more than we need to prove completeness, opening an unexpected gap.

Finally, I would like to acknowledge my gratitude to Prof. M. Fattorosi-Barnaba for the many helpful conversations I had with him about the topic of the present work.

via di Bravetta, Italy.

REFERENCES

[1] C. Cerrato. Modal sequents for normal modal logics. *Math. Logic Quarterly*, **39**, 231–240, 1993.
[2] C. Cerrato. Cut-free modal sequents for normal modal logics. *Notre Dame J. Formal Logic*, **34**, 564–582, 1993.
[3] C. Cerrato. Natural deduction based upon strict implication for normal modal logics. *Notre Dame J. Formal Logic*, **35**, 471–495, 1994.
[4] C. Cerrato. Modal tree-sequents. *Math. Logic Quarterly*. To appear.
[5] B. F. Chellas. *Modal Logic; An Introduction*. Cambridge Univ. Press, 1980.
[6] K. Došen. Sequent-systems for modal logic. *J. Symbolic Logic*, **50**, 149–168, 1985.
[7] M. C. Fitting. *Proof Methods for Modal and Intuitionistic Logics*. Reidel, Dordrecht, 1983.
[8] J.-Y. Girard. *Proof Theory and Logical Complexity*, Vol. I, Bibliopolis, 1987.
[9] R. Goré. Cut-free Sequents and Tableau Systems for Propositional Diodorean Modal Logics. *Studia Logica*, **53**, 433–457, 1994.
[10] R. Goré. Communication to C. Cerrato, Manchester, 1993.
[11] S. C. Kleene. *Introduction to Metamathematics*. North-Holland, Amsterdam, 1952.
[12] S. C. Kleene. A completeness theorem in modal logic. *J. Symbolic Logic*, **24**, 1–14, 1959.
[13] M. Ohnishi, K. Matsumoto. Gentzen Method in Modal Calculi. *Osaka Math. J.* , **9**, 113–130, 1957.

[14] M. Ohnishi, K. Matsumoto. Gentzen Method in Modal Calculi II. *Osaka Math. J.* , **11**, 115–120, 1959.
[15] K. Schütte. Schlussweisen-Kalküle der rädikatenlogik. *Math. Annalen* , **122**, 47–65, 1951.
[16] G. Sundholm. Systems of deduction. In: D. M. Gabbay, F. Guenthner (eds.), *Handbook of Philosophical Logic*, Vol I, pp. 133–188, Reidel, Dordrecht, 1983.
[17] G. Takeuti. *Proof Theory*. North-Holland, Amsterdam, 1975.
[18] M. Takano. Subformula property as a substitute for cut-elimination in modal propositional logics. *Mathematica Japonica* , **37**, 1129–1145, 1992.
[19] M. Takano. Communication to C. Cerrato, Niigata, 1993.

KOSTA DOŠEN AND ZORAN PETRIĆ

MODAL FUNCTIONAL COMPLETENESS

INTRODUCTION

This paper is a companion to [3], where it was shown how a modal version of the deduction theorem induces modal extensions of substructural logics. Here we shall prove some of the results announced in the concluding section of [3]. Namely, we shall prove modal functional completeness for categories corresponding to the main modal substructural propositional logics. This modal functional completeness is related to the modal version of the deduction theorem of [3] as functional completeness for bicartesian closed categories is related to the deduction theorem for intuitionistic propositional logic. Roughly speaking, the deduction theorem says that the system is strong enough to express its own deductive metatheory. Functional completeness says that the deductive metatheory can be embedded in the system.

Functional completeness for categories is a property of the same kind as combinatorial completeness for systems of combinators. Combinatorial completeness permits us to define functional abstraction and demonstrate the equivalence with systems of lambda terms. Functional completeness permits us similarly to find systems of typed lambda terms as internal languages of closed categories. A well-known use of typed lambda terms in proof theory, which goes by the name of the Curry–Howard correspondence, is to serve as codes of natural-deduction proofs. If categories are conceived as logical systems with an equivalence relation among proofs, which is induced by normalization, this coding and extracting the internal language boil down to the same thing.

We learned about these matters from [6] and refer to that book for a demonstration of the importance of functional completeness. We shall try to follow the style of [6], so that our results may be compared with the standard functional completeness results proved there for cartesian, cartesian closed and bicartesian closed categories. An acquaintance with [6] may also help to grasp the proof-theoretical import of what we will present here—a matter about which we don't have space to say much. Further logical motivation may be found in the aforementioned paper [3] and in [4]. We refer to these works only for motivation. Otherwise, our paper will be rather self-contained.

H. Wansing (ed.), Proof Theory of Modal Logic, 167–211.

For our functional completeness results we concentrate on categories corresponding to what we take to be the minimal substructural logic: namely, Lambek's nonassociative calculus. The modal postulates assumed for these categories are those of S4 plus modalized versions of the missing structural rules—the same kind of modal postulates one finds in usual presentations of linear logic. The modal functional completeness theorem proved for these categories can easily be extended to categories corresponding to better-known substructural logics, like linear and relevant logic, as we indicate in Section 6. Working with the minimal substructural logic has technical advantage and should help dispelling the impression that this sort of theorem, as well as some others, is exclusively tied to a particular substructural logic. Results about linear logic are often presented in a fashion that fosters such misleading impressions. For example, modal translation results (which are not foreign to the matter we are treating) are presented as if they had to do with linear logic specifically, though analogous results may be obtained as easily for other substructural logics.

We shall devote a section to a transformation of our proof of modal functional completeness into a proof of ordinary, nonmodal, functional completeness for cartesian, cartesian closed and bicartesian closed categories axiomatized in a nonstandard manner. Namely, instead of having as primitives projection arrows and a pairing operation on arrows, we have arrows corresponding to structural rules, and we show how with arrows corresponding to the structural rules of thinning and contraction, Mac Lane's axiomatization of symmetric monoidal categories, which among other things has arrows corresponding to association and permutation, can be extended to an axiomatization of cartesian categories. This means putting cartesian categories in a substructural perspective, where we envisage rejecting structural rules.

We produce such nonmodal substructural categories when in a separate section, at the end, we consider restricted nonmodal functional completeness for them. The restrictions in question are obtained by having special notions of what in the terminology of [6] is called polynomials in polynomial categories. This is related to having restricted classes of typed lambda terms as codes of proofs in substructural logics. We leave for the future a more thorough investigation of this and some other matters mentioned in the concluding section.

We also devote a section to justifying the assumptions made for our categories, and, in particular, assumptions related to structural rules. We shall show for a number of them, and in particular the more abstruse, that they are not only sufficient, but also necessary for functional completeness. Most of these assumptions are quite well known and can be found in Mac Lane's book [7]. There they are motivated differently: they appear in connection with coherence problems (cf. the very end of our paper). We motivate the same assumptions by functional completeness. Though coherence may also be related to logic, we believe the sort of motivation we provide does more justice to the logical character of our categories. Anyway, it should be better than just taking over these assumptions from category theory, as it is sometimes done when categories are presented as 'models' for logics.

However, we didn't find in the literature all of these assumptions: some of those tied to the structural rule of contraction may be new. Among them we single out something we shall call *octagonal equation*, a principle we shall use quite often, which is comparable to Mac Lane's pentagonal and hexagonal diagrams of natural associativity and commutativity.

To end this introduction, let us mention a notational and terminological matter. Though we shall in general imitate the style of [6], we shall diverge somewhat in notation. In particular, we write arrows in categories with a turnstile, as $f : A \vdash B$, instead of $f : A \rightarrow B$. This we do because we reserve $\rightarrow$ for implication, an operation on objects, for which category theory uses an exponential notation or square brackets. We prefer to stay close to logical notation because our motivation is in logic. We want to suggest that objects in categories correspond to formulae, operations on objects to connectives, arrows to sequents, special arrows to axioms, operations on arrows to rules of inference, and equations between arrows to conversions of proofs, like those we make in normalization. However, we dont want to diverge from category theory also in terminology. So we call $f : A \vdash B$ an *arrow*, rather than a *sequent*. This hybrid between logical notation and categorial terminology may be slightly awkward, both from the point of view of logic and of category theory, but it just reflects the nature of our work. We use categories to say something about logical systems.

Because of its connection with the deduction theorem and proof theory, functional completeness could perhaps be called *deductive completeness*. But, to follow the policy exposed in the previous paragraph, and to prevent misunderstanding, we prefer to stick to the established terminology.

1 NL CATEGORIES

A *graph* consists of a class of *arrows* and a class of *objects*, together with the functions that to every arrow assign the objects that are its *source* and *target*. We use $f, g, h, \ldots$, possibly with indices, for arrows, and $A, B, C, \ldots$, possibly with indices, for objects. We write $f : A \vdash B$ to say that A is the source of f and B the target of f. For such an f we say that it is of type $A \vdash B$.

A *deductive system* is a graph in which we have a special arrow for every object

A:

$$1_A : A \vdash A$$

and the binary (partial) operation of composition of arrows:

$$\frac{f : A \vdash B \quad g : B \vdash C}{gf : A \vdash C}$$

A *category* is a deductive system in which the following equations between arrows are satisfied:

(cat 1) For $f : A \vdash B$, $\quad 1_B f = f, \quad f 1_A = f.$

(cat 2) For $f : A \vdash B, g : B \vdash C$ and $h : C \vdash D$, $\quad h(gf) = (hg)f.$

An NL deductive system ('NL' stands for 'nonassociative Lambek') has the following in addition to what every deductive system must have:

binary operations on objects: $\bullet, \rightarrow, \leftarrow, \wedge, \vee$

special objects: $\mathrm{I}, \top, \bot$

special arrows for every object A and B:

$$\begin{array}{lcl}
\sigma_A : \mathrm{I} \bullet A \vdash A & & \delta_A : A \bullet \mathrm{I} \vdash A \\
\sigma^i_A : A \vdash \mathrm{I} \bullet A & & \delta^i_A : A \vdash A \bullet \mathrm{I} \\
\varepsilon^{\rightarrow}_{A,B} : A \bullet (A \rightarrow B) \vdash B & & \varepsilon^{\leftarrow}_{A,B} : (B \leftarrow A) \bullet A \vdash B \\
\pi_{A,B} : A \wedge B \vdash A & & \pi'_{a,B} : A \wedge B \vdash B \\
& \tau_A : A \vdash \top & \\
\kappa_{A,B} : A \vdash A \vee B & & \kappa'_{A,B} : B \vdash A \vee B \\
& \iota_A : \bot \vdash A &
\end{array}$$

('σ' is to be associated with 'sinister', 'δ' with 'dexter', 'τ' with 'terminal', 'ι' with 'initial', and the superscript 'i' stands for 'inverse' or, maybe, 'introduction'; the remaining notation for arrows and operations on them is modelled after [6]) *operations on arrows:*

$$\frac{f : A \vdash B \quad g : C \vdash D}{f \bullet g : A \bullet C \vdash B \bullet D}$$

$$\frac{f : A \bullet C \vdash B}{{}^{*}f : C \vdash A \rightarrow B} \qquad \frac{f : C \bullet A \vdash B}{f^{*} : C \vdash B \leftarrow A}$$

$$\frac{f : C \vdash A \quad g : C \vdash B}{\langle f, g \rangle : C \vdash A \wedge B}$$

$$\frac{f : A \vdash C \quad g : B \vdash C}{[f, g] : A \vee B \vdash C}$$

The minimal NL deductive system is an extension of Lambek's nonassociative calculus with the propositional constant I, the lattice connectives $\wedge, \vee, \top$ and $\bot$ (lattice connectives are called 'additive' in the terminology of linear logic), and the associated arrows and operations on arrows. Note that the extension with I and the σ, σ^i, δ and δ^i arrows amounts to adding some structural rules; namely, rules for dealing with the empty collection of premises. Hence, it shouldn't be surprising that this extension is not conservative (it enables us to prove, for example, $(A \rightarrow A) \rightarrow B \vdash B$). We shall call the σ, σ^i, δ and δ^i arrows *$\sigma\delta$ arrows.*

An NL category is an NL deductive system that is a category in which the following equations between arrows are satisfied:

- equations

  ($\bullet$) For $f_1 : A_1 \vdash B_1, g_1 : B_1 \vdash C_1, f_2 : A_2 \vdash B_2$ and $g_2 : B_2 \vdash C_2$, $(g_1 \bullet g_2)(f_1 \bullet f_2) = (g_1 f_1) \bullet (g_2 f_2)$.

  ($\bullet 1$) $1_A \bullet 1_B = 1_{A \bullet B}$

  $\sigma\delta$ equations

  (σ) For $f : A \vdash B$, $f\sigma_A = \sigma_B(1_{\mathrm{I}} \bullet f)$.

  (δ) For $f : A \vdash B$, $f\delta_A = \delta_B(f \bullet 1_{\mathrm{I}})$.

  ($\sigma\sigma^i$) $\sigma_A\sigma^i_A = 1_A$, $\sigma^i_A\sigma_A = 1_{\mathrm{I} \bullet A}$

  ($\delta\delta^i$) $\delta_A\delta^i_A = 1_A$, $\delta^i_A\delta_A - 1_{A \bullet \mathrm{I}}$

  ($\sigma\delta$) $\sigma_{\mathrm{I}} = \delta_{\mathrm{I}}$

  closure equations

$(\rightarrow\beta)$ For $f : A \bullet C \vdash B$, $\varepsilon^{\rightarrow}_{A,B}(1_A \bullet {}^*f) = f$.
$(\rightarrow\eta)$ For $g : C \vdash A \rightarrow B$, ${}^*(\varepsilon^{\rightarrow}_{A,B}(1_A \bullet g)) = g$.
$(\leftarrow\beta)$ For $f : C \bullet A \vdash B$, $\varepsilon^{\leftarrow}_{A,B}(f^* \bullet 1_A) = f$.
$(\leftarrow\eta)$ For $g : C \vdash B \leftarrow A$, $(\varepsilon^{\leftarrow}_{A,B}(g \bullet 1_A))^* = g$.

bicartesian equations

$(\wedge\beta)$ For $f : C \vdash A$ and $g : C \vdash B, \pi_{A,B}\langle f, g\rangle = f$, $\pi'_{A,B}\langle f, g\rangle = g$.
$(\wedge\eta)$ For $h : C \vdash A \wedge B$, $\langle \pi_{A,B}h, \pi'_{A,B}h\rangle = h$.
$(\top)$ For $f : A \vdash \top$, $\tau_A = f$.
$(\vee\beta)$ For $f : A \vdash C$ and $g : B \vdash C, [f, g]\kappa_{A,B} = f$, $[f, g]\kappa'_{A,B} = g$.
$(\vee\eta)$ For $h : A \vee B \vdash C$, $[h\kappa_{A,B}, h\kappa'_{A,B}] = h$.
$(\bot)$ For $f : \bot \vdash A$, $\iota_A = f$.

With the help of either $(\rightarrow\beta)$ or $(\leftarrow\beta)$ we can derive $(\bullet 1)$: we have

$$\begin{aligned} 1_A \bullet 1_B &= \varepsilon^{\rightarrow}_{A,A\bullet B}(1_A \bullet {}^*1_{A\bullet B})(1_A \bullet 1_B), \text{ with (cat 1) and } (\rightarrow\beta) \\ &= 1_{A\bullet B}, \text{ with } (\bullet), \text{ (cat 1) and } (\rightarrow\beta) \end{aligned}$$

and can proceed analogously with $(\leftarrow\beta)$. However, we have preferred to include $(\bullet 1)$ among the $\bullet$ equations so that we may have it even in the absence of $\rightarrow$ and $\leftarrow$ (see Section 4).

With the definitions

$$D \bullet f =_{\mathrm{df}} 1_D \bullet f$$
$$f \bullet D =_{\mathrm{df}} f \bullet 1_D$$

we can derive the following equations by using (cat 1) and $(\bullet)$:

$(\bullet 2)$ $D \bullet (gf) = (D \bullet g)(D \bullet f)$, $(gf) \bullet D = (g \bullet D)(f \bullet D)$
$(\bullet\text{bifunctor})$ $(f_1 \bullet B_2)(A_1 \bullet f_2) = (B_1 \bullet f_2)(f_1 \bullet A_2)$

Conversely, if the unary operations on arrows $D \bullet \_$ and $\_ \bullet D$ are primitive instead of the binary operation on arrows $\_ \bullet \_$, we can define the latter by either of the following two definitions

$$f_1 \bullet f_2 =_{\mathrm{df}} (f_1 \bullet B_2)(A_1 \bullet f_2)$$
$$f_1 \bullet f_2 =_{\mathrm{df}} (B_1 \bullet f_2)(f_1 \bullet A_2)$$

and derive (•) by using (•2) and (•bifunctor). With these unary operations primitive, (•1)is replaced by

$$D \bullet 1_B = 1_{D \bullet B}, \qquad 1_A \bullet D = 1_{A \bullet D}.$$

When it is more convenient to work with $D \bullet \_$ and $\_ \bullet D$ instead of $\_ \bullet \_$, we may freely avail ourselves of this opportunity.

The first six $\sigma\delta$ equations assert that σ and δ are natural isomorphisms. They can be replaced by the four equations

$$\begin{array}{lll} (\mathrm{I}\beta) & \sigma_B(1_{\mathrm{I}} \bullet f)\sigma^i_A = f, & \delta_B(f \bullet 1_{\mathrm{I}})\delta^i_A = f \\ (\mathrm{I}\eta) & \sigma^i_B f \sigma_A = 1_{\mathrm{I}} \bullet f, & \delta^i_B f \delta_A = f \bullet 1_{\mathrm{I}} \end{array}$$

which are more parallel with the closure and bicartesian equations. The letters 'β' and 'η' in the names of all these equations should point towards the analogy with β and η conversion in the lambda calculus, which are themselves analogous to the conversions of two types of detours in natural deduction: introduction followed by elimination and elimination followed by introduction. However, the $\sigma\delta$ equations as we have given them are more transparent. These equations are assumed by Mac Lane [7, VII.1] for monoidal categories. The difference is that for • in NL categories we need not have a natural associativity isomorphism.

For an NL category $\mathcal{C}$, the operations $\_ \bullet \_$ on objects and arrows determine a functor from $\mathcal{C} \times \mathcal{C}$ to $\mathcal{C}$, i.e. a bifunctor, whereas $D \bullet \_$ and $\_ \bullet D$ are functors from $\mathcal{C}$ to $\mathcal{C}$. With the definitions

$$\begin{array}{l} D \to f =_{\mathrm{df}} {}^*(f \varepsilon_{\overrightarrow{D},A}) \\ f \leftarrow D =_{\mathrm{df}} (f \varepsilon_{\overleftarrow{D},A})^* \end{array}$$

we obtain the functors $D \to \_$ and $\_ \leftarrow D$ from $\mathcal{C}$ to $\mathcal{C}$. We also have the definitions

$$\begin{array}{ll} \varepsilon_{\overrightarrow{D}}(A) =_{\mathrm{df}} \varepsilon_{\overrightarrow{D},A} & \varepsilon_{\overleftarrow{D}}(A) =_{\mathrm{df}} \varepsilon_{\overleftarrow{D},A} \\ \eta_{\overrightarrow{D}}(A) =_{\mathrm{df}} {}^*1_{D \bullet A} & \eta_{\overleftarrow{D}}(A) =_{\mathrm{df}} 1_{A \bullet D}{}^* \end{array}$$

(conversely, we can define ${}^*f$ as $(D \to f)\eta_{\overrightarrow{D}}(A)$, and f^* as $(f \leftarrow D)\eta_{\overleftarrow{D}}(A)$).

The • and closure equations amount to asserting that $\_ \bullet \_$ is a bifunctor, that for every object D the functors $D \bullet \_$ and $D \to \_$ are adjoined, the first being left adjoined and the second right adjoined, with the natural transformations $\varepsilon_{\overrightarrow{D}}$ and $\eta_{\overrightarrow{D}}$ being the counit and unit of the adjunction, and analogously with the functors $\_ \bullet D$

and $\_ \leftarrow D$ and the natural transformations $\varepsilon^{\leftarrow}_D$ and $\eta^{\leftarrow}_D$. All this may be expressed by saying that NL categories are biclosed. The bicartesian equations amount to asserting that NL categories are bicartesian with respect to the operations $\wedge$ and $\vee$, their terminal object being $\top$ and their initial object $\bot$.

The equations $(\rightarrow \eta)$ and $(\leftarrow \eta)$ can be replaced by the equations

$$(^*)\quad \text{For } f : A \bullet D \vdash B \text{ and } g : C \vdash D,\quad {}^*(f(1_A \bullet g)) = ({}^*f)g.$$
$$\text{For } f : D \bullet A \vdash B \text{ and } g : C \vdash D,\quad (f(g \bullet 1_A))^* = (f^*)g.$$

$$({}^*\varepsilon)\quad {}^*\varepsilon^{\rightarrow}_{A,B} = 1_{A\rightarrow B},\qquad \varepsilon^{\leftarrow}_{A,B}{}^* = 1_{B\leftarrow A}.$$

The style of these equations is comparable to the style of the equation (σ) and the second $(\sigma\sigma^i)$ equation (or (δ) and the second $(\delta\delta^i)$ equation): the equations $(^*)$ say how the operations * permute with composition, whereas the equations $(*\varepsilon)$ exhibit the result of eliminating and then reintroducing an implication. Similarly, the equations $(\wedge\eta)$ and $(\vee\eta)$ can be replaced by the equations:

$$\langle fh, gh\rangle = \langle f, g\rangle h,\quad \langle \pi_{A,B}, \pi'_{A,B}\rangle = 1_{A\wedge B}$$
$$[hf, hg] = h[f, g],\quad [\kappa_{A,B}, \kappa'_{A,B}] = 1_{A\vee B}.$$

The equations $(\top)$ and $(\bot)$ can be replaced by the equations:

$$\text{For } f : A \vdash B,\quad \tau_A = \tau_B f,\quad \tau_\top = 1_\top.$$
$$\text{For } f : B \vdash A,\quad \iota_A = f\iota_B,\quad \iota_\bot = 1_\bot.$$

2 NL□ CATEGORIES

To define the modal NL□ deductive systems and NL□ categories we need the following piece of terminology. First we define inductively the *factors* of an object in a deductive system that has the binary operation $\bullet$ on objects: A is a factor of A; if $B \bullet C$ is a factor of A, then B and C are factors of A. An *atomic factor* of A is a factor of A that has no factors save itself. Let us consider deductive systems that have the binary operation $\bullet$ and a unary operation $\Box$ on objects, and also the special object I. We shall say that an object $\Box A$ in such a deductive system is *boxed*. An object is *modalized* if and only if each of its atomic factors is either boxed or else it is I.

An NL□ deductive system is an NL deductive system that, moreover, has the following:

unary operation on objects: $\Box$

special arrows:

$\mathbf{r}_A : \Box A \vdash A,$	for every object A,
$\mathbf{b}^{\rightarrow}_{A,B,C} : A \bullet (B \bullet C) \vdash (A \bullet B) \bullet C,$	provided A or B or C is modalized,
$\mathbf{b}^{\leftarrow}_{A,B,C} : (A \bullet B) \bullet C \vdash A \bullet (B \bullet C),$	provided A or B or C is modalized,
$\mathbf{c}_{A,B} : A \bullet B \vdash B \bullet A,$	provided A or B is modalized.
$\mathbf{k}_A : A \vdash \mathrm{I},$	provided A is modalized,
$\mathbf{w}_A : A \vdash A \bullet A,$	provided A is modalized,

operation on arrows:

$$\frac{f : B \vdash A}{f^{\Box} : B \vdash \Box A} \qquad \text{provided } B \text{ is modalized.}$$

Note immediately that for modalized objects A we have the arrows $1^{\Box}_A$: $A \vdash \Box A$. These and the $\mathbf{r}$ arrows would enable us to formulate the provisos for the $\mathbf{b}^{\rightarrow}, \mathbf{b}^{\leftarrow}, \mathbf{c}, \mathbf{k}$ and $\mathbf{w}$ arrows by restricting ourselves to boxed objects A, B or C, rather than any modalized objects. However, for technical reasons, it is more convenient to have the provisos in the equivalent form above.

The $\mathbf{b}^{\rightarrow}, \mathbf{b}^{\leftarrow}, \mathbf{c}, \mathbf{k}$ and $\mathbf{w}$ arrows are related to the combinators usually named with the corresponding capital letters. A $\mathbf{b}^{\rightarrow}$ arrow is used to obtain the arrow

$$^{**}(\varepsilon^{\rightarrow}_{A,B}(\varepsilon^{\rightarrow}_{C,A} \bullet 1_{A\rightarrow B})\mathbf{b}^{\rightarrow}_{C,C\rightarrow A,A\rightarrow B}) : A \rightarrow B \vdash (C \rightarrow A) \rightarrow (C \rightarrow B)$$

which corresponds to the functional type of the combinator $\mathbf{B}$ taking arguments on the left-hand side, whereas $\mathbf{b}^{\leftarrow}$ is used for

$$(\varepsilon^{\leftarrow}_{A,B}(1_{B\leftarrow A} \bullet \varepsilon^{\leftarrow}_{C,A})\mathbf{b}^{\leftarrow}_{B\leftarrow A,A\leftarrow C,C})^{**} : B \leftarrow A \vdash (B \leftarrow C) \leftarrow (A \leftarrow C)$$

which corresponds to the functional type of $\mathbf{B}$ taking arguments on the right-hand side. This explains the upper indices of $\mathbf{b}^{\rightarrow}$ and $\mathbf{b}^{\leftarrow}$. The $\mathbf{b}^{\rightarrow}$ and $\mathbf{b}^{\leftarrow}$ arrows will be called **b** *arrows*.

The **b**, **c**, **k** and **w** arrows are related to structural rules, too: **b** arrows to association, **c** arrows to permutation, **k** arrows to thinning and **w** arrows to contraction (to give the full force of thinning, **k** arrows have to cooperate with $\sigma\delta$ arrows). So we shall call these arrows *structural arrows*. We have said in the previous section that the $\sigma\delta$ arrows may be taken as structural. Such are also the arrows 1_A, which correspond to the combinator **I**. However, when we say here *structural arrows*, we mean the **b**, **c**, **k** and **w** arrows.

An NL$\Box$ deductive system has modalized forms of the structural rules missing from NL deductive systems; moreover, it has S4 modal postulates. We can replace the operation on arrows $^\Box$ by special arrows of types

$$\begin{array}{l} \Box(A \rightarrow B) \vdash \Box A \rightarrow \Box B \\ \Box A \vdash \Box\Box A \end{array}$$

and the restricted form of $^\Box$ where B is I; this form of $^\Box$ corresponds to the modal rule of necessitation. An alternative is to replace $^\Box$ by special arrows of types

$$\begin{array}{l} \Box A \bullet \Box B \vdash \Box(A \bullet B) \\ \mathrm{I} \vdash \Box\mathrm{I} \end{array}$$

and the restricted form of $^\Box$ where B is boxed (cf. the derivation of $^\Box$ in Section 5). Still another alternative is to replace $^\Box$ by special arrows of the last two types plus $\Box A \vdash \Box\Box A$ and the operation on arrows

$$\frac{f : A \vdash B}{\Box f : \Box A \vdash \Box B}$$

An NL$\Box$ category is an NL$\Box$ deductive system that is an NL category in which the following equations between arrows are satisfied:

$\Box$ *equations*:

$$\begin{array}{lll} (\Box\beta) & \text{For } f : B \vdash A, & \mathbf{r}_A f^\Box = f. \\ (\Box\eta) & \text{For } f : B \vdash \Box A, & (\mathbf{r}_A f)^\Box = f. \end{array}$$

b *equations*:

(b) For $f : A \vdash D, g : B \vdash E$ and $h : C \vdash F$ $((f \bullet g) \bullet h)\mathbf{b}^{\rightarrow}_{A,B,C} = \mathbf{b}^{\rightarrow}_{D,E,F}(f \bullet (g \bullet h))$.

(bb) $\mathbf{b}^{\rightarrow}_{A,B,C}\mathbf{b}^{\leftarrow}_{A,B,C} = 1_{(A\bullet B)\bullet C}$, $\mathbf{b}^{\leftarrow}_{A,B,C}\mathbf{b}^{\rightarrow}_{A,B,C} = 1_{A\bullet(B\bullet C)}$

($\sigma\delta$**b**) $(\delta_A \bullet 1_B)\mathbf{b}^{\rightarrow}_{A,\mathrm{I},B} = 1_A \bullet \sigma_B$

(b5) $\mathbf{b}^{\rightarrow}_{A\bullet B,C,D}\mathbf{b}^{\rightarrow}_{A,B,C\bullet D} = (\mathbf{b}^{\rightarrow}_{A,B,C} \bullet 1_D)\mathbf{b}^{\rightarrow}_{A,B\bullet C,D}(1_A \bullet \mathbf{b}^{\rightarrow}_{B,C,D})$

c *equations:*

(c) For $f : A \vdash C$ and $g : B \vdash D$, $(g \bullet f)\mathbf{c}_{A,B} = \mathbf{c}_{C,D}(f \bullet g)$.

(**cc**) $\mathbf{c}_{B,A}\mathbf{c}_{A,B} = 1_{A\bullet B}$

($\sigma\delta$**c**) $\sigma_A\mathbf{c}_{A,\mathrm{I}} = \delta_A$

(**bc6**) $\mathbf{b}^{\rightarrow}_{C,A,B}\mathbf{c}_{A\bullet B,C}\mathbf{b}^{\rightarrow}_{A,B,C} = (\mathbf{c}_{A,C} \bullet 1_B)\mathbf{b}^{\rightarrow}_{A,C,B}(1_A \bullet \mathbf{c}_{B,C})$

k *equations*:

(**k**) For $f : A \vdash B$, $\mathbf{k}_A = \mathbf{k}_B f$.

(1**k**) $\mathbf{k}_\mathrm{I} = 1_\mathrm{I}$

w *equations*:

(**w**) For $f : A \vdash B$, $(f \bullet f)\mathbf{w}_A = \mathbf{w}_B f$.

($\sigma\delta$**w**) $\sigma_\mathrm{I}\mathbf{w}_\mathrm{I} = 1_\mathrm{I}$

(**bw**) $\mathbf{b}^{\rightarrow}_{A,A,A}(1_A \bullet \mathbf{w}_A)\mathbf{w}_A = (\mathbf{w}_A \bullet 1_A)\mathbf{w}_A$

(**cw**) $\mathbf{c}_{A,A}\mathbf{w}_A = \mathbf{w}_A$

(**bcw8**) If $\mathbf{c}^m_{A,B,C,D} =_{\mathrm{df}} \mathbf{b}^{\rightarrow}_{A,C,B\bullet D}(1_A\bullet(\mathbf{b}^{\leftarrow}_{C,B,D}(\mathbf{c}_{B,C}\bullet 1_D)\mathbf{b}^{\rightarrow}_{B,C,D}))\mathbf{b}^{\leftarrow}_{A,B,C\bullet D}$,
$\mathbf{c}^m_{A,B,A,B}\mathbf{w}_{A\bullet B} = \mathbf{w}_A \bullet \mathbf{w}_B$.

(σ**kw**) $\sigma_A(\mathbf{k}_A \bullet 1_A)\mathbf{w}_A = 1_A$ (δ**kw**) $\delta_A(1_A \bullet \mathbf{k}_A)\mathbf{w}_A = 1_A$

Of course, the arrows f in ($\Box\beta$) and ($\Box\eta$) must have B modalized, and, likewise, the other equations involve arrows with provisos for modalized objects. The equation ($\Box\eta$) can be replaced by the two equations

$$(^\Box) \quad (fg)^\Box = f^\Box g$$
$$(^\Box\mathbf{r}) \quad \mathbf{r}^\Box_A = 1_{\Box A}$$

which is quite parallel to replacing $(\rightarrow\eta)$ and $(\leftarrow\eta)$ by $(^*)$ and $(^*\varepsilon)$; the $\mathbf{r}$ arrows are analogous to the ε arrows, and the operation on arrows $^\Box$ is analogous to *. For the sake of example, let us derive $(^\Box)$:

$$\begin{array}{l} fg = fg \\ \mathbf{r}_A(fg)^\Box = \mathbf{r}_A f^\Box g, \text{ with } (\Box\beta) \\ (\mathbf{r}_A(fg)^\Box)^\Box = (\mathbf{r}_A f^\Box g)^\Box \\ (fg)^\Box = f^\Box g, \text{ with } (\Box\eta). \end{array}$$

It follows immediately from $(\Box\beta)$, $(^\Box)$ and $(^\Box\mathbf{r})$ that for modalized A the arrow $1_A^\Box : A \vdash \Box A$ is an isomorphism, its inverse being $\mathbf{r}_A$.

If for $f : A \vdash B$ we define $\Box f$ as $(f\mathbf{r}_A)^\Box$ and $\mathbf{t}_A$ as $1^\Box_{\Box A}$, then the functor $\Box\_$, the $\mathbf{r}$ arrows and the $\mathbf{t}$ arrows make a comonad (or cotriple). It is easy to check that the $\mathbf{r}$ and $\mathbf{t}$ arrows are natural transformations and that we have equations corresponding to the three commutative diagrams of [7, VI.1, p. 135] ($\mathbf{r}$ corresponds to Mac Lane's ε, and $\mathbf{t}$ to Mac Lane's δ). However, the $\mathbf{r}$ and $\mathbf{t}$ arrows and the operation on arrows $\Box\_$ don't suffice to define our operation on arrows $\_^\Box$. As we have already noted, we need moreover arrows of types $\Box A \bullet \Box B \vdash \Box(A \bullet B)$ and $\mathrm{I} \vdash \Box\mathrm{I}$ (which in the axiomatization of [8, 9.7, pp. 87–90] must be recuperated in a roundabout way, via isomorphisms between $\Box A \bullet \Box B$ and $\Box(A \wedge B)$, and between I and $\Box\top$; in the absence of $\mathbf{k}$ and $\mathbf{w}$ arrows, which are recuperated similarly, such an axiomatization becomes impracticable).

The equations (**b**) and (**bb**) can be replaced by the two equations

$$\begin{array}{l} \mathbf{b}^{\rightarrow}_{D,E,F}(f \bullet (g \bullet h))\mathbf{b}^{\leftarrow}_{A,B,C} = (f \bullet g) \bullet h \\ \mathbf{b}^{\leftarrow}_{D,E,F}((f \bullet g) \bullet h)\mathbf{b}^{\rightarrow}_{A,B,C} = f \bullet (g \bullet h) \end{array}$$

(analogous to $(\mathrm{I}\beta)$ and$(\mathrm{I}\eta)$ from Section 1), but (**b**) and (**bb**) make it more transparent we are dealing with a natural isomorphism. Similarly, (**c**) and (**cc**) can be replaced by

$$\mathbf{c}_{D,C}(g \bullet f)\mathbf{c}_{A,B} = f \bullet g$$

but, again, (**c**) and (**cc**) make it more transparent we are dealing with a natural isomorphism. Of course, from (**b**) and (**bb**) we obtain immediately

$$(f \bullet (g \bullet h))\mathbf{b}^{\leftarrow}_{A,B,C} = \mathbf{b}^{\leftarrow}_{D,E,F}((f \bullet g) \bullet h)$$

and this equation could replace (**b**).

The equation (**b**) can be replaced by the three equations

$$
\begin{array}{ll}
(\mathbf{b}_1) & ((f \bullet B) \bullet C)\mathbf{b}^{\rightarrow}_{A,B,C} = \mathbf{b}^{\rightarrow}_{D,B,C}(f \bullet (B \bullet C)) \\
(\mathbf{b}_2) & ((A \bullet g) \bullet C)\mathbf{b}^{\rightarrow}_{A,B,C} = \mathbf{b}^{\rightarrow}_{A,E,C}(A \bullet (g \bullet C)) \\
(\mathbf{b}_3) & ((A \bullet B) \bullet h)\mathbf{b}^{\rightarrow}_{A,B,C} = \mathbf{b}^{\rightarrow}_{A,B,F}(A \bullet (B \bullet h))
\end{array}
$$

where $D \bullet f$ is $1_D \bullet f$ and$f \bullet D$ is $f \bullet 1_D$, as we have defined them in Section 1. Similarly, (**c**) can be replaced by either of the two equations

$$
\begin{array}{ll}
(\mathbf{c}_1) & (B \bullet f)\mathbf{c}_{A,B} = \mathbf{c}_{C,B}(f \bullet B) \\
(\mathbf{c}_2) & (g \bullet A)\mathbf{c}_{A,B} = \mathbf{c}_{A,D}(A \bullet g).
\end{array}
$$

It is natural to assume these substitute equations if the unary operations on arrows $D \bullet \_$ and $\_ \bullet D$ are primitive instead of the binary operation on arrows $\_ \bullet \_$, though (**b**) and (**c**) would do as well. The equation (**w**) then reads

$$(f \bullet B)(A \bullet f)\mathbf{w}_A = \mathbf{w}_B f.$$

From $(\sigma\delta\mathbf{b})$, which amounts to

$$\mathbf{b}^{\rightarrow}_{A,\mathrm{I},B} = \delta^i_A \bullet \sigma_B$$

we can derive (without using **c** arrows) the following two analogous equations:

$$
\begin{array}{ll}
(\sigma\mathbf{b}) & (\sigma_A \bullet 1_B)\mathbf{b}^{\rightarrow}_{\mathrm{I},A,B} = \sigma_{A \bullet B} \\
(\delta\mathbf{b}) & \delta_{A \bullet B}\mathbf{b}^{\rightarrow}_{A,B,\mathrm{I}} = 1_A \bullet \delta_B
\end{array}
$$

[7, VII.1, p. 161, Exercise 1]. We call these three equations, and those derived from them with $(\sigma\sigma^i)$, $(\delta\delta^i)$ and (**bb**), *triangular equations*, because Mac Lane assumes $(\sigma\delta\mathbf{b})$ for monoidal categories as a triangular commutative diagram [7, VII.1, p. 159].

The equation (**b5**) is Mac Lane's pentagonal diagram for monoidal categories [7, VII.1, p. 158]. We call this equation, and equations derived from it with (**bb**), *pentagonal equations.*

The equation $(\sigma\delta\mathbf{c})$, assumed by Mac Lane for symmetric monoidal categories [7, VII.7, p. 180], enables us to define the δ and δ^i arrows in terms of the σ and σ^i arrows, or the other way round. With this equation, it is superfluous to assume in Section 1 the equations (δ) and $(\delta\delta^i)$, or (σ) and $(\sigma\sigma^i)$. From $(\sigma\delta\mathbf{c})$ and $(\sigma\sigma^i)$ it follows that $\mathbf{c}_{A,\mathrm{I}} = \sigma^i_A \delta_A$, which with $(\sigma\delta)$ and $(\sigma\sigma^i)$ yields $\mathbf{c}_{\mathrm{I},\mathrm{I}} = \mathbf{1}_{\mathrm{I}\bullet\mathrm{I}}$. This last equation with $(\sigma\delta\mathbf{c})$ immediately yields $(\sigma\delta)$.

The equation $(\mathbf{bc}6)$ is Mac Lane's hexagonal diagram for symmetric monoidal categories [7, VII.7, p. 180]. It says, intuitively, that permutation of products like $A \bullet B$ in $\mathbf{c}_{A\bullet B,C}$ can be replaced by permutation of the factors A and B in $\mathbf{c}_{A,C}$ and $\mathbf{c}_{B,C}$. We call (**bc**6), and equations derived from it with (**bb**) and (**cc**), *hexagonal equations*.

If we forget about the provisos for modalized objects, the **b** equations together with the $\bullet$ and $\sigma\delta$ equations of Section 1 axiomatize monoidal categories. If to that we add the **c** equations, we obtain symmetric monoidal categories. The closure equations of Section 1 transform the former into monoidal biclosed categories and the latter into symmetric monoidal closed categories (cf. Section 6).

The **k** equations can be replaced by the single equation:

$$\text{For } f : A \vdash \mathrm{I}, \quad \mathbf{k}_A = f.$$

At the end of Section 1 we have noted something quite analogous concerning the equation $(\top)$. The **k** equations say that I is a terminal object when we restrict ourselves to arrows from modalized objects. Formulating these equations as we did makes clearer the parallelism with the **b**, **c** and **w** equations. The equation (**k**) corresponds to (**b**), (**c**) and (**w**): when written $\mathbf{1}_\mathrm{I}\mathbf{k}_A = \mathbf{k}_B f$, it says that **k** is a natural transformation from the identity functor to the constant functor that maps objects into I and arrows into $\mathbf{1}_\mathrm{I}$. Of course, **k** need not be an isomorphism as **b** and **c** are. The equation $(\mathbf{1k})$ should be compared with

$\mathbf{b}^{\rightarrow}_{A,\mathrm{I},B} = \delta^i_A \bullet \sigma_B$ and $\mathbf{b}^{\rightarrow}_{\mathrm{I},\mathrm{I},\mathrm{I}} = \sigma^i_\mathrm{I} \bullet \sigma_\mathrm{I}$, which are related to $(\sigma\delta\mathbf{b})$ and $(\sigma\delta)$,

$\mathbf{c}_{A,\mathrm{I}} = \sigma^i_A \delta_A$ and $\mathbf{c}_{\mathrm{I},\mathrm{I}} = \mathbf{1}_{\mathrm{I}\bullet\mathrm{I}}$, which are related to $(\sigma\delta\mathbf{c})$ and $(\sigma\delta)$,

$\mathbf{w}_\mathrm{I} = \sigma^i_\mathrm{I}$ and $\mathbf{w}_\mathrm{I} = \delta^i_\mathrm{I}$, which are related to $(\sigma\delta\mathbf{w})$ and $(\sigma\delta)$,

but it may also be compared with (**bb**) and (**cc**).

The equation (**w**) says that **w** is a natural transformation from the identity functor to the square functor that maps objects A into $A \bullet A$ and arrows f into $f \bullet f$. However, **w** need not be an isomorphism.

The equation $(\sigma\delta\mathbf{w})$ has 'δ' in its name because through $(\sigma\delta)$ it involves δ_I. The equations $(\sigma\delta\mathbf{w})$, $(\mathbf{bw})$, $(\mathbf{cw})$, $(\sigma\mathbf{kw})$ and $(\delta\mathbf{kw})$ say, intuitively, how σ, δ, **b**, **c** and **k** arrows behave if they are composed with **w** arrows. Though $\mathbf{c}_{\mathrm{I},\mathrm{I}}$ equals $\mathbf{1}_{\mathrm{I}\bullet\mathrm{I}}$, the arrow $\mathbf{c}_{A,A}$ need not be equal to $\mathbf{1}_{A\bullet A}$; but it behaves like $\mathbf{1}_{A\bullet A}$ if composed with $\mathbf{w}_A$. (We can derive $\mathbf{c}_{\mathrm{I},\mathrm{I}} = 1_{\mathrm{I}\bullet\mathrm{I}}$ from (**cw**), ($\sigma\delta$**w**) and ($\sigma\sigma^i$), which is different from the derivation we gave above.) Similarly, $\mathbf{b}^{\rightarrow}_{A,A,A}$ need not be equal to $\mathbf{c}_{A,A\bullet A}$, but from (**bw**), (**cw**) and (**c**) it follows that

$$\mathbf{b}^{\rightarrow}_{A,A,A}(1_A \bullet \mathbf{w}_A)\mathbf{w}_A = \mathbf{c}_{A,A\bullet A}(1_A \bullet \mathbf{w}_A)\mathbf{w}_A.$$

From that with ($\sigma\delta$**w**) and ($\sigma\sigma^i$) we obtain, however, $\mathbf{b}^{\rightarrow}_{\mathrm{I},\mathrm{I},\mathrm{I}} = \mathbf{c}_{\mathrm{I},\mathrm{I}\bullet\mathrm{I}}$ (we have said already that $\mathbf{b}^{\rightarrow}_{\mathrm{I},\mathrm{I},\mathrm{I}}$ is equal to $\sigma^i_{\mathrm{I}} \bullet \sigma_{\mathrm{I}}$). The equation (**bw**) is obtained by reversing the arrows of the first diagram for monoids in [7, VII.3, p. 166, diagram (1)].

The arrow

$$\mathbf{c}^m_{A,B,C,D} : (A \bullet B) \bullet (C \bullet D) \vdash (A \bullet C) \bullet (B \bullet D)$$

for which, with ($\bullet$), (**b**) and (**c**), we can prove

$$((f \bullet h) \bullet (g \bullet j))\mathbf{c}^m_{A,B,C,D} = \mathbf{c}^m_{F,G,H,J}((f \bullet g) \bullet (h \bullet j))$$

is, like $\mathbf{c}_{A,B}$, a natural isomorphism (the upper index 'm' stands for 'middle'). In the equation (**bcw**8) it enables us, intuitively, to replace contraction of products like $A \bullet B$ in $\mathbf{w}_{A\bullet B}$ by contraction of the factors A and B in $\mathbf{w}_A$ and $\mathbf{w}_B$. In this sense, this equation is analogous to Mac Lane's hexagonal equations. We shall call (**bcw**8), and equations derived from it with (**bb**) and (**cc**), *octagonal equations*, because the commutative diagram corresponding to

$$\mathbf{b}^{\rightarrow}_{A,A,B\bullet B}(1_A \bullet \mathbf{b}^{\leftarrow}_{A,B,B})(1_A \bullet (\mathbf{c}_{B,A} \bullet 1_B))(1_A \bullet \mathbf{b}^{\rightarrow}_{B,A,B})\mathbf{b}^{\leftarrow}_{A,B,A\bullet B}\mathbf{w}_{A\bullet B} = (\mathbf{w}_A \bullet 1_{B\bullet B})(1_A \bullet \mathbf{w}_B)$$

has eight sides. There are two main octagonal equations: (**bcw**8) and

$$\mathbf{c}^m_{A,A,B,B}(\mathbf{w}_A \bullet \mathbf{w}_B) = \mathbf{w}_{A\bullet B}.$$

As (**bw**) is obtained by reversing the arrows of a diagram for monoids, so the equations (σ**kw**) and (δ**kw**) are obtained from the remaining diagrams for monoids in [7, VII.3, p. 166, diagram (2)] by the same procedure (which involves replacing σ by σ^i and δ by δ^i). These equations can be replaced by the equations:

$$\text{For} f : A \vdash B, \quad \sigma_B(\mathbf{k}_A \bullet f)\mathbf{w}_A = f, \quad \delta_B(f \bullet \mathbf{k}_A)\mathbf{w}_A = f.$$

If $\mathbf{p}_{A,B} : A \bullet B \vdash A$ and $\mathbf{p}'_{A,B} : A \bullet B \vdash A$ are defined by

$$\mathbf{p}_{A,B} =_{\mathrm{df}} \delta_A(1_A \bullet \mathbf{k}_B), \quad \mathbf{p}'_{A,B} =_{\mathrm{df}} \sigma_B(\mathbf{k}_A \bullet 1_B)$$

then $(\sigma\mathbf{kw})$ becomes $\mathbf{p}'_{A,A}\mathbf{w}_A = 1_A$ and $(\delta\mathbf{kw})$ becomes $\mathbf{p}_{A,A}\mathbf{w}_A = 1_A$. It follows from $(1\mathbf{k})$ that $\sigma_A = \mathbf{p}'_{\mathrm{I},A}$ and $\delta_A = \mathbf{p}_{A,\mathrm{I}}$ (cf. Section 4). In the presence of **c** arrows, $(\sigma\mathbf{kw})$ and $(\delta\mathbf{kw})$ are not independent: one can be derived from the other.

The equation $(\sigma\delta)$, i.e. $\sigma_{\mathrm{I}} = \delta_{\mathrm{I}}$, becomes superfluous in the presence of **(k)** and $(1\mathbf{k})$. Likewise, $(\sigma\delta\mathbf{w})$ follows from $(1\mathbf{k})$ and $(\sigma\mathbf{kw})$. However, these superfluous equations may be needed even if we dont have **k** arrows (see Section 6), and because of that we keep them as primitive. In the same spirit, we have assumptions for both σ and δ arrows, though these assumptions are not mutually independent in the presence of the **c** arrows, as we have noted above. From the beginning, we also have the superfluous equation $(\bullet 1)$.

3 MODAL FUNCTIONAL COMPLETENESS FOR NL□ CATEGORIES

Given an NL□ category $\mathcal{C}$ and an object A of $\mathcal{C}$, we build the *polynomial* NL□ category $\mathcal{C}[x]$ with an 'indeterminate' arrow $x : \mathrm{I} \vdash A$ by a procedure described in [6, I.5, p. 57]. Namely, we add a new arrow $x : \mathrm{I} \vdash A$ to the underlying graph of $\mathcal{C}$ and then build the NL□ category freely generated by the new graph. That this will succeed is guaranteed by the fact that NL□ categories are equationally presented, i.e. that all our assumptions about arrows are equations. We shall not rehearse here this procedure, which is explained in sufficient detail in [6]. It is also covered by theorems in universal algebra concerning equationally presented algebras with partial operations [5, Section 5, p. 124, corollary to Lemma 3], and [1, Section 7, p. 129, Corollary 1 to Proposition 18]. The name 'polynomial category' is explained by thinking about the arrows of $\mathcal{C}[x]$ as polynomials in x.

We can now state the central result of this paper:

MODAL FUNCTIONAL COMPLETENESS THEOREM.
For every arrow $\varphi : B \vdash C$ of the polynomial NL□ category $\mathcal{C}[x]$ built over the NL□ category $\mathcal{C}$ with $x : \mathrm{I} \vdash A$, there is a unique arrow $f : \Box A \bullet B \vdash C$ of $\mathcal{C}$ such that $f(x^{\Box} \bullet 1_B)\sigma^i_B = \varphi$ holds in $\mathcal{C}[x]$.

We also have the following:

COROLLARY
For every arrow $\varphi : \mathrm{I} \vdash C$ of the polynomial NL□ category $\mathcal{C}[x]$ built over the NL□ category $\mathcal{C}$ with $x : \mathrm{I} \vdash A$, there is a unique arrow $g : \Box A \vdash C$ of $\mathcal{C}$ such that $gx^{\Box} = \varphi$ holds in $\mathcal{C}[x]$.

If B is I in the Modal Functional Completeness Theorem, then in $\mathcal{C}[x]$ we have

$$f(x^{\Box} \bullet 1_{\mathrm{I}})\sigma^i_{\mathrm{I}} = \varphi$$
$$f\delta^i_{\Box A}x^{\Box} = \varphi, \text{ with } (\sigma\delta), (\delta) \text{ and } (\delta\delta^i).$$

So, in the Corollary, we shall take g to be $f\delta^i_{\Box A}$. Actually, as with cartesian closed categories in [6, I.6, p. 61], the Corollary entails the Theorem, since the arrows $\varphi : B \vdash C$ of $\mathcal{C}[x]$ are in one-one correspondence with the arrows $^*(\varphi\delta_B) : \mathrm{I} \vdash B \to C$ (or $(\varphi\sigma_B)^* : \mathrm{I} \vdash C \leftarrow B$). This one–one correspondence also entails that our restriction of x to arrows of type $\mathrm{I} \vdash A$ is surmountable (cf. [6, I.2, p. 52, Exercise 1]). However, the Theorem has a better form than the Corollary for the proof we are going to present. The remainder of this section will be devoted to this proof of the Modal Functional Completeness Theorem.

If φ ranges over arrows of the polynomial NL□ category $\mathcal{C}[x]$ built over the NL□ category $\mathcal{C}$ with $x : \mathrm{I} \vdash A$, whereas f ranges over the arrows of type $\Box A \bullet B \vdash C$ of $\mathcal{C}$, for some objects B and C, and

$$f\text{`}x =_{\mathrm{df}} f(x^{\Box} \bullet 1_B)\sigma^i_B$$

then the Modal Functional Completeness Theorem asserts that 'x is an *onto* and *one-one* function from the f arrows to the φ arrows (from the definition of 'x it is clear that it is a function). Our proof of the Modal Functional Completeness Theorem (inspired by the proof of functional completeness for cartesian closed categories in [6, I.6]) will proceed by defining a function μ_x from the φ arrows to the f arrows and showing that μ_x is the inverse of 'x. Applying μ_x to φ is related to applying the functional abstraction operator λx. Syntactically, μ_x binds the variables x that may occur in the polynomial φ. Semantically, it extracts from φ a function that may be applied to x in the sense of ': if μ_x is like functional abstraction, 'x is like application to x. The analogue of μ_x in [6, I.2, I.6] is $\kappa_{x\in A}$, whereas the analogue of 'x is $\langle x\bigcirc_B, 1_B\rangle$.

We define the function μ_x by the following inductive clauses, which cover all possible forms an arrow of $\mathcal{C}[x]$ may have:

(μ0.1) $\mu_x x = \mathbf{r}_A\delta_{\Box A}$

(μ0.2) For $h : D \vdash E$ an arrow of $\mathcal{C}$,

$$\begin{aligned}\mu_x h &= h\sigma_D(\mathbf{k}_{\Box A} \bullet 1_D)\\ &= h\mathbf{p}'_{\Box A,D}.\end{aligned}$$

(μ1) For $\psi : D \vdash E$ and $\xi : E \vdash F$,

$$\mu_x(\xi\psi) = \mu_x\xi(1_{\Box A} \bullet \mu_x\psi)\mathbf{b}^{\leftarrow}_{\Box A,\Box A,D}(\mathbf{w}_{\Box A} \bullet 1_D).$$

(μ2) For $\psi : D \vdash E$ and $\xi : F \vdash G$,

$$\mu_x(\psi \bullet \xi) = (\mu_x\psi \bullet \mu_x\xi)\mathbf{c}^m_{\Box A,\Box A,D,F}(\mathbf{w}_{\Box A} \bullet 1_{D\bullet F}).$$

(μ3.1) For $\psi : E \bullet D \vdash F$,

$$\mu_x({}^*\psi) = {}^*(\mu_x\psi\mathbf{b}^{\leftarrow}_{\Box A,E,D}(\mathbf{c}_{E,\Box A} \bullet 1_D)\mathbf{b}^{\rightarrow}_{E,\Box A,D}).$$

(μ3.2) For $\psi : D \bullet E \vdash F$,

$$\mu_x(\psi^*) = (\mu_x\psi\mathbf{b}^{\leftarrow}_{\Box A,D,D})^*.$$

(μ4) For $\psi : D \vdash E$ and $\xi : D \vdash F$,

$$\mu_x\langle\psi, \xi\rangle = \langle\mu_x\psi, \mu_x\xi\rangle.$$

(μ5) For $\psi : E \vdash D, \xi : F \vdash D$ and $\mathbf{d}_{G,E,F} : G \bullet (E \vee F) \vdash (G \bullet E) \vee (G \bullet F)$ defined by $\mathbf{d}_{G,E,F} =_{\text{df}} \varepsilon^{\rightarrow}_{G,(G\bullet E)\vee(G\bullet F)}(1_G \bullet [{}^*\kappa_{G\bullet E,G\bullet F}, {}^*\kappa'_{G\bullet E,G\bullet F}])$,

$$\mu_x[\psi, \xi] = [\mu_x\psi, \mu_x\xi]\mathbf{d}_{\Box A,E,F}.$$

(μ6) For $\psi : D \vdash E$ with D modalized,

$$\mu_x(\psi^{\Box}) = (\mu_x\psi)^{\Box}.$$

Let us first deduce that the following equations hold in $\mathcal{C}$:

(μ1.1) For $\psi : D \vdash E$ and $h : E \vdash F$ an arrow of $\mathcal{C}$,

$$\mu_x(h\psi) = h\mu_x\psi.$$

(μ1.2) For $h : D \vdash E$ an arrow of $\mathcal{C}$ and $\xi : E \vdash F$,

$$\mu_x(\xi h) = \mu_x\xi(1_{\Box A} \bullet h).$$

We could as well have taken $(\mu 1.1)$ and $(\mu 1.2)$ as clauses in the definition of μ_x, but then we would have to show that they are compatible with $(\mu 1)$ and $(\mu 0.2)$, and this compatibility is demonstrated by deducing them from $(\mu 1)$ and $(\mu 0.2)$. For $(\mu 1.1)$ we have

$$\begin{aligned}
\mu_x(h\psi) &= h\sigma_E(\mathbf{k}_{\Box A} \bullet 1_E)(1_{\Box A} \bullet \mu_x\psi)\mathbf{b}^{\leftarrow}_{\Box A,\Box A,D}(\mathbf{w}_{\Box A} \bullet 1_D), && \text{with } (\mu 1) \text{ and } (\mu 0.2)\\
&= h\sigma_E(1_{\mathrm{I}} \bullet \mu_x\psi)(\mathbf{k}_{\Box A} \bullet 1_{\Box A \bullet D})\mathbf{b}^{\leftarrow}_{\Box A,\Box A,D}(\mathbf{w}_{\Box A} \bullet 1_D), && \text{with } (\bullet)\\
&= h\mu_x\psi\sigma_{\Box A \bullet D}\mathbf{b}^{\leftarrow}_{\mathrm{I},\Box A,D}((\mathbf{k}_{\Box A} \bullet 1_{\Box A}) \bullet 1_D)(\mathbf{w}_{\Box A} \bullet 1_D), && \text{with } (\sigma) \text{ and } (\mathbf{b})\\
&= h\mu_x\psi((\sigma_{\Box A}(\mathbf{k}_{\Box A} \bullet 1_{\Box A})\mathbf{w}_{\Box A}) \bullet 1_D), && \text{with } (\sigma\mathbf{b}), (\mathbf{bb}) \text{ and } (\bullet).
\end{aligned}$$

Then we apply $(\sigma\mathbf{kw})$ and $(\bullet 1)$. For $(\mu 1.2)$ we have

$$\begin{aligned}
\mu_x(\xi h) &= \mu_x\xi(1_{\Box A} \bullet (h\sigma_D(\mathbf{k}_{\Box A} \bullet 1_D)))\mathbf{b}^{\leftarrow}_{\Box A,\Box A,D}(\mathbf{w}_{\Box A} \bullet 1_D), && \text{with } (\mu 1) \text{ and } (\mu 0.2)\\
&= \mu_x\xi(1_{\Box A} \bullet h)(1_{\Box A} \bullet \sigma_D)\mathbf{b}^{\leftarrow}_{\Box A,\mathrm{I},D}((1_{\Box A} \bullet \mathbf{k}_{\Box A}) \bullet 1_D)(\mathbf{w}_{\Box A} \bullet 1_D), && \text{with } (\bullet) \text{ and } (\mathbf{b})\\
&= \mu_x\xi(1_{\Box A} \bullet h)((\delta_{\Box A}(1_{\Box A} \bullet \mathbf{k}_{\Box A})\mathbf{w}_{\Box A}) \bullet 1_D), && \text{with } (\sigma\delta\mathbf{b}), (\mathbf{bb}) \text{ and } (\bullet).
\end{aligned}$$

It remains to apply $(\delta\mathbf{kw})$ and $(\bullet 1)$.

In a rather similar manner, with the help of $(\mu 0.2)$, $(\sigma\mathbf{kw})$ and $(\delta\mathbf{kw})$, we derive from $(\mu 2)$:

$(\mu 2.1)$ For $\psi : D \vdash F$,

$$\mu_x(E \bullet \psi) = (E \bullet \mu_x\psi)\mathbf{b}^{\leftarrow}_{E,\Box A,D}(\mathbf{c}_{\Box A,E} \bullet D)\mathbf{b}^{\rightarrow}_{\Box A,E,D}.$$

$(\mu 2.2)$ For $\psi : D \vdash F$,

$$\mu_x(\psi \bullet E) = (\mu_x\psi \bullet E)\mathbf{b}^{\rightarrow}_{\Box A,D,E}.$$

In these clauses we write $D \bullet f$ for $1_D \bullet f$ and $f \bullet D$ for $f \bullet 1_D$. If the unary operations on arrows $D \bullet \_$ and $\_ \bullet D$ are primitive instead of the binary operation on arrows $\_ \bullet \_$, then we can derive $(\mu 2)$ from $(\mu 2.1)$ and $(\mu 2.2)$. These substitute clauses, which are often simpler to work with than $(\mu 2)$, will also serve for the results of Sections 5 and 6.

Then we have to check that $\mu_x h$ is well-defined for arrows h in $\mathcal{C}$. For example, we must check for $h = ts$, $s : D \vdash E$ and $t : E \vdash F$, that an equation corresponding to clause $(\mu 1)$, namely,

$$h\mathbf{p}'_{\Box A,D} = t\mathbf{p}'_{\Box A,E}(1_{\Box A} \bullet (s\mathbf{p}'_{\Box A,D}))\mathbf{b}^{\leftarrow}_{\Box A,\Box A,D}(\mathbf{w}_{\Box A} \bullet 1_D)$$

holds in $\mathcal{C}$. This amounts to the deduction of $(\mu 1.1)$. We have to check similar equations corresponding to clauses $(\mu 2)$–$(\mu 6)$. We shall not go into the details of this lengthy, but rather straightforward, exercise. However, let us note as a hint that it may be easier to check such equations with $D \bullet \_$ and $\_ \bullet D$, rather than with $\_ \bullet \_$. In that case, instead of $(\mu 2)$ we use $(\mu 2.1)$ and $(\mu 2.2)$. Let us note as another hint that when we check the equation corresponding to $(\mu 5)$, we use the fact that the distribution arrow $\mathbf{d}_{G,E,F}$ is an isomorphism (actually, a natural isomorphism), its inverse being $[1_G \bullet \kappa_{E,F}, 1_G \bullet \kappa'_{E,F}]$.

Next we check that μ_x is indeed a function:

LEMMA 1. *If $\varphi = \psi$ holds in $\mathcal{C}[x]$, then $\mu_x \varphi = \mu_x \psi$ holds in $\mathcal{C}$.*

Proof. From the inductive definition of μ_x it follows immediately that if $\mu_x \varphi = \mu_x \psi$ holds in $\mathcal{C}$, then $\mu_x(\xi\varphi) = \mu_x(\xi\psi)$ and $\mu_x(\varphi\xi) = \mu_x(\psi\xi)$ hold in $\mathcal{C}$. We have analogous implications for the other operations on arrows of NL$\Box$ categories.

If φ and ψ are arrows of $\mathcal{C}$ and $\varphi = \psi$ holds in $\mathcal{C}[x]$, then $\varphi = \psi$ holds in $\mathcal{C}$. So, we have in $\mathcal{C}$

$$\begin{array}{l} \varphi\mu_x 1_B = \psi\mu_x 1_B \\ \mu_x\varphi = \mu_x\psi, \text{ by } (\mu 1.1) \text{ and (cat 1)} \end{array}$$

(we could as well have used $(\mu 1.2)$).

It remains to check that for all the equations $\varphi = \psi$ we have assumed for NL$\Box$ categories, in which arrows of $\mathcal{C}[x]$ not in $\mathcal{C}$ may occur, $\mu_x\varphi = \mu_x\psi$ holds in $\mathcal{C}$.

The equations of (cat 1) are covered by $(\mu 1.1)$ and $(\mu 1.2)$. For (cat 2) we have

$$\begin{array}{l} \mu_x(\xi(\psi\varphi)) \\ = \mu_x\xi(1_{\Box A} \bullet (\mu_x\psi(1_{\Box A} \bullet \mu_x\varphi)\overleftarrow{\mathbf{b}}_{\Box A,\Box A,B}(\mathbf{w}_{\Box A} \bullet 1_B)))\overleftarrow{\mathbf{b}}_{\Box A,\Box A,B}(\mathbf{w}_{\Box A} \bullet 1_B) \\ = \mu_x\xi(1_{\Box A} \bullet \mu_x\psi)(1_{\Box A} \bullet (1_{\Box A} \bullet \mu_x\varphi))(1_{\Box A} \bullet \overleftarrow{\mathbf{b}}_{\Box A,\Box A,B})\overleftarrow{\mathbf{b}}_{\Box A,\Box A \bullet \Box A,B} \\ \qquad ((1_{\Box A} \bullet \mathbf{w}_{\Box A}) \bullet 1_B)(\mathbf{w}_{\Box A} \bullet 1_B), \text{ with } (\bullet) \text{ and } (\mathbf{b}); \end{array}$$

then we have the pentagonal equation

$$(1_{\Box A} \bullet \overleftarrow{\mathbf{b}}_{\Box A,\Box A,B})\overleftarrow{\mathbf{b}}_{\Box A,\Box A \bullet \Box A,B} = \overleftarrow{\mathbf{b}}_{\Box A,\Box A,\Box A \bullet B}\overleftarrow{\mathbf{b}}_{\Box A \bullet \Box A,\Box A,B}(\overrightarrow{\mathbf{b}}_{\Box A,\Box A,\Box A} \bullet 1_B)$$

which with the help of $(\mathbf{bw})$ and $(\mathbf{b})$ yields

$$\begin{array}{ll} \mu_x(\xi(\psi\varphi)) & = \mu_x\xi(1_{\Box A} \bullet \mu_x\psi)\overleftarrow{\mathbf{b}}_{\Box A,\Box A,C}(1_{\Box A \bullet \Box A} \bullet \mu_x\varphi)(\mathbf{w}_{\Box A} \bullet 1_{\Box A \bullet B}) \\ & \qquad \overleftarrow{\mathbf{b}}_{\Box A,\Box A,B}(\mathbf{w}_{\Box A} \bullet 1_B) \\ & = \mu_x((\xi\psi)\varphi), \text{ with } (\bullet). \end{array}$$

For $(\bullet)$ it is again easier (and more instructive for what we do in Sections 5 and 6) to work with $D \bullet \_$ and $\_ \bullet D$ instead of $\_ \bullet \_$, i.e. with $(\mu 2.1)$ and $(\mu 2.2)$ instead of $(\mu 2)$, and check $(\bullet 2)$ and ($\bullet$ bifunctor). For the first equation of $(\bullet 2)$ we need pentagonal and hexagonal equations, whereas for the second we need only a pentagonal equation. For ($\bullet$ bifunctor) we need (**cw**) besides pentagonal and hexagonal equations in order to check that

$$\mathbf{c}^m_{\Box A,A_1,\Box A,A_2}\mathbf{b}^{\leftarrow}_{\Box A\bullet A_1,\Box A,A_2}(\mathbf{c}_{\Box A,\Box A\bullet A_1} \bullet A_2)\mathbf{b}^{\rightarrow}_{\Box A,\Box A\bullet A_1 A_2}(\Box A \bullet \mathbf{b}^{\rightarrow}_{\Box A,A_1,A_2})$$
$$\mathbf{b}^{\leftarrow}_{\Box A,\Box A,A_1\bullet A_2}(\mathbf{w}_{\Box A} \bullet (A_1 \bullet A_2)) = \mathbf{w}_{\Box A} \bullet (A_1 \bullet A_2).$$

For (σ) and (δ) we apply $(\mu 1.1)$ and $(\mu 1.2)$. Among the remaining equations for NL categories that need checking, namely, the closure and bicartesian equations, let us consider $(\bot)$.

For $\varphi : \bot \vdash B$, we have to show $\mu_x \iota_B = \mu_x \varphi$. With $(\bot)$ we have

$$\begin{aligned}
{}^*(\mu_x \iota_B) &= \iota_{\Box A \to B} \\
{}^*(\mu_x \iota_B) &= {}^*(\mu_x \varphi) \\
\varepsilon^{\rightarrow}_{\Box A,\Box A\to B}(\mathbf{1}_{\Box A} \bullet {}^*(\mu_x \iota_B)) &= \varepsilon^{\rightarrow}_{\Box A,\Box A \to B}(\mathbf{1}_{\Box A} \bullet {}^*(\mu_x \varphi)) \\
\mu_x \iota_B &= \mu_x \varphi, \text{ with } (\to \beta).
\end{aligned}$$

A similar trick involving * is applied when we show that the distribution arrow **d** is an isomorphism, and this we need when we check the lemma for $(\vee\eta)$.

It remains to check $(\Box\beta)$, $(\Box\eta)$, (**b**), (**c**), (**k**) and (**w**). The cases with $(\Box\beta)$ and $(\Box\eta)$ follow readily by applying $(\mu 1.1)$ and $(\mu 6)$. For the case with (**b**) it is easier and more instructive to break the checking into checking $(\mathbf{b}_1)$, $(\mathbf{b}_2)$ and $(\mathbf{b}_3)$ with the clauses $(\mu 2.1)$ and $(\mu 2.2)$. Then for $(\mathbf{b}_1)$ we need the pentagonal equation (**b5**) (see the justification of pentagonal equations in Section 5). For $(\mathbf{b}_2)$ we use two pentagonal equations, and for $(\mathbf{b}_3)$ one hexagonal equation besides three pentagonal equations. To check $(\mathbf{c}_1)$ or $(\mathbf{c}_2)$ we need a hexagonal equation (see the justification of hexagonal equations in Section 5). For (**k**) we just use the fact that the object I is terminal for arrows from modalized objects. Finally, for (**w**) we have

$$\begin{aligned}
\mu_x((\varphi \bullet \varphi)\mathbf{w}_B) &= (\mu_x\varphi \bullet \mu_x\varphi)\mathbf{c}^m_{\Box A \Box A,B,B}(\mathbf{w}_{\Box A} \bullet \mathbf{w}_B), \\
&\qquad \text{with } (\mu 2) \text{ and } (\mu 1.2) \\
&= (\mu_x\varphi \bullet \mu_x\varphi)\mathbf{w}_{\Box A \bullet B}, \text{ with an octagonal equation} \\
&= \mu_x(\mathbf{w}_C \varphi), \text{ with } (\mathbf{w}) \text{ and } (\mu 1.1).
\end{aligned}$$

∎

The following two lemmata assert that μ_x is the inverse of ${}^{\prime}x$. As μ_x corresponds to functional abstraction and ${}^{\prime}x$ to application, so Lemma 2 corresponds to

β-conversion and Lemma 3 to η-conversion.

LEMMA 2. *For* $\varphi : B \vdash C$ *an arrow of* $\mathcal{C}[x]$, *the equation* $(\mu_x\varphi)`x = \varphi$ *holds in* $\mathcal{C}[x]$.

Proof. We proceed by induction on the complexity of φ. For the basis we have:

(0.1) If φ is x, then

$$\begin{aligned}(\mu_x x)`x &= \mathbf{r}_A\delta_{\Box A}(x^{\Box} \bullet \mathbf{1}_\mathrm{I})\sigma^i_\mathrm{I}, \text{ by definition}\\ &= \mathbf{r}_A x^{\Box}\delta_\mathrm{I}\sigma^i_\mathrm{I}, \text{ with } (\delta)\\ &= x, \text{ with } (\Box\beta), (\sigma\delta) \text{ and } (\sigma\sigma^i)\end{aligned}$$

(0.2) If φ is an arrow h of $\mathcal{C}$, then, by definition,

$$(\mu_x h)`x = h\sigma_B(\mathbf{k}_{\Box A} \bullet \mathbf{1}_B)(x^{\Box} \bullet \mathbf{1}_B)\sigma^i_B$$

and we use $(\bullet)$, $\mathbf{k}_{\Box A}x^{\Box} = \mathbf{1}_\mathrm{I}$ and $(\sigma\sigma^i)$ to show that the right-hand side is equal to h.

In the induction step we shall only consider the following case as an example:

(1) If φ is $\xi\psi$, with $\psi : B \vdash E$ and $\xi : E \vdash C$, then by the induction hypothesis we have $(\mu_x\psi)`x = \psi$ and $(\mu_x\xi)`x = \xi$. We need to prove $(\mu_x(\xi\psi))`x = \xi\psi$, which amounts to $\mu_x\xi(\mathbf{1}_{\Box A}\bullet\mu_x\psi)\mathbf{b}^{\leftarrow}_{\Box A,\Box A,B}(\mathbf{w}_{\Box A}\bullet\mathbf{1}_B)(x^{\Box}\bullet\mathbf{1}_B)\sigma^i_B = \mu_x\xi(x^{\Box} \bullet \mathbf{1}_E)\sigma^i_E\mu_x\psi(x^{\Box} \bullet \mathbf{1}_B)\sigma^i_B$. For the left-hand side (lhs) of this equation we have

$$\begin{aligned}\text{lhs} &= \mu_x\xi(\mathbf{1}_{\Box A} \bullet \mu_x\psi)\mathbf{b}^{\leftarrow}_{\Box A,\Box A,B}((x^{\Box} \bullet x^{\Box}) \bullet \mathbf{1}_B)(\mathbf{w}_\mathrm{I} \bullet \mathbf{1}_B)\sigma^i_B,\\ &\qquad\qquad \text{with } (\bullet) \text{ and } (\mathbf{w})\\ &= \mu_x\xi(\mathbf{1}_{\Box A} \bullet \mu_x\psi)(x^{\Box} \bullet (x^{\Box} \bullet \mathbf{1}_B))\mathbf{b}^{\leftarrow}_{\mathrm{I,I},B}(\mathbf{w}_\mathrm{I} \bullet \mathbf{1}_B)\sigma^i_B, \text{ with } (\mathbf{b})\\ &= \mu_x\xi(x^{\Box} \bullet \mathbf{1}_E)(\mathbf{1}_\mathrm{I} \bullet (\mu_x\psi(x^{\Box} \bullet \mathbf{1}_B)))\mathbf{b}^{\leftarrow}_{\mathrm{I,I},B}(\sigma^i_\mathrm{I} \bullet \mathbf{1}_B)\sigma^i_B,\\ &\qquad\qquad \text{with } (\bullet), (\sigma\delta\mathbf{w}) \text{ and } (\sigma\sigma^i)\end{aligned}$$

which, with (σ) and a triangular equation derived from $(\sigma\mathbf{b})$ with $(\sigma\sigma^i)$ and $(\mathbf{bb})$, is equal to the right- hand side.

As a lengthy exercise, it remains to check cases corresponding to clauses $(\mu2)$–$(\mu6)$ (for ease, it may be preferable to work with $(\mu2.1)$ and $(\mu2.2)$ instead of $(\mu2)$; see the justification of $(\sigma\mathbf{b})$ and $(\sigma\delta\mathbf{b})$ in Section 5). ∎

LEMMA 3. *For* $f : \Box A \bullet B \vdash C$ *an arrow of* $\mathcal{C}$, *the equation* $\mu_x(f`x) = f$ *holds in* $\mathcal{C}$.

Proof. By using $(\mu1.1), (\mu1.2), (\mu2.2), (\mu6)$ and $(\mu0.1)$ we obtain

$$\begin{aligned}\mu_x(f`x) &= f((\mathbf{r}_A\delta_{\Box A})^{\Box} \bullet 1_B)\mathbf{b}^{\rightarrow}_{\Box A,\mathrm{I},B}(1_{\Box A} \bullet \sigma^i_B)\\ &= f(\delta_{\Box A} \bullet 1_B)\mathbf{b}^{\rightarrow}_{\Box A,\mathrm{I},B}(1_{\Box A} \bullet \sigma^i_B), \text{ with } (\Box\eta);\end{aligned}$$

with the triangular equation $(\sigma\delta\mathbf{b}), (\bullet)$ and $(\sigma\sigma^i)$, the right-hand side is equal to f. ■

As a corollary of Lemma 2 we can easily obtain a more general statement; namely, for every arrow $a : \mathrm{I} \vdash A$ of $\mathcal{C}[x]$ the equation $(\mu_x\varphi)`a = \varphi^x_a$ holds in $\mathcal{C}[x]$. In this equation, which corresponds exactly to $\beta$-conversion, $f`a$ is defined analogously to $f`x$, and φ^x_a is obtained from φ by uniformly substituting a for x. We shall not define this substitution with more precision since we don't need the corollary here. Lemma 2 suffices; i.e., Lemmata 1–3 immediately give the Modal Functional Completeness Theorem.

4 FUNCTIONAL COMPLETENESS FOR CARTESIAN, CARTESIAN CLOSED AND BICARTESIAN CLOSED CATEGORIES

An NL□ category in which every object A is isomorphic with $\Box A$ is a cartesian category with respect to $\bullet$ and I, a cartesian closed category with respect to $\bullet, \rightarrow$ and I, and a bicartesian closed category with respect to $\bullet, \rightarrow, \mathrm{I}, \vee$ and $\bot$. From our axiomatization of NL□ categories in Sections 1 and 2 one can easily extract non-standard axiomatizations of these sorts of category by selecting the assumptions tied with the mentioned operations and objects (of course, we always assume that we have the 1_A arrows, composition, (cat 1) and (cat 2); i.e. that we are in a category).

As a matter of fact, to axiomatize bicartesian closed categories in such a non-standard way, all we have to do is forget about □ in Section 2, without making any selection among our assumptions. Namely, let us assume whatever is assumed in Section 1 for NL categories, and let us, moreover, assume the structural arrows **b**, **c**, **k** and **w** *without* provisos concerning modalized objects. We forget about the **r** arrows and the operation on arrows $^{\Box}$: the arrow $\mathbf{r}_A$ may be identified with 1_A and $f^{\Box}$ is just f. For the structural arrows without provisos we assume the equations of Section 2.

Remember that we have defined the **p** and **p**$'$ arrows at the end of Section 2 by

$$\mathbf{p}_{A,B} =_{\mathrm{df}} \delta_A(1_A \bullet \mathbf{k}_B), \qquad \mathbf{p}'_{A,B} =_{\mathrm{df}} \sigma_B(\mathbf{k}_A \bullet 1_B).$$

In the same spirit, we define the binary operation on arrows

$$\frac{f : C \vdash A \quad g : C \vdash B}{\{f, g\} : C \vdash A \bullet B}$$

by

$$\{f, g\} =_{\text{df}} (f \bullet g)\mathbf{w}_C.$$

Then we can show first that $\langle \mathbf{p}_{A,B}, \mathbf{p}'_{a,B} \rangle$ is a natural isomorphism from $A \bullet B$ to $A \wedge B$, its inverse being $\{\pi_{A,B}, \pi'_{A,B}\}$. The essential step in checking

$$\{\pi_{A,B}, \pi'_{A,B}\}\langle \mathbf{p}_{A,B}, \mathbf{p}'_{A,B} \rangle = \mathbf{1}_{A \bullet B}$$

is to show $\{\mathbf{p}_{A,B}, \mathbf{p}'_{A,B}\} = \mathbf{1}_{A \bullet B}$, i.e.

$$(\sigma\delta\mathbf{kw}) \quad ((\delta_A(\mathbf{1}_A \bullet \mathbf{k}_B)) \bullet (\sigma_B(\mathbf{k}_A \bullet \mathbf{1}_B)))\mathbf{w}_{A \bullet B} = \mathbf{1}_{A \bullet B}.$$

In this equation is hidden the octagonal principle: with an octagonal equation, the left-hand side is equal to

$$(\delta_A \bullet \sigma_B)\mathbf{c}^m_{A,\mathrm{I},\mathrm{I},B}((\mathbf{1}_A \bullet \mathbf{k}_A) \bullet (\mathbf{k}_B \bullet \mathbf{1}_B))(\mathbf{w}_A \bullet \mathbf{w}_B)$$

which with $\mathbf{c}^m_{A,\mathrm{I},\mathrm{I},B} = \mathbf{1}_{(A \bullet \mathrm{I}) \bullet (\mathrm{I} \bullet B)}$ (an equation related to $\mathbf{c}_{\mathrm{I},\mathrm{I}} = \mathbf{1}_{\mathrm{I} \bullet \mathrm{I}}$, mentioned in Section 2), $(\delta\mathbf{kw})$, $(\sigma\mathbf{kw})$, $(\bullet)$ and $(\bullet 1)$ is equal to $\mathbf{1}_{A \bullet B}$.

We can also prove equations that correspond exactly to $(\wedge\beta)$ and $(\wedge\eta)$:

$$\begin{array}{lll} (\bullet\beta) & \mathbf{p}_{A,B}\{f, g\} = f, & \mathbf{p}'_{A,B}\{f, g\} = g \\ (\bullet\eta) & \text{For } h : C \vdash A \bullet B, & \{\mathbf{p}_{A,B}h, \mathbf{p}'_{A,B}h\} = h. \end{array}$$

For $(\bullet\eta)$ we use $(\sigma\delta\mathbf{kw})$. Note that, with $(\bullet)$, the equation $(\sigma\delta\mathbf{kw})$ yields also

$$\{f\mathbf{p}_{A,B}, g\mathbf{p}'_{A,B}\} = f \bullet g.$$

To show all that we don't need $\rightarrow, \leftarrow, \top, \vee, \bot$ and the assumptions tied with them. The equation last displayed permits us to give $(\leftarrow \beta)$ and $(\leftarrow \eta)$ the form they have in [6, I.3, p. 53, E4a, E4b]. Of course, we can analogously rewrite $(\rightarrow \beta)$ and $(\rightarrow \eta)$.

Next we show that τ_{I} is a natural isomorphism from I to $\top$, its inverse being $\mathbf{k}_{\top}$. For that we need only $(\top)$ and the **k** equations.

Finally, we show that we can keep $\rightarrow$ and reject $\leftarrow$; actually, it does not matter which one of the two implications we keep while rejecting the other. This follows from the fact that $(\varepsilon^{\rightarrow}_{a,B}\mathbf{c}_{A\rightarrow B,A})^*$ is a natural isomorphism from $A \rightarrow B$ to $B \leftarrow A$, its inverse being ${}^*(\varepsilon^{\leftarrow}_{A,B}\mathbf{c}_{A,B\leftarrow A})$. To demonstrate that, we need only $(\bullet)$, closure equations, (**c**) and (**cc**).

So we have indeed an axiomatization of cartesian closed categories, which, given that we also have the assumptions for $\vee$ and $\bot$, amounts to an axiomatization of bicartesian closed categories. We can forget in this nonstandard axiomatization about $\leftarrow$, $\wedge$, $\top$ and the assumptions tied with them.

Our proof of the Modal Functional Completeness Theorem then yields an alternative proof of ordinary, nonmodal, functional completeness for bicartesian closed categories. We only have to forget about $\Box$. That means that in the statement of the Functional Completeness Theorem we shall have $f : A{\bullet}B \vdash C$ and $f(x{\bullet}1_B)\sigma^i_B = \varphi$, i.e. $f`x$ is $f(x \bullet 1_B)\sigma^i_B$ (cf. the Substructural Functional Completeness Theorem in Section 6). In the definition of μ_x, in clause $(\mu 0.1)$ we shall have

$$\begin{aligned} \mu_x x &= 1_A\delta_A \\ &= \delta_A \end{aligned}$$

whereas in the other clauses we just delete $\Box$ wherever it occurs (clause $(\mu 6)$ is omitted). Then it may be checked that our new μ_x may be identified with $\kappa_{x\in A}$ of [6, I.2, p. 51]. For example, since

$$\kappa_{x\in A}(\xi\psi) = \kappa_{x\in A}\xi\langle\pi_{A,D}, \kappa_{x\in A}\psi\rangle$$

the connection with clause $(\mu 1)$ in the new definition of μ_x is achieved by verifying

$$(1_A \bullet \mu_x\psi)\mathbf{b}^{\leftarrow}_{A,A,D}(\mathbf{w}_A \bullet 1_D) = \{\mathbf{p}_{A,D}, \mu_x\psi\}$$

which is done with the help of an octagonal equation.

It should be clear that we also have an alternative proof of ordinary functional completeness for cartesian and cartesian closed categories. As we have already remarked, from our nonstandard axiomatization of bicartesian closed categories we obtain a nonstandard axiomatization of cartesian closed categories by rejecting the assumptions tied with the bicartesian equations. For cartesian categories we reject moreover the assumptions tied with the closure equations. In our nonstandard axiomatization of bicartesian closed categories, the essentially nonstandard part is the cartesian part, which is given by the operation $\bullet$ on objects, the object I, the arrows 1_A, the $\sigma\delta$ arrows, the structural arrows **b**, **c**, **k** and **w** without provisos for modalized objects, the operations composition and $\bullet$ on arrows, and the equations between arrows: (cat 1), (cat 2), the $\bullet$ equations, the $\sigma\delta$ equations and the **b**, **c**, **k** and **w** equations.

In the same style, we can make as nonstandard the axiomatization of the part involving the coproduct $\vee$ and the initial object $\bot$ in bicartesian categories. Namely, we would have special arrows analogous to the $\sigma\delta$, **b** and **c** arrows in which $\bullet$ is replaced everywhere by $\vee$ and I by $\bot$. Instead of **k** and **w** arrows we would have $\iota_A : \bot \vdash A$ and $\mathbf{w}^{\vee}_A : A \vee A \vdash A$. We would assume for these arrows equations analogous to the $\bullet$ equations, the $\sigma\delta$ equations and the **b**, **c**, **k** and **w** equations. In the equations involving $\mathbf{w}^{\vee}$ and ι the order would have to be reversed and the analogues of σ and δ replaced by the analogues of σ^i and δ^i (the equations corresponding to the **k** equations would state that $\bot$ is an initial object; we have written down these equations at the very end of Section 1).

In our nonstandard axiomatization of cartesian categories there are redundancies among the special arrows (there are also redundancies among the equations, as we have noted at the end of Section 2; however, ($\bullet$1) is not redundant now: we use it to derive ($\sigma\delta\mathbf{kw}$)). Actually, since the definition of **p** is given in terms of δ and **k**, the definition of $\mathbf{p}'$ in terms of σ and **k**, and the definition of the curly brackets in terms of **w**, we can keep only the special arrows $1_A, \sigma, \delta$, **k** and **w**, and define all the others with $\mathbf{p}, \mathbf{p}'$ and the curly brackets, as this is done with the standard axiomatization of cartesian categories. Namely, we would have

$$
\begin{aligned}
&\sigma^i_A =_{\text{df}} \{\mathbf{k}_A, 1_A\}, \quad \delta^i_A =_{\text{df}} \{1_A, \mathbf{k}_A\} \\
&\mathbf{b}^{\rightarrow}_{A,B,C} =_{\text{df}} \{\{\mathbf{p}_{A,B\bullet C}, \mathbf{p}_{B,C}\mathbf{p}'_{A,B\bullet C}\}, \mathbf{p}'_{B,C}\mathbf{p}'_{A,B\bullet C}\} \\
&\mathbf{b}^{\leftarrow}_{A,B,C} =_{\text{df}} \{\mathbf{p}_{A,B}\mathbf{p}_{A\bullet B,C}, \{\mathbf{p}'_{A,B}\mathbf{p}_{A\bullet B,C}, \mathbf{p}'_{A\bullet B,C}\}\} \\
&\mathbf{c}_{A,B} =_{\text{df}} \{\mathbf{p}'_{A,B}, \mathbf{p}_{A,B}\}.
\end{aligned}
$$

The equations behind these definitions all hold in our nonstandard axiomatization of cartesian categories (for the first two we use ($\sigma\mathbf{kw}$) and ($\delta\mathbf{kw}$), whereas for the last three we use octagonal equations). Note that the last definition says that in Gentzen's sequent systems we can derive permutation from thinning and contraction, provided we are allowed to contract sequences of formulae, rather than single formulae only. Here is a permutation obtained by two thinnings followed by a contraction:

$$\dfrac{\dfrac{\dfrac{B, A \vdash C}{A, B, A \vdash C}}{A, B, A, B \vdash C}}{A, B \vdash C}$$

To obtain cartesian categories with this reduced stock of special arrows we have to assume (cat 1), (cat 2), ($\bullet$), (σ), (δ), (**k**), (1**k**), (**w**), (σ**kw**), (δ**kw**) and ($\sigma\delta$**kw**). The octagonal equations enter into this axiomatization via ($\sigma\delta$**kw**). By the way, the $\mathbf{c}^m$ arrows have the following symmetrical definition in terms of $\mathbf{p}, \mathbf{p}'$ and the curly brackets:

$$\mathbf{c}^m_{A,B,C,D} =_{\mathrm{df}}$$
$$\{\{\mathbf{p}_{A,B}\mathbf{p}_{A\bullet B,C\bullet D}, \mathbf{p}_{C,D}\mathbf{p}'_{A\bullet B,C\bullet D}\}, \{\mathbf{p}'_{A,B}\mathbf{p}_{A\bullet B,C\bullet D}, \mathbf{p}'_{C,D}\mathbf{p}'_{A\bullet B,C\bullet D}\}\}.$$

We cannot economize similarly on $\sigma^i, \delta^i, \mathbf{b}$ and $\mathbf{c}$ arrows with the restricted, modalized, versions of the arrows **k** and **w** in Section 2. With the definitions of $\sigma^i, \delta^i, \mathbf{b}$ and $\mathbf{c}$ given above, σ^i_A and δ^i_A would be lacking if A is not modalized, whereas $\mathbf{b}^{\rightarrow}_{A,B,C}$, $\mathbf{b}^{\leftarrow}_{A,B,C}$ and $\mathbf{c}_{A,B}$ would be lacking if A, B and C are not *all* modalized. For example, $\mathbf{w}_{A\bullet B}$ hidden in the curly brackets of $\{\mathbf{p}'_{A,B}, \mathbf{p}_{A,B}\}$ is not available if A and B are not *both* modalized, whereas in Sections 2 and 3 we need $\mathbf{c}_{A,B}$ even in cases where only A or only B is modalized, and similarly with **b** arrows (the arrows σ^i_B with B not necessarily modalized are involved in the formulation of the Modal Functional Completeness Theorem).

Another axiomatization of cartesian categories may be obtained by taking the special arrows **p** or **p**$'$ as primitive instead of **k**. With **p** primitive, we define $\mathbf{k}_A$ as $\mathbf{p}_{\mathrm{I},A}\sigma^i_A$ and $\mathbf{p}'_{A,B}$ as $\mathbf{p}_{B,A}\mathbf{c}_{A,B}$. A further possibility is to define δ_A as $\mathbf{p}_{A,\mathrm{I}}$ and σ_A as $\mathbf{p}_{A,\mathrm{I}}\mathbf{c}_{\mathrm{I},A}$. We shall not investigate here what reshuffling of our equations this change of primitives would require. Let us only note that in Section 2 we may have taken $\mathbf{p}_{A,B}$ as primitive instead of $\mathbf{k}_A$ *provided* B is modalized. Still another possibility is to take the curly brackets operation on arrows as primitive instead of the **w** arrows and define $\mathbf{w}_A$ as $\{1_A, 1_A\}$. In Section 2, we would have to require that with $f : C \vdash A$ and $g : C \vdash B$ we have $\{f, g\}$ only if C is modalized. However, as we have explained in the previous paragraph, having **p** and these restricted curly brackets primitive would not permit us to economize on **b** and **c** arrows in Section 2.

5 NECESSITY OF ASSUMPTIONS FOR NL□ CATEGORIES

We have seen that what we have assumed for NL□ categories is sufficient to demonstrate modal functional completeness. We want now to address the question whether these assumptions are also necessary. This is not a question we can answer in an absolute sense, but only relatively to some presuppositions. These presuppositions are contained in the particular notion of category without functional completeness that we extend to obtain a notion of category with functional completeness, as our notion of NL category was extended to the notion of NL□ category. But, foremost, the exact form of the functional completeness theorem carries presuppositions about what are polynomial arrows φ and about the type of the arrow f (the arrow f is of type $\Box A \bullet B \vdash C$ rather than $B \bullet \Box A \vdash C$, or some other type). Moreover, we require that a particular function from the f arrows to the φ arrows be *onto* and *one–one*. We shall see in the next section that we may understand functional completeness in different ways by restricting the notion of polynomial (still another way to restrict functional completeness is briefly mentioned in the concluding section).

For the time being we assume we have NL categories, axiomatized as in Section 1, and we shall try to see to what extent the additional assumptions for NL□ categories, in Section 2, are necessary for proving modal functional completeness as this is done in Section 3. After that we shall try to see whether at least some assumptions about NL categories in Section 1 are necessary in the same sense.

That in NL□ categories we must have the special arrows **r**, the modalized structural arrows **b**, **c**, **k** and **w**, and the operation on arrows $^{\Box}$, follows, as explained in [3], from the *Modal Deduction Theorem* and its converse, which are both consequences of the Modal Functional Completeness Theorem. The Modal Deduction Theorem says:

($\Box\downarrow$) For every arrow $\varphi : B \vdash C$ of the polynomial NL□ category $\mathcal{C}[x]$ built over the NL□ category $\mathcal{C}$ with $x : \mathrm{I} \vdash A$, there is an arrow $f : \Box A \bullet B \vdash C$ of $\mathcal{C}$.

It is clear that to prove ($\Box\downarrow$) it is enough to take for f the arrow $\mu_x\varphi$. The converse of the Modal Deduction Theorem says:

($\Box\uparrow$) For every arrow $f : \Box A \bullet B \vdash C$ of the NL□ category $\mathcal{C}$, there is an arrow $\varphi : B \vdash C$ of the polynomial NL□ category $\mathcal{C}[x]$ built over $\mathcal{C}$ with $x : \mathrm{I} \vdash A$.

It is clear that to prove ($\Box\uparrow$) it is enough to take for φ the arrow $f`x$. It is also clear that though we may infer ($\Box\downarrow$) and ($\Box\uparrow$) from the Modal Functional Completeness Theorem, this theorem does not follow from ($\Box\downarrow$) and ($\Box\uparrow$) alone. In ($\Box\downarrow$) and ($\Box\uparrow$) nothing is said about μ_x and ${}^{\backprime}x$, and their functional character.

Although this is something proven in [2, 3], we shall show again in more detail, and in a way adapted to the present context, how ($\Box\downarrow$) and ($\Box\uparrow$) deliver the special modal arrows and the operation on arrows $^\Box$ of NL$\Box$ categories. For that it will be useful to have the abbreviations:

$$\text{For } g : C \vdash A \to B, \quad {}^{\circ}g =_{\text{df}} \varepsilon^{\rightarrow}_{A,B}(1_A \bullet g).$$
$$\text{For } g : C \vdash A \leftarrow B, \quad g^{\circ} =_{\text{df}} \varepsilon^{\leftarrow}_{A,B}(1_A \bullet g).$$

r) Since in $\mathcal{C}[x]$ we have $x : \mathrm{I} \vdash A$, by ($\Box\downarrow$) we must have $f : \Box A \bullet \mathrm{I} \vdash A$ in $\mathcal{C}$, and we take $f\delta^{i}_{\Box A}$ to be $\mathbf{r}_A$.

b) Since in $\mathcal{C}$ we have

$$1_{(\Box A \bullet B)\bullet C}{}^{*} : \Box A \bullet B \vdash ((\Box A \bullet B) \bullet C) \leftarrow C$$

by ($\Box\uparrow$) we must have in $\mathcal{C}[x]$, where $x : \mathrm{I} \vdash A$,

$$\varphi : B \vdash ((\Box A \bullet B) \bullet C) \leftarrow C$$

and hence also $\varphi^{\circ} : B \bullet C \vdash (\Box A \bullet B) \bullet C$. Then, by ($\Box\downarrow$), we must have in $\mathcal{C}$

$$f : \Box A \bullet (B \bullet C) \vdash (\Box A \bullet B) \bullet C$$

and we take this f to be $\mathbf{b}^{\rightarrow}_{\Box A,B,C}$.

Since in $\mathcal{C}$ we have

$${}^{*}(1_{(A\bullet\Box B)\bullet C}{}^{*})\delta_{\Box B} : \Box B \bullet \mathrm{I} \vdash A \to (((A \bullet \Box B) \bullet C) \leftarrow C)$$

by ($\Box\uparrow$) we must have in $\mathcal{C}[x]$, where $x : \mathrm{I} \vdash B$,

$$\varphi : \mathrm{I} \vdash A \to (((A \bullet \Box B) \bullet C) \leftarrow C)$$

and hence also ${}^{*}(({}^{\circ}\varphi\delta^{i}_{A})^{\circ}) : C \vdash A \to ((A \bullet \Box B) \bullet C)$. Then, by ($\Box\downarrow$), we must have in $\mathcal{C}$

$$f : \Box B \bullet C \vdash A \to ((A \bullet \Box B) \bullet C)$$

and we take ${}^{\circ}f$ to be $\mathbf{b}^{\rightarrow}_{A,\Box B,C}$.

Since in $\mathcal{C}$ we have

$$^{*}1_{(A\bullet B)\bullet\Box C}\delta_{\Box C} : \Box C \bullet \mathrm{I} \vdash (A \bullet B) \rightarrow ((A \bullet B) \bullet \Box C)$$

by ($\Box\uparrow$) we must have in $\mathcal{C}[x]$, where $x : \mathrm{I} \vdash C$,

$$\varphi : \mathrm{I} \vdash (A \bullet B) \rightarrow ((A \bullet B) \bullet \Box C)$$

and hence also ${}^{*}({}^{*}({}^{\circ}\varphi\delta^{i}_{A\bullet B})\delta_B) : \mathrm{I} \vdash B \rightarrow (A \rightarrow ((A \bullet B) \bullet \Box C))$. Then, by ($\Box\downarrow$), we must have in $\mathcal{C}$

$$f : \Box C \bullet \mathrm{I} \vdash B \rightarrow (A \rightarrow ((A \bullet B) \bullet \Box C))$$

and we take ${}^{\circ\circ}(f\delta^{i}_{\Box C})$ to be $\mathbf{b}^{\rightarrow}_{A,B,\Box C}$. We proceed analogously for the $\mathbf{b}^{\leftarrow}$ arrows.

c) Since in $\mathcal{C}$ we have

$$^{*}1_{B\bullet\Box A}\delta_{\Box A} : \Box A \bullet \mathrm{I} \vdash B \rightarrow (B \bullet \Box A)$$

by ($\Box\uparrow$) we must have in $\mathcal{C}[x]$, where $x : \mathrm{I} \vdash A$,

$$\varphi : \mathrm{I} \vdash B \rightarrow (B \bullet \Box A)$$

and hence also $({}^{\circ}\varphi\delta^{i}_{B}\sigma_B)^{*} : \mathrm{I} \vdash (B \bullet \Box A) \leftarrow B$. Then, by ($\Box\downarrow$), we must have in $\mathcal{C}$

$$f : \Box A \bullet \mathrm{I} \vdash (B \bullet \Box A) \leftarrow B$$

and we take $(f\delta^{i}_{\Box A})^{\circ}$ to be $\mathbf{c}_{A,\Box B}$. We proceed analogously for $\mathbf{c}_{A,\Box B}$.

k) Since in $\mathcal{C}[x]$ we have 1_{I}, by ($\Box\downarrow$) in $\mathcal{C}$ we must have $f : \Box A \bullet \mathrm{I} \vdash \mathrm{I}$. We take $f\delta^{i}_{\Box A}$ to be $\mathbf{k}_{\Box A}$.

w) Since in $\mathcal{C}$ we have $\delta_{\Box A}$, by ($\Box\uparrow$) we must have $\varphi : \mathrm{I} \vdash \Box A$ in $\mathcal{C}[x]$, where $x : \mathrm{I} \vdash A$. Then in $\mathcal{C}[x]$ we have $(\varphi \bullet \varphi)\sigma^{i}_{\mathrm{I}} : \mathrm{I} \vdash (\Box A \bullet \Box A)$, and by ($\Box\downarrow$), in $\mathcal{C}$ we must have $f : \Box A \bullet \mathrm{I} \vdash (\Box A \bullet \Box A)$. We take $f\delta^{i}_{\Box A}$ to be $\mathbf{w}_{\Box A}$.

$^{\Box}$) Since in $\mathcal{C}$ we have $\delta_{\Box A}$, by ($\Box\uparrow$) we must have $\varphi : \mathrm{I} \vdash \Box A$ in $\mathcal{C}[x]$, where $x : \mathrm{I} \vdash A$. We take this φ to be $x^{\Box}$. So we have $^{\Box}$ applying to arrows of type $\mathrm{I} \vdash A$.

We want to show that we have also $^{\Box}$ applying to arrows of type $B \vdash A$ where B is any modalized object.

Since in the polynomial NL$\Box$ category $\mathcal{C}[x][y]$ built with $y : \mathrm{I} \vdash B$ over the polynomial NL$\Box$ category $\mathcal{C}[x]$, which was itself built over the NL$\Box$ category $\mathcal{C}$ with $x : \mathrm{I} \vdash A$, we have

$$((x \bullet y)\sigma^i_{\mathrm{I}})^{\Box} : \mathrm{I} \vdash \Box(A \bullet B)$$

by ($\Box \downarrow$) we must have $g : \Box B \bullet \mathrm{I} \vdash \Box(A \bullet B)$ in $\mathcal{C}[x]$, and hence also

$$g\delta^i_{\Box B} : \Box B \vdash \Box(A \bullet B).$$

Then, again by ($\Box \downarrow$), we must have in $\mathcal{C}$

$$f : \Box A \bullet \Box B \vdash \Box(A \bullet B).$$

We also have $1^{\Box}_{\mathrm{I}}$ in $\mathcal{C}$.

Suppose that we have in $\mathcal{C}$ an arrow $h : \Box C \vdash A$, and hence also $h\delta_{\Box C}$. Then by ($\Box \uparrow$) we must have $\varphi : \mathrm{I} \vdash A$ in $\mathcal{C}[x]$, where $x : \mathrm{I} \vdash C$. Hence we have $\varphi^{\Box}$ in $\mathcal{C}[x]$, and by ($\Box \downarrow$) we must have $f : \Box C \bullet \mathrm{I} \vdash \Box A$ and $f\delta^i_{\Box C} : \Box C \vdash \Box A$ in $\mathcal{C}$. Taking $f\delta^i_{\Box C}$ to be $h^{\Box}$, we have $^{\Box}$ applying to arrows of type $\Box C \vdash A$. To have full $^{\Box}$, applying to arrows of type $B \vdash A$ where B is any modalized object, it remains to show that for modalized B we have in $\mathcal{C}$ an arrow of type $B \vdash \Box B$, and this we do by induction on the complexity of B, using the arrows and operations on arrows we have already secured.

The fact that for modalized A we have an arrow of type $A \vdash \Box A$, as well as $\mathbf{r}_A$, gives us the structural arrows **b**, **c**, **k** and **w** with the provisos involving any modalized objects, as we have assumed them in Section 2, and not only boxed objects, as we have inferred them above.

Now we shall justify to a certain extent the equations of Section 2. In this justification we presuppose the equations (**b**), (**bb**), (**c**), (**cc**), (**k**) and (**w**); i.e., we presuppose that the **b** and **c** arrows are natural isomorphisms and that the **k** and **w** arrows are natural transformations. We also presuppose the equation ($^{\Box}$), which is of the same sort as (**b**), (**c**), (**k**) and (**w**). Proof- theoretically, it amounts to permuting the rule $^{\Box}$ with cut. We presuppose that modalized objects A are isomorphic with $\Box A$; i.e., for modalized A we presuppose the equations

$$\mathbf{r}_A 1^{\Box}_A = 1_A, \qquad 1^{\Box}_A \mathbf{r}_A = 1_{\Box A}.$$

These equations are of the same sort as (**bb**) and (**cc**). They amount to the conversion of some detours in proofs. Furthermore, we presuppose that the definitions of $`x$ and $\mu_x$ are given and that $\mu_x$ satisfies also clauses $(\mu 1.1), (\mu 1.2)$, $(\mu 2.1)$ and $(\mu 2.2)$. Finally, we presuppose Lemmata 1–3, i.e. that $\mu_x$ is a function and that it is the inverse of $`x$. Our derivation of the remaining equations of Section 2 from modal functional completeness will depend on all these presuppositions.

In some cases our derivation will fall short of obtaining the required equation in full generality in which it holds. This will happen with the pentagonal equations, the hexagonal equations, the triangular equation $(\sigma\delta\mathbf{b})$ and the equation $(\sigma\delta\mathbf{c})$. For them we shall push our derivation only up to instances of these equations where particular indices are modalized. We shall indicate at what places other instances are required, if such is the case.

First we derive $(\Box\beta)$. As a consequence of Lemma 2 (see case (0.1) in the proof) we have

$$(\Box\beta x) \quad \mathbf{r}_A x^{\Box} = x.$$

Then for $f : B \vdash A$ with B modalized and $y : \mathrm{I} \vdash B$ we have

$$\begin{array}{l} \mathbf{r}_A f^{\Box} y = \mathbf{r}_A (fy)^{\Box}, \text{ with } (^{\Box}) \\ \mathbf{f}_A f^{\Box} y = fy, \text{ with } (\Box\beta x) \\ \mu_y(\mathbf{r}_A f^{\Box} y) = \mu_y(fy), \text{ by Lemma 1} \\ \mathbf{r}_A f^{\Box} \mathbf{r}_B \delta_{\Box B} = f\mathbf{r}_b \delta_{\Box B}, \text{ with } (\mu 1.1) \text{ and } (\mu 0.1) \end{array}$$

which with $(\delta\delta^i)$ and $\mathbf{r}_B 1_B^{\Box} = 1_B$ yields $(\Box\beta)$. The equation $(\Box\eta)$ amounts to $(^{\Box})$ and $1_A^{\Box}\mathbf{r}_A = 1_{\Box A}$, which we have presupposed.

Next we justify the pentagonal equations. From Lemma 1 and from the equation (**b**), more precisely ($\mathbf{b}_1$), it follows that we must have

$$\mu_x(((\psi \bullet 1_C) \bullet 1_D)\mathbf{b}^{\rightarrow}_{B,C,D}) = \mu_x(\mathbf{b}^{\rightarrow}_{E,C,D}(\psi \bullet (1_C \bullet 1_D)))$$

which by using $(\mu 1.1), (\mu 1.2), (\mu 2.2)$ and (**b**) gives

$$\begin{array}{l} ((\mu_x\psi \bullet 1_C) \bullet 1_D)(\mathbf{b}^{\rightarrow}_{\Box A,B,C} \bullet 1_D)\mathbf{b}^{\rightarrow}_{\Box A,B\bullet C,D}(1_{\Box A} \bullet \mathbf{b}^{\rightarrow}_{B,C,D}) = \\ \qquad ((\mu_x\psi \bullet 1_C) \bullet 1_D)\mathbf{b}^{\rightarrow}_{\Box A\bullet B,C,D}\mathbf{b}^{\rightarrow}_{\Box A,B,C\bullet D}. \end{array}$$

Then we substitute $1_{\Box A\bullet B}`x$ for ψ and, by Lemma 3, obtain the pentagonal equation (**b5**) with A boxed. Since a modalized object A is isomorphic with $\Box A$, we have the pentagonal equation (**b**5) for modalized A, too. This is not the full pentagonal equation (**b**5), because, in the full one, A need not be modalized provided two objects among B, C and D are modalized. Other forms of pentagonal equations, not covered by (**b**5) with A modalized, are involved in the proof of Lemma 1 when we derive the equations obtained by prefixing μ_x to the two sides of $(\mathbf{b}_2)$ and $(\mathbf{b}_3)$, instead of $(\mathbf{b}_1)$ as above.

At this place we can make the following remark. By going carefully over our proof of the Modal Functional Completeness Theorem, one finds that we never need the arrows $\mathbf{b}^{\rightarrow}_{A,B,C}$ and $\mathbf{b}^{\leftarrow}_{A,B,C}$ where C is modalized, but only where A or B is modalized, except when in the proof of Lemma 1 we derive the equation obtained by prefixing μ_x to the two sides of $(\mathbf{b}_3)$. This has to do with the fact that $\mu_x\varphi$ is taken to be of type $\Box A \bullet B \vdash C$, rather than $B \bullet \Box A \vdash C$. We might as well have defined an analogous $\mu_x\varphi$ of this other type. Then for the first index of the **b** arrows and the equation $(\mathbf{b}_1)$ we would have the same thing that we have now for the third index and $(\mathbf{b}_3)$. However, with our definition of μ_x, we could not exclude the **b** arrows in which only the object in the third index is modalized because these are definable as follows in terms of **c** arrows and the remaining **b** arrows:

$$\mathbf{b}^{\rightarrow}_{A,B,C} =_{\text{df}} \mathbf{c}_{C,A\bullet B}\mathbf{b}^{\leftarrow}_{C,A,B}(\mathbf{c}_{A,C} \bullet \mathbf{1}_B)\mathbf{b}^{\rightarrow}_{A,C,B}(\mathbf{1}_A \bullet \mathbf{c}_{B,C})$$

where only C is modalized. The equation installed by this definition is a hexagonal equation.

Next we justify the hexagonal equations. From Lemma 1 and the equation (**c**), more precisely $(\mathbf{c}_1)$, it follows that we must have

$$\mu_x((\psi \bullet \mathbf{1}_C)\mathbf{c}_{C,B}) = \mu_x(\mathbf{c}_{C,D}(\mathbf{1}_C \bullet \psi))$$

which by using $(\mu 1.1)$, $(\mu 1.2)$, $(\mu 2.1)$, $(\mu 2.2)$ and (**c**) gives

$$(\mu_x\psi \bullet \mathbf{1}_C)\mathbf{b}^{\rightarrow}_{\Box A,B,C}(\mathbf{1}_{\Box A} \bullet \mathbf{c}_{C,B}) = (\mu_x\psi \bullet \mathbf{1}_C)\mathbf{c}_{C,\Box A\bullet B}\mathbf{b}^{\leftarrow}_{C,\Box A,B}(\mathbf{c}_{\Box A,C} \bullet \mathbf{1}_B)\mathbf{b}^{\rightarrow}_{\Box A,C,B}.$$

Then we substitute $\mathbf{1}_{\Box A\bullet B}$'$x$ for ψ and, by Lemma 3, obtain a hexagonal equation. The same hexagonal equation is induced by prefixing μ_x to the two sides of $(\mathbf{c}_2)$. As before, $\Box A$ can be replaced by modalized A. This doesn't yet amount to the full hexagonal equation (**bc**6), because, there, A need not be modalized if C is (it is not enough that only B be modalized). A hexagonal equation where only C is modalized is involved in the proof of Lemma 1 when we derive the equation obtained by prefixing μ_x to the two sides of $(\mathbf{b}_3)$. Such an equation is also installed by the definition of $\mathbf{b}^{\rightarrow}_{A,B,C}$ where only C is modalized, which we gave in the preceding paragraph.

To derive the octagonal equations it is enough to consider the case with **(w)** in the proof of Lemma 1 and proceed as for the pentagonal and hexagonal equations. The octagonal equations are completely justified by that (unlike the pentagonal and hexagonal equations, whose justification we have pushed only up to a point).

Let us now justify the triangular equations and $(\sigma\delta\mathbf{c})$. As a consequence of Lemma 2 we have that

$$\mu_x(\psi \bullet 1_E)`x = (\mu_x \psi`x) \bullet 1_E$$

which, by substituting $1_{\Box A \bullet D}`x$ for ψ, and by using $(\mu 2.2)$ and Lemma 3, reduces to

$$\mathbf{b}^{\rightarrow}_{\Box A,D,E}(x^{\Box} \bullet 1_{D\bullet E})\sigma^{i}_{D\bullet E} = ((x^{\Box} \bullet 1_D)\sigma^{i}_{D}) \bullet 1_E$$
$$((x^{\Box} \bullet 1_D) \bullet 1_E)\mathbf{b}^{\rightarrow}_{\mathrm{I},D,E}\sigma^{i}_{D\bullet E} = ((x^{\Box} \bullet 1_D) \bullet 1_E)(\sigma^{i}_{D} \bullet 1_E), \text{ with } \mathbf{(b)} \text{ and } (\bullet).$$

By prefixing μ_x to the two sides of the last equation, as Lemma 1 allows, with $(\mu 1.2), (\mu 2.2), (\mu 6), (\mu 0.1)$ and $(\Box\eta)$ we obtain

$$(((\delta_{\Box A} \bullet 1_D)\mathbf{b}^{\rightarrow}_{\Box A,\mathrm{I},D}) \bullet 1_E)\mathbf{b}^{\rightarrow}_{\Box A,\mathrm{I}\bullet D,E}(1_{\Box A} \bullet (\mathbf{b}^{\rightarrow}_{\mathrm{I},D,E}\sigma^{i}_{D\bullet E})) =$$
$$(((\delta_{\Box A} \bullet 1_D)\mathbf{b}^{\rightarrow}_{\Box A,\mathrm{I},D}) \bullet 1_E)\mathbf{b}^{\rightarrow}_{\Box A,\mathrm{I}\bullet D,E}(1_{\Box A} \bullet (\sigma^{i}_{D} \bullet 1_E)).$$

With $(\delta\delta^{i})$, **(bb)** and $(\bullet)$ this reduces to

$$1_{\Box A} \bullet (\mathbf{b}^{\rightarrow}_{\mathrm{I},D,E}\sigma^{i}_{D\bullet E}) = 1_{\Box A} \bullet (\sigma^{i}_{D} \bullet 1_E)$$

from which, by taking A to be I and by using $(\bullet)$ and $\mathbf{r}_{\mathrm{I}} 1_{\Box \mathrm{I}} 1^{\Box}_{\mathrm{I}} = 1_{\mathrm{I}}$, we obtain

$$1_{\mathrm{I}} \bullet (\mathbf{b}^{\rightarrow}_{\mathrm{I},D,E}\sigma^{i}_{D\bullet E}) = 1_{\mathrm{I}} \bullet (\sigma^{i}_{D} \bullet 1_E).$$

By prefixing $\sigma_{(\mathrm{I}\bullet D)\bullet E}$ to both sides, with (σ) and $(\sigma\sigma^{i})$ we derive the triangular equation

$$(\sigma^{i}\mathbf{b}) \qquad \mathbf{b}^{\rightarrow}_{\mathrm{I},D,E}\sigma^{i}_{D\bullet E} = \sigma^{i}_{D} \bullet 1_E$$

which, with $(\sigma\sigma^{i})$ and $(\bullet)$, amounts to $(\sigma\mathbf{b})$. (The triangular equation $(\delta\mathbf{b})$, in which, however, A is modalized, is derivable by prefixing μ_x to the two sides of (δ).)

As a consequence of Lemma 2 we also have

$$\mu_x(1_E \bullet \psi)`x = 1_E \bullet (\mu_x \psi`x)$$

which, by substituting $1_{\Box \mathrm{I} \bullet D}`x$ for ψ, and by using $(\mu 2.1)$ and Lemma 3, reduces to

$$\mathbf{b}^{\leftarrow}_{E,\Box \mathrm{I},D}(\mathbf{c}_{\Box \mathrm{I},E} \bullet 1_D)\mathbf{b}^{\rightarrow}_{\Box \mathrm{I},E,D}(x^{\Box} \bullet 1_{E \bullet D})\sigma^i_{E \bullet D} = 1_E \bullet ((x^{\Box} \bullet 1_D)\sigma^i_D)$$
$$(1_E \bullet (x^{\Box} \bullet 1_D))\mathbf{b}^{\leftarrow}_{E,\mathrm{I},D}(\mathbf{c}_{\mathrm{I},E} \bullet 1_D)\mathbf{b}^{\rightarrow}_{\mathrm{I},E,D}\sigma^i_{E \bullet D} = (1_E \bullet (x^{\Box} \bullet 1_D))(1_E \bullet \sigma^i_D),$$
with (**b**), (**c**), and (•).

By prefixing μ_x to both sides and proceeding as in the last paragraph, with $(\sigma^i\mathbf{b})$ and (•) we obtain

$$\mathbf{b}^{\leftarrow}_{E,\mathrm{I},D}((\mathbf{c}_{\mathrm{I},E}\sigma^i_E) \bullet 1_D) = 1_E \bullet \sigma^i_D$$

which, with $(\sigma\sigma^i)$, (**bb**), (**cc**) and (•), amounts to

$$(\sigma\mathbf{c}\sigma\mathbf{b}) \qquad ((\sigma_E \mathbf{c}_{E,\mathrm{I}}) \bullet 1_D)\mathbf{b}^{\rightarrow}_{E,\mathrm{I},D} = 1_E \bullet \sigma_D.$$

This would be the triangular equation $(\sigma\delta\mathbf{b})$ if we had $\sigma_E \mathbf{c}_{E,\mathrm{I}} = \delta_E$, i.e. $(\sigma\delta\mathbf{c})$.

Now, let us see what we can do for $(\sigma\delta\mathbf{c})$. From Lemma 3 (see the proof of the lemma) we can derive

$$(\delta_{\Box E} \bullet 1_D)\mathbf{b}^{\rightarrow}_{\Box E,\mathrm{I},D} = 1_{\Box E} \bullet \sigma_D$$
$$\delta_{\Box E} \bullet 1_D = (\sigma_{\Box E}\mathbf{c}_{\Box E,\mathrm{I}}) \bullet 1_D, \text{ with } (\sigma\mathbf{c}\sigma\mathbf{b}) \text{ and } (\mathbf{bb}).$$

By taking D to be I and prefixing $\delta_{\Box E}$ to both sides, with (δ) and $(\delta\delta^i)$ we obtain

$$\delta_{\Box E} = \sigma_{\Box E}\mathbf{c}_{\Box E,\mathrm{I}}.$$

So, we have $(\sigma\delta\mathbf{c})$ for modalized A.

In our proof of modal functional completeness, we actually dont need $(\sigma\delta\mathbf{b})$ and $(\sigma\delta\mathbf{c})$ except when A is modalized, provided we have assumed other triangular and pseudotriangular equations, like $(\sigma\mathbf{b})$ and $(\sigma\mathbf{c}\sigma\mathbf{b})$. However, such a reduced stock of triangular and pseudotriangular equations works only because we have taken $\mu_x\varphi$ to be of type $\Box A \bullet B \vdash C$, rather than $B \bullet \Box A \vdash C$. With this other type we would need other equations. (With the first type, σ is preponderant over δ; with the second, it is the other way round.) Our tidier, more symmetric, axiomatization of NL$\Box$ categories, in which $(\sigma\delta\mathbf{b})$ and $(\sigma\delta\mathbf{c})$ hold for nonmodalized A too, is insensitive to this change of type for $\mu_x\varphi$. It works equally well with either type. This slight generalization has repercussions on the underlying nonmodal part of NL$\Box$ categories, i.e. the NL part, because σ and δ, which were not linked in it, are now linked through **c**. Formerly independent assumptions for σ and δ can now be derived from each other, as we noted in Section 2.

To derive (1**k**), we have as a consequence of Lemma 2 (see case (0.2) in the proof)

$$1_{\mathrm{I}}\sigma_{\mathrm{I}}(\mathbf{k}_{\Box A} \bullet 1_{\mathrm{I}})(x^{\Box} \bullet 1_{\mathrm{I}})\sigma_{\mathrm{I}}^{i} = 1_{\mathrm{I}}$$

and then we use $(\sigma\delta)$, (δ), $(\sigma\sigma^{i})$ and (**k**).

It remains to derive $(\sigma\delta\mathbf{w})$, (**bw**), (**cw**), $(\sigma\mathbf{kw})$ and $(\delta\mathbf{kw})$. We derive $(\sigma\delta\mathbf{w})$ from the case we have considered in the induction step of the proof of Lemma 2 (case (1)), together with equations we already have, in particular $(\sigma\mathbf{b})$, which we have derived in full generality. We derive (**bw**) by prefixing μ_x to the two sides of (cat 2) and using equations we already have, in particular a pentagonal equation with A modalized (see the proof of Lemma 1). Similarly, we derive (**cw**) by prefixing μ_x to the two sides of (•bifunctor) and using equations we already have, in particular pentagonal and hexagonal equations with A modalized. Finally, to derive $(\sigma\mathbf{kw})$, take ψ to be $1_{\Box A\bullet\mathrm{I}}$‘$x$ and h to be $1_{\Box A\bullet\mathrm{I}}$ in the last equation displayed in our derivation of $(\mu1.1)$ in Section 3, so as to obtain

$$(\sigma_{\Box A}(\mathbf{k}_{\Box A} \bullet 1_{\Box A})\mathbf{w}_{\Box A}) \bullet 1_{\mathrm{I}} = 1_{\Box A} \bullet 1_{\mathrm{I}}.$$

Then we prefix $\delta_{\Box A}$ to the two sides of this equation; with (δ) and $(\delta\delta^{i})$ this gives $(\sigma\mathbf{kw})$. (We could as well have taken ψ to be $\delta_{\Box A}$‘x). For $(\delta\mathbf{kw})$ we proceed analogously using the last equation displayed in the derivation of $(\mu1.2)$. Note that we already have all the equations used in the derivations of $(\mu1.1)$ and $(\mu1.2)$ up to the last displayed equations; in particular, we have $(\sigma\mathbf{b})$ and $(\sigma\delta\mathbf{b})$ with A modalized. (We use $(\sigma\mathbf{kw})$ and $(\delta\mathbf{kw})$ also in the derivation of $(\mu2.1)$ and $(\mu2.2)$ from $(\mu2)$.) With these derivations, (1**k**), $(\sigma\delta\mathbf{w})$, (**bw**), (**cw**), $(\sigma\mathbf{kw})$ and $(\delta\mathbf{kw})$ are completely justified, and we have accomplished our partial justification of the assumptions made for NL$\Box$ categories in Section 2.

Can we justify in a similar manner the assumptions of Section 1? The existence of an operation • on objects and arrows follows from the formulation of the Modal Functional Completeness Theorem. That there is a functor behind these operations is something we have to presuppose, as we presupposed about the structural arrows that they are natural isomorphisms or natural transformations (proof-theoretically, the equation ($\bullet$) amounts to permuting the rule $\bullet$ with cut). The assumptions concerning $\bullet$ can be understood as structural assumptions, too. Such are also the assumptions about the arrows 1_A and composition. (The equations (cat 1) imply that the arrows 1_A are natural isomorphisms from the identity functor to the identity functor; proof-theoretically, these equations amount to eliminating some cuts, whereas (cat 2) amounts to permuting cut with itself.)

Since they involve the modalized object I, and are related to structural rules, the $\sigma\delta$ arrows might be understood as modal structural arrows, on a par with **b**, **c**, **k** and **w** arrows. So, perhaps the assumptions concerning $\sigma\delta$ arrows could be shifted to Section 2, and would need to be justified as much as the modal assumptions of that section. (We put, however, $\sigma\delta$ arrows in Section 1 because they don't involve $\Box$, and because I makes sense in the absence of $\Box$, too.) The arrows σ^i are needed for the definition of the function $`x$ (they are in the formulation of the Modal Functional Completeness Theorem), whereas σ and δ arrows are needed for the definition of μ_x: the first for (μ0.2) and the second for (μ0.1). The δ^i arrows are not absolutely needed (as we have remarked above, σ is preponderant over δ), but this is only because we take $\mu_x\varphi$ to be of type $\Box A \bullet B \vdash C$, rather than $B \bullet \Box A \vdash C$. The equations ($\sigma$) and ($\delta$) are comparable to (**b**), (**c**), (**k**) and (**w**), whereas ($\sigma\sigma^i$) and ($\delta\delta^i$) are comparable to (**bb**) and (**cc**). They have the same proof-theoretical import. We presuppose these $\sigma\delta$ equations as we presupposed the corresponding **b**, **c**, **k** and **w** equations earlier in this section.

We can, however, derive ($\sigma\delta$) from modal functional completeness. As a consequence of Lemma 2 (see case (0.1) in the proof) we have

$$\mathbf{r}_{\mathrm{I}}\delta_{\Box\mathrm{I}}(x^{\Box} \bullet 1_{\mathrm{I}})\sigma^i_{\mathrm{I}} = x$$
$$\mathbf{r}_{\mathrm{I}}x^{\Box}\delta_{\mathrm{I}}\sigma^i_{\mathrm{I}} = x, \text{ with } (\delta)$$
$$\mu_x(\mathbf{r}_{\mathrm{I}}x^{\Box}\delta_{\mathrm{I}}\sigma^i_{\mathrm{I}}) = \mu_x x, \text{ by Lemma 1}$$
$$\mathbf{r}_{\mathrm{I}}(\mathbf{r}_{\mathrm{I}}\delta_{\Box\mathrm{I}})^{\Box}(1_{\Box\mathrm{I}} \bullet (\delta_{\mathrm{I}}\sigma^i_{\mathrm{I}})) = \mathbf{r}_{\mathrm{I}}\delta_{\Box\mathrm{I}}, \text{ with } (\mu 1.1), (\mu 1.2), (\mu 6) \text{ and } (\mu 0.1)$$
$$\delta_{\Box\mathrm{I}}(1_{\Box\mathrm{I}} \bullet (\delta_{\mathrm{I}}\sigma^i_{\mathrm{I}})) = \delta_{\Box\mathrm{I}}, \text{ with } 1^{\Box}_{\mathrm{I}}\mathbf{r}_{\mathrm{I}} = 1_{\Box\mathrm{I}} \text{ and } (^{\Box})$$

which, with ($\delta\delta^i$), ($\bullet$), $\mathbf{r}_{\mathrm{I}}1_{\Box\mathrm{I}}1^{\Box}_{\mathrm{I}} = 1_{\mathrm{I}}$, ($\sigma$) and ($\sigma\sigma^i$) yields ($\sigma\delta$). (Note that we have used ($\sigma\delta$) in the justification of ($\Box\beta$); here we dont use ($\Box\beta$), but only the isomorphism between I and $\Box$I.)

The assumptions concerning $\rightarrow, \leftarrow, \wedge, \top, \vee$ and $\bot$ are obviously independent from modal functional completeness. Note, however, that we wouldn't have functional completeness with $\vee$ if we didnt have $\rightarrow$, or without a primitive distribution arrow **d** (with $\mu_x\varphi$ of type $B \bullet \Box A \vdash C$ instead of $\Box A \bullet B \vdash C$, we would need $\leftarrow$, or a distribution arrow of type $(E \vee F) \bullet G \vdash (E \bullet G) \vee (F \bullet G)$). A related fact (about which we heard from Djordje Čubrić) is that a bicartesian category is functionally complete in the ordinary sense if and only if it is distributive; in that case $\wedge$, usually written $\times$, plays the role of $\bullet$ for formulating functional completeness.

To conclude this section, let us note that for a weakened form of modal functional completeness where we would be happy with asserting that 'x is *onto*, without necessarily being *one–one*, we would need far less assumptions for our categories. In the justification above we could appeal only to Lemma 2, and for proving this lemma we dont need $1_A^{\Box}\mathbf{r}_A = 1_{\Box A}$, the pentagonal, hexagonal and octagonal equations, (**bw**), (**cw**), $(\sigma\mathbf{kw})$ and $(\delta\mathbf{kw})$.

6 SUBSTRUCTURAL FUNCTIONAL COMPLETENESS

Let us now take NL deductive systems with the unary operations on arrows $D\bullet\_$ and $\_\bullet D$ primitive instead of the binary operation on arrows $\_\bullet\_$, and let us consider the following hierarchy of nonmodal deductive systems, obtained by assuming, in addition to what we have for NL deductive systems, the structural arrows mentioned in parentheses *without* provisos concerning modalized objects:

AL *deductive systems* (**b** arrows)
M *deductive systems* (**b** and **c** arrows)
BCK *deductive systems* (**b**, **c** and **k** arrows)
R *deductive systems* (**b**, **c** and **w** arrows)
H *deductive systems* (**b**, **c**, **k** and **w** arrows).

For easier comparison, we use the labels introduced in [3]. The label 'AL' stands for 'associative Lambek', 'M' for 'multiset' (in the light of latter-day developments, it would be more intelligible if we called these systems *linear*—more precisely, *intuitionistic linear*), 'R' for 'relevant' (intuitionistic and without distribution of $\wedge$ over $\vee$) and 'H' for 'Heyting'. The label 'BCK' is pretty standard (the linear logic trade has recently produced some ersatz names for BCK systems; the BCK systems we consider here are of an intuitionistic sort).

Let us write S for NL, AL, M, BCK, R or H (the variable 'S' stands for 'substructural') and let S *categories* be S deductive systems that are NL categories and satisfy moreover the equations between arrows of Section 2, which, of course, apply only if the arrows in question are present in the deductive system. So, AL categories must satisfy **b** equations, M categories **b** and **c** equations, BCK categories **b**, **c** and **k** equations, R categories **b**, **c** and **w** equations except $(\sigma\mathbf{kw})$ and $(\delta\mathbf{kw})$, and H categories all the **b**, **c**, **k** and **w** equations.

For all S categories except NL and AL categories, we have that $A \rightarrow B$ is naturally isomorphic to $B \leftarrow A$, as in Section 4. However, for none except H categories, which are the bicartesian closed categories of Section 4, we need have that $A \bullet B$ is isomorphic to $A \wedge B$. Only for BCK and H categories I must be naturally isomorphic to $\top$. The AL categories are monoidal biclosed with respect to $\bullet$, I, $\rightarrow$ and $\leftarrow$, and M categories are symmetric monoidal closed with respect to $\bullet$, I and $\rightarrow$. However, they are all also bicartesian with respect to $\wedge$ and $\vee$, their terminal object being $\top$ and their initial object $\perp$.

We can extend the axiomatization of S categories to the axiomatization of the corresponding S$\Box$ categories as we extended the axiomatization of NL categories to the axiomatization of NL$\Box$ categories: we just add the missing modal assumptions from Section 2. For example, the axiomatization of M categories is extended to the axiomatization of M$\Box$ categories by adding the operation on objects $\Box$, the **r** arrows, the operation on arrows $^{\Box}$ and the modal structural arrows $\mathbf{k}_A$ and $\mathbf{w}_A$, *with* the proviso for modalized A, together with the $\Box$, **k** and **w** equations (the **b** and **c** equations are already assumed for the unrestricted structural arrows of M categories). The M$\Box$ categories correspond to intuitionistic modal linear propositional logic. For them we can prove modal functional completeness as we did for NL$\Box$ categories, and similarly with other S$\Box$ categories. For H$\Box$ categories we have to add just the **r** arrows, the operation on arrows $^{\Box}$ and the $\Box$ equations, which gives categories corresponding to intuitionistic S4 propositional logic. For these categories we can prove modal functional completeness, but ordinary, nonmodal, functional completeness fails, though we have it for H categories, as shown in Section 4 (otherwise we would always have arrows of type $A \vdash \Box A$ in H$\Box$ categories). We can also infer the necessity of assumptions for our modal categories as we did in Section 5.

We want now to state a general functional completeness theorem, which as a special case covers ordinary functional completeness for H categories, i.e. bicartesian closed categories, and in other cases yields restricted functional completeness for other S categories. To state this theorem we need special notions of polynomials in polynomial categories, which we proceed to define.

Given an S category $\mathcal{C}$, we build the polynomial S category $\mathcal{C}[x]$ with $x : \mathrm{I} \vdash A$ as before. This we can always do because, for every S, the S categories are equationally presented. If all arrows of $\mathcal{C}[x]$ are called *polynomials*, then our notion of polynomial satisfies the following clauses (parallel with the inductive clauses for μ_x in Section 3):

(P0.1) The arrow x is a polynomial.

(P0.2) Every arrow of $\mathcal{C}$ is a polynomial.

(P1) If $\psi : D \vdash E$ and $\xi : E \vdash F$ are polynomials, then $\xi\psi$ is a polynomial.

(P1.1) If $\psi : D \vdash E$ is a polynomial and $h : E \vdash F$ is an arrow of $\mathcal{C}$, then $h\psi$ is a polynomial.

(P1.2) If $h : D \vdash E$ is an arrow of $\mathcal{C}$ and $\xi : E \vdash F$ is a polynomial, then ξh is a polynomial.

(P2.1) If ψ is a polynomial, then $E \bullet \psi$ is a polynomial.

(P2.2) If ψ is a polynomial, then $\psi \bullet E$ is a polynomial.

(P3.1) If $\psi : E \bullet D \vdash F$ is a polynomial, then ${}^*\psi$ is a polynomial.

(P3.2) If $\psi : D \bullet E \vdash F$ is a polynomial, then ψ^* is a polynomial.

(P4) If $\psi : D \vdash E$ and $\xi : D \vdash F$ are polynomials, then $\langle \psi, \xi \rangle$ is a polynomial.

(P5) If $\psi : E \vdash D$ and $\xi : F \vdash D$ are polynomials, then $[\psi, \xi]$ is a polynomial.

Clauses (P1.1) and (P1.2) are redundant in the presence of (P0.2) and (P1), but we have listed them nevertheless because we need them for the inductive definitions of restricted notions of polynomials. These notion are obtained by assuming the clauses mentioned in parentheses:

M *polynomial* (all clauses save (P0.2) and (P1))
BCK *polynomial* (all clauses save (P1))
R *polynomial* (all clauses save (P0.2))
H *polynomial* (all clauses)

(NL and AL polynomials will be considered below). Of course, H polynomials are not restricted: they coincide with all the arrows of $\mathcal{C}[x]$. (As we have just noted above, we may omit clauses (P1.1) and (P1.2) from their definition.) However, the other notions of polynomial reject some arrows of $\mathcal{C}[x]$.

Before looking into that, let us make a point concerning the nature of the arrow x in $\mathcal{C}[x]$. This arrow must be *new* to $\mathcal{C}$ (otherwise, with NL□ categories we would need the equation $x^{\Box}\mathbf{k}_{\Box A} = 1_{\Box A}$ to make (μ0.1) and (μ0.2) match; with H categories we would need $x\mathbf{k}_A = 1_A$, which doesn't hold necessarily in cartesian categories). However, nothing prevents us from introducing a new arrow x of type $\mathrm{I} \vdash \mathrm{I}$, which in the course of constructing $\mathcal{C}[x]$ will be identified with 1_{I} for some categories (in NL□ categories we have $1_{\mathrm{I}}^{\Box}\mathbf{k}_{\Box\mathrm{I}} = 1_{\Box\mathrm{I}}$ because $\mathbf{k}_{\Box\mathrm{I}} = \mathbf{r}_{\mathrm{I}}$; in BCK and H categories, and NL□ categories as well, we have $\mathbf{k}_{\mathrm{I}} = 1_{\mathrm{I}}$). So, in some S categories, an arrow of $\mathcal{C}[x]$ may qualify as a polynomial on more than one ground:

in BCK, and hence also H, categories $x : \mathrm{I} \vdash \mathrm{I}$ will be a polynomial both by (P0.1) and (P0.2).

It can easily be checked that if we exclude the operations on arrows mentioned in clauses (P4) and (P5), an M polynomial is an arrow of $\mathcal{C}[x]$ in whose construction x occurs *exactly* once, a BCK polynomial an arrow of $\mathcal{C}[x]$ in whose construction x occurs *at most* once, and an R polynomial an arrow of $\mathcal{C}[x]$ in whose construction x occurs *at least* once.

Note that for R and H polynomials we obtain as a derived clause:

(P2) If ψ and ξ are polynomials, then $\psi \bullet \xi$ is a polynomial.

This is because $\psi \bullet \xi$ is equal by definition to $(\psi \bullet E)(F \bullet \xi)$ in all S categories (we have taken the unary operations on arrows $D \bullet \_$ and $\_ \bullet D$ as primitive), and the arrow $(\psi \bullet E)(F \bullet \xi)$ is an R polynomial by (P2.1), (P2.2) and (P1). Since we lack (P1) for M and BCK polynomials, we shall also lack (P2). This is parallel with the fact that if we replace clause $(\mu 2)$ by $(\mu 2.1)$ and $(\mu 2.2)$, the **w** arrows enter into the definition of μ_x only via clause $(\mu 1)$. The **k** arrows enter into this definition only via clause $(\mu 0.2)$.

Let now the function ‘x be defined as it was defined in Section 4:

$$\text{For } x : \mathrm{I} \vdash A \text{ and } f : A \bullet B \vdash C, \quad f`x =_{\text{df}} f(x \bullet 1_B)\sigma^i_B.$$

For μ_x as in Section 4, we replace clause $(\mu 0.1)$ by

$$\mu_x x = \delta_A$$

and assume $(\mu 1.1), (\mu 1.2), (\mu 2.1), (\mu 2.2), (\mu 3.1), (\mu 3.2), (\mu 4)$ and $(\mu 5)$ with $\Box$ deleted everywhere. Clause $(\mu 0.2)$ with $\Box$ deleted will be assumed only in the presence of **k** arrows, which in the present context means only for S being BCK or H. Similarly, clause $(\mu 1)$ with $\Box$ deleted will be assumed only in the presence of **b** and **w** arrows, which in the present context means only for S being R or H. When we refer to the μ clauses from now on, we assume these are the newly introduced, nonmodal, clauses. If S is R, clause $(\mu 1)$ is independent from $(\mu 1.1)$ and $(\mu 1.2)$, because arrows of $\mathcal{C}$ dont qualify as polynomials. If S is H, we have to show this clause is compatible with $(\mu 1.1)$, $(\mu 1.2)$ and $(\mu 0.2)$; this we do by deriving $(\mu 1.1)$ and $(\mu 1.2)$ from $(\mu 1)$ and $(\mu 0.2)$, quite analogously to what we did in Section 3. Since $D \bullet \_$ and $\_ \bullet D$ are primitive, instead of $\_ \bullet \_$, we shall have clause $(\mu 2)$ with $\Box$ deleted only as a derived clause when **b**, **c** and **w** arrows are present, and, again, this will happen only for S being R or H. That $(\mu 2)$ can actually replace $(\mu 2.1)$

and $(\mu 2.2)$ will be the case only for H (we use $(\mu 0.2)$, $(\sigma \mathbf{kw})$ and $(\delta \mathbf{kw})$ to derive $(\mu 2.1)$ and $(\mu 2.2)$ from $(\mu 2)$).

We can now state our general functional completeness theorem:

Substructural Functional Completeness Theorem If S is M, BCK, R or H, then for every S polynomial $\varphi : B \vdash C$ of the polynomial S category $\mathcal{C}[x]$ built over the S category $\mathcal{C}$ with $x : \mathrm{I} \vdash A$, there is a unique arrow $f : A \bullet B \vdash C$ of $\mathcal{C}$ such that $f(x \bullet 1_B)\sigma^i_B = \varphi$ holds in $\mathcal{C}[x]$.

In other words, the function 'x is an *onto* and *one–one* function from the arrows $f : A \bullet B \vdash C$ of $\mathcal{C}$ to the S polynomials of $\mathcal{C}[x]$.

We can prove this theorem by a straightforward adaptation of the argument in Section 3. For example, for M categories, we have to check that for an M polynomial φ the **k** and **w** arrows are not involved in $\mu_x\varphi$, and that Lemmata 1–3 can be demonstrated for M polynomials φ and ψ, and M categories $\mathcal{C}$. For other S categories covered by the theorem, we similarly have to check that for an S polynomial φ the structural arrows rejected in S categories are not involved in $\mu_x\varphi$, and that the proofs of Lemmata 1–3 work.

In analogy with the demonstration of necessity of Section 5, we can show that structural arrows and equations we have assumed for them are necessary. This is now simpler than in Section 5, since the complications involving modalized objects are eschewed. In particular, we can completely justify pentagonal, hexagonal and triangular equations.

In [3], our restrictions concerning polynomials are matched by restrictions concerning structural rules in the deductive metalogic, and it is demonstrated that if the metalogic is appropriately restricted, BCK logic is minimal in the presence of $\wedge$ for proving the deduction theorem and its converse. The discrepancy between that and the Substructural Functional Completeness Theorem, which covers also M and R, is explained by the fact that [3] always allows the polynomials $\langle \psi, h \rangle$ and $\langle h, \psi \rangle$ where ψ is a polynomial of $\mathcal{C}[x]$ and h an arrow of $\mathcal{C}$. Here, we allow that for BCK and H polynomials, but not for M and R polynomials.

The Substructural Functional Completeness Theorem does not cover NL and AL categories. The corresponding notions of polynomial should presumably be

NL *polynomial*	(all clauses save (P0.2), (P1), (P2.1), (P2.2), (P3.1) and (P3.2))
AL *polynomial*	(all clauses save (P0.2), (P1), (P2.1) and (P3.1))

because, without ($\mu 2$), the **b** arrows enter into the definition of μ_x via clauses ($\mu 1$), ($\mu 2.1$), ($\mu 2.2$), ($\mu 3.1$) and ($\mu 3.2$), and the **c** arrows via clauses ($\mu 2.1$) and ($\mu 3.1$). If for $\varphi : B \vdash C$, the arrow $\mu_x\varphi$ were redefined so as to be of type $B \bullet A \vdash C$ instead of $A \bullet B \vdash C$, then for AL polynomials we would reject (P2.2) and (P3.2) instead of (P2.1) and (P3.1). With NL and AL polynomials we shall run into trouble in the proof of Lemma 1, because, for example, the NL and AL polynomial $x\sigma_{\mathrm{I}}$ is equal to $\sigma_A(\mathrm{I} \bullet x)$, by ($\sigma$), and $\mu_x(\sigma_A(\mathrm{I} \bullet x))$ is, by ($\mu 1.1$), ($\mu 2.1$) and ($\mu 0.1$), equal to

$$\sigma_A(\mathrm{I} \bullet \delta_A)\mathbf{b}^{\leftarrow}_{\mathrm{I},A,\mathrm{I}}(\mathbf{c}_{A,\mathrm{I}} \bullet \mathrm{I})\mathbf{b}^{\rightarrow}_{A,\mathrm{I},\mathrm{I}}$$

which involves both **b** and **c** arrows. The problem is that for S being NL or AL, the class of S polynomials considered as arrows in an S deductive system is not closed under equality of arrows in S categories (since we lack (P2.1), the arrow $\sigma_A(\mathrm{I} \bullet x)$ is neither an NL nor an AL polynomial), whereas for S being M, BCK, R or H, this is the case. The proof of Lemma 3, too, does not work for NL categories, because of an essential use of **b** arrows: $f`x$ is not an NL polynomial. However, the proof of Lemma 2 works, and we can demonstrate a weakened version of the Substructural Functional Completeness Theorem with NL or AL substituted for S if we dont require that $f$ be unique; i.e. $`x$ is *onto*, but not necessarily *one–one*.

7 CONCLUSION

We conclude this paper by brief indications about matters related to our results that we intend to treat in the future.

There is another way of restricting functional completeness, different from the way of the Substructural Functional Completeness Theorem. We may redefine μ_x so that for an arbitrary polynomial $\varphi : B \vdash C$ (i.e. an H polynomial), the arrow $\mu_x\varphi$ is not necessarily of type $A \bullet B \vdash C$ (nor $B \bullet A \vdash C$), but of some type $B[A] \vdash C$, where $B[A]$ is obtained from B by replacing factors D of B by $A \bullet D$ or $D \bullet A$, there are as many A's in $B[A]$ as there are x's in φ, and these A's are distributed in $B[A]$ in a way matching the distribution of x's in φ. This distribution requirement becomes unnecessary in the presence of **b** and **c** arrows, whereas there can be more A's in $B[A]$ than x's in φ in the presence of **k** arrows, and less in the presence of **w** arrows. With such a μ_x we may also be able to prove restricted functional completeness for various S categories.

Next, as functional completeness for cartesian closed categories enables us to extract systems of typed lambda terms as the internal languages of these categories, so modal functional completeness should lead to a kind of system of lambda terms with modalized types, such a system being the internal language of an S□ category. On the other hand, the restricted functional completeness of Section 6 leads to systems of typed lambda terms with restricted functional abstraction (for example, we may bind with a lambda operator exactly one variable, or not more than one, or at least one).

At the end of [3] it is supposed that freely extending a nonmodal S category to an S□ category might result in the former being a full subcategory of the latter. We suppose this could be demonstrated by a normalization technique, perhaps inspired by Gentzen's methods, or lambda conversion.

Another matter we will try to consider is the relationship between functional completeness and coherence. It is remarkable that assumptions about categories that Mac Lane needed to prove coherence for monoidal and symmetric monoidal categories should reappear as necessary for proving functional completeness. We conjecture that some sort of equivalence between coherence and functional completeness could be established. The reason for this equivalence should be that in functional completeness we transform a polynomial $\varphi : B \vdash C$ into $\mu_x\varphi : A \bullet B \vdash C$ irrespectively of where x occurs in φ. This requires that certain diagrams whose nodes are obtained from B by replacing factors D of B by $A \bullet D$ of $D \bullet A$ should commute. And the commuting of these diagrams should be sufficient for functional completeness.

ACKNOWLEDGEMENTS

The first-mentioned among us is very grateful to the Department of Mathematics of the University of Montpellier III and especially Anne Preller for their exceptional hospitality and the support received for this work. He is also grateful to Heinrich Wansing for the invitation to the conference 'Proof Theory of Modal Logic' in Hamburg, in November 1993, and to Josep Maria Font for the invitation to the 'Meeting on Mathematical Logic' in Barcelona, in January 1994, at which he gave talks based on this paper. The Alexander von Humboldt Foundation has generously provided support for his participation at the first conference, while the second was organized under the auspices of the Centre for Mathematical Research of the Institute of Catalan Studies and marked by the hospitality of the logicians of Barcelona. We are both grateful to the Mathematical Institute in Belgrade for supporting our work through Grant 0401A of the Science Fund of Serbia.

NOTE ADDED IN PRINT

The proof of Lemma 1 is incomplete as it stands, because it is not necessarily the case, as claimed in the second paragraph of that proof, that if φ and ψ are arrows of $\mathcal{C}$ and $\varphi = \psi$ holds in $\mathcal{C}\{x\}$, then $\varphi = \psi$ holds in $\mathcal{C}$. When this is not the case, the proof should be phrased as the corresponding part of the proof of Proposition 6.1 in [6, chapter I.6]. The calculations of the proof of Lemma 1 are sufficient for this rephrasing.

In the comments about the bijection 'x one should bear in mind that if $\Box A \bullet \_$ is not one-one on objects, the bijection is only local; i.e., it exists only between the arrows $f : \Box A \bullet B \vdash C$ and the polynomials $\psi : B \vdash C$ for B given in advance (and the same with $A \bullet \_$ instead of $\Box A \bullet \_$).

Kosta Došen
Université Paul Sabatier, France, and
Mathematical Institute of Belgrade, Yugoslavia

Zoran Petrić
University of Belgrade, Yugoslavia

REFERENCES

[1] G. Birkhoff, J. D. Lipson. Heterogeneous algebras. *Journal of Combinatorial Theory*, **8**, 115–133, 1970.
[2] K. Došen. Sequent systems for modal logic. *The Journal of Symbolic Logic*, **50**, 149–168, 1985.
[3] K. Došen. Modal logic as metalogic. *Journal of Logic, Language and Information*, **1**, 173–201, 1992.
[4] K. Došen, P. Schroeder-Heister (eds.), *Substructural Logics*. Oxford University Press, Oxford, 1993.
[5] P. J. Higgins. Algebras with a scheme of operators. *Mathematische Nachrichten*, **27**, 115–132, 1963.
[6] J. Lambek, P. J. Scott. *Introduction to Higher-Order Categorical Logic*, Cambridge University Press, Cambridge, 1986.
[7] S. Mac Lane. *Categories for the Working Mathematician*. Springer-Verlag, Berlin, 1971.
[8] A. S. Troelstra. *Lectures on Linear Logic*. Center for the Study of Language and Information, Stanford, 1992.

SIMONE MARTINI AND ANDREA MASINI

A COMPUTATIONAL INTERPRETATION OF MODAL PROOFS

1 INTRODUCTION

Proof theory of modal logics, though largely studied since the fifties, has always been a delicate subject, the main reason being the apparent impossibility to obtain elegant, natural systems for intensional operators (with the excellent exception of intuitionistic logic). For example Segerberg, not earlier than 1984 [5], observed that the Gentzen format, which works so well for truth functional and intuitionistic operators, cannot be *a priori* expected to remain valid for modal logics; carrying to the limit this observation one could even assert that 'logics with no good proof theory are *unnatural*'. In such a way we should mark as 'unnatural' all modal logics (with great delight of a large number of logicians!).

One of the main drawbacks of modal proof theory is well exemplified, in the context of the logics K, KT, and KD, by Scott's rule:

$$\frac{\Gamma \vdash \sigma}{\Box\Gamma \vdash \Box\sigma}.$$

This rule, though allowing the proof of a cut-elimination theorem, is neither a left nor a right rule, thus destroying the deep symmetries of the sequent calculus. The situation is even worse with the natural deduction formulation of the same logics, where either there are no explicit modal introduction and elimination rules, as in [5], or the modal rule is formulated as

$$\frac{\Box\Gamma \qquad \begin{matrix}[\Gamma] \\ \vdots \\ \sigma\end{matrix}}{\Box\sigma}\,\Box K$$

which is neither an introduction nor an elimination rule and in which the same formula(s) appear in the premise twice, once boxed and once unboxed.

The situation is somehow better for S4, with reasonable sequent rules,

$$\frac{\Gamma, \tau \vdash \sigma}{\Gamma, \Box\tau \vdash \sigma}\,\Box\vdash \qquad \frac{\Box\Gamma \vdash \sigma}{\Box\Gamma \vdash \Box\sigma}\,\vdash\Box$$

H. Wansing (ed.), Proof Theory of Modal Logic, 213–241.

but, once again, problematic natural deduction. The sequent rules appear strictly related to those for the universal quantifier, and following this analogy Prawitz [14] attempted the definition of a natural deduction system styled after the first order system. This naive approach, however, does not work, and in order to obtain normalization, Prawitz was forced to introduce more elaborated concepts and to formulate rules whose application depends on global constraints on the deduction (see Section 4.1).

In the last few years, however, it has been discovered by several authors that the difficulty was in a too strict interpretation of the 'Gentzen format' and that several extensions [6, 3, 7, 11, 12] allow a good and 'natural' proof theory for modal logics.

We will focus here on the proposal by the second author for the logic KD, whose main idea is a two-dimensional generalization of the notion of sequent. Instead of asserting provability ($\vdash$) between two sequences of formulas, provability is asserted between two-dimensional sequences of formulas.

Developed in [11] as a sequent calculus for classical KD, in [12] the approach is tailored to the intuitionistic framework, for which it is also given an equivalent *natural deduction system.*

The goal of the present paper is to study the computational properties of the system in [12] and of several extensions to other logics, namely the positive fragments of K4, KT, and S4.

To this aim, we introduce first a complete natural deduction system for each of the logics at hand. All the rules of these systems act only on the conclusions and/or the premises of the deductions they are applied to, no introduction rule has a premise containing the introduced connective (as it is case for rule $\Box K$, for instance), and the modal rules strictly match the standard rules for an universal quantifier, the side conditions on variables becoming a side condition on levels.

One the main features of the approach is its *scalability*, the various logics differring only in one parameter of the elimination rule for $\Box$. This allows both a compact treatment and the study of the 'fine structure' of the $\Box$ modality (that is, how several sublogics interact to obtain a larger logic, in this case S4).

By defining a suitable notion of reduction on the terms representing proofs (extending standard β-reduction for λ-terms), we obtain a natural computational interpretation for proofs, which is proved strongly normalizing and confluent.

Finally, this local, two dimensional approach enforces remarkable *global properties* on the resulting deductions. Though this is not the main subject of the paper, in Section 4.1 we will show that all deductions in our system for S4, in fact satisfy Prawitz's global requirements on proofs, thus giving an immediate forgetful translation of our system in Prawitz's one.

It should be noted that Kripke's work on semantics of modalities [10] already contains Gentzen systems based on indexed formulas. Indexes represent paths in an abstract tree, whose obvious interpretation is the accessibility relation of a (Kripke) model. Formally, our approach is a particular case of Kripke's, where indexes are just linearly ordered (the level structure). The intended interpretation, however, is not (and cannot be) a reflection of the accessibility relation into the syntax. The approach is motivated only by a 'purist' approach to modal proof-theory. It searches for local, geometrical properties of sequents in order to validate the introduction of modal quantifiers.

2 SYSTEMS

Before introducing formally the systems we will deal with, we briefly, and informally recall the natural deduction approach taken in [12]. Let us denote formulas with lowercase Greek letters $\alpha, \beta, \sigma, \ldots$, and sequences of formulas with uppercase Gothic letters $\mathfrak{S}, \mathfrak{B}, \ldots$, while ε will be the empty sequence. An *intuitionistic 2-sequent* is a two dimensional expression of the form

$$(1) \qquad \begin{array}{ccc} \mathfrak{S}_1 & & \varepsilon \\ \mathfrak{S}_2 & & \varepsilon \\ \vdots & \vdash & \vdots \\ \mathfrak{S}_k & & \sigma \end{array}$$

whose intended meaning is the formula

$$\bigwedge \mathfrak{S}_1 \supset \Box(\bigwedge \mathfrak{S}_2 \supset \ldots \Box(\bigwedge \mathfrak{S}_k \supset \sigma) \ldots).$$

Note that the formula σ, the *conclusion* of the deduction, lies at a *level*, k, which is greater than, or equal to, the level of any assumption it depends on (any of the $\mathfrak{S}_j$'s may be empty, of course). The propositional rules act over these two dimensional stuctures in the expected way, just 'respecting the levels'. The rules for $\supset$, for instance, can be expressed as:

$$\dfrac{\begin{array}{ccc} \mathfrak{S}_1 & & \varepsilon \\ \mathfrak{S}_2 & & \varepsilon \\ \vdots & \vdash & \vdots \\ \mathfrak{S}_k, \sigma & & \tau \end{array}}{\begin{array}{ccc} \mathfrak{S}_1 & & \varepsilon \\ \mathfrak{S}_2 & & \varepsilon \\ \vdots & \vdash & \vdots \\ \mathfrak{S}_k & & \sigma \supset \tau \end{array}} \supset\mathcal{I} \qquad \dfrac{\begin{array}{ccc} \mathfrak{S}_1 & & \varepsilon \\ \mathfrak{S}_2 & & \varepsilon \\ \vdots & \vdash & \vdots \\ \mathfrak{S}_k & & \sigma \supset \tau \end{array} \quad \begin{array}{ccc} \mathfrak{B}_1 & & \varepsilon \\ \mathfrak{B}_2 & & \varepsilon \\ \vdots & \vdash & \vdots \\ \mathfrak{B}_k & & \sigma \end{array}}{\begin{array}{ccc} \mathfrak{S}_1, \mathfrak{B}_1 & & \varepsilon \\ \mathfrak{S}_2, \mathfrak{B}_2 & & \varepsilon \\ \vdots & \vdash & \vdots \\ \mathfrak{S}_k, \mathfrak{B}_k & & \tau \end{array}} \supset\mathcal{E}$$

Formulas may change their level only by means of modal rules:

$$
\frac{\begin{array}{lcl} \mathfrak{S}_1 & & \varepsilon \\ \mathfrak{S}_2 & & \varepsilon \\ \vdots & \vdash & \vdots \\ \mathfrak{S}_k & & \varepsilon \\ \varepsilon & & \sigma \end{array}}{\begin{array}{lcl} \mathfrak{S}_1 & & \varepsilon \\ \mathfrak{S}_2 & & \varepsilon \\ \vdots & \vdash & \vdots \\ \mathfrak{S}_k & & \Box\sigma \end{array}} \Box\mathcal{I}
\qquad\qquad
\frac{\begin{array}{lcl} \mathfrak{S}_1 & & \varepsilon \\ \mathfrak{S}_2 & & \varepsilon \\ \vdots & \vdash & \vdots \\ \mathfrak{S}_k & & \Box\sigma \end{array}}{\begin{array}{lcl} \mathfrak{S}_1 & & \varepsilon \\ \mathfrak{S}_2 & & \varepsilon \\ \vdots & \vdash & \vdots \\ \mathfrak{S}_k & & \varepsilon \\ \varepsilon & & \sigma \end{array}} \Box\mathcal{E}
$$

Thus, the only way to introduce a modality on a formula occurrence σ at level k is that σ be the only formula present at that level. As a result of the rule, the introduced formula is lifted one level up. Vice versa, the elimination rule pushes a formula down one level (but there is no restriction on its premise). The levels thus represent in the calculus a notion of modal dependence: the conclusion σ at level k modally depends on the assumptions at the same level. If there is not any such assumption, then we are allowed to assert that σ is necessary.

As proved in [12], the given rules for $\Box$ characterize the minimal normal modal logic K.[1] In this paper we will extend this approach to a class of modal logics based on K (KT, K4, and S4), focusing on the computational aspects of the resulting proofs.

Before going into the details of the systems, we adopt a more compact representation for 2-sequents. Instead of writing two-dimensional judgements, we will denote each formula σ at level k with σ^k and write the judgment (2) as $\Gamma \vdash \sigma^k$, where

$$
\Gamma = \begin{array}{l} \mathfrak{S}_1 \\ \mathfrak{S}_2 \\ \vdots \\ \mathfrak{S}_k \end{array}
$$

will be seen as a multiset $\{\tau_1^{i_1}, \ldots, \tau_n^{i_n}\}$. The usefulness of this representation will be especially clear in Section 5. The reader should always bear in mind, however, that the indexes on formulas are only a metatheoretical notation for two dimensional structures.

[1]More precisely, [12] introduces a system for the minimal deontic normal modal logic KD, that in absence of negation and $\Diamond$ is equivalent to K.

2.1 *Basic definitions*

Formally, formulas are built out of *atoms* (ranged over by p); compound formulas are obtained with the connectives: $\Box$ (unary), $\wedge$ and $\supset$ (binary). Any formula of the calculus will be marked with a *level index*, varying in $\mathbb{N}^+$; an indexed formula σ of level i will be written σ^i.

The following definition will introduce a calculus of *terms* and *formulas*, such that to any term M there will correspond a unique indexed formula σ^i, called *its type*; this fact will be denoted with $M : \sigma^i$. For any indexed formula σ^i we assume the existence of a numberable set $\{x_1^i, x_2^i, \ldots\}$ of variables of type σ^i. A *set of assumptions* (called sometimes also a *basis*, or a *context*) is a set $\Gamma = \{x_1^{i_1} : \sigma_1^{i_1}, \ldots, x_n^{i_n} : \sigma_n^{i_n}\}$, where all the variables are different: if $x^k : \sigma^k \in \Gamma$ and $x^h : \tau^h \in \Gamma$, then $\sigma^k = \tau^h$ (and in particular $k = h$). No variable name, hence, can appear in a set of assumptions with two different levels (this is the *level variable convention*). For such a set of assumptions Γ, define $\#\Gamma = \max\{k_j \mid x_j^{k_j} : \sigma_j^{k_j} \in \Gamma\}$; $\#\Gamma = 0$ when Γ is empty. Moreover, $\Gamma^{<i} = \{x^k : \sigma^k \in \Gamma \mid k < i\}$; the set $\Gamma^{\geq i}$ is defined analogously. Finally, for $n \in \mathbb{Z}, \Gamma(n) = \{x^{k+n} : \sigma^{k+n} \mid x^k : \sigma^k \in \Gamma\}$.

DEFINITION 1 *[Calculus λ^ℓ]* Terms *and* derivations *are inductively defined by the following rules.*

$$x^k : \sigma^k$$

$$\frac{\begin{matrix}\Gamma \\ \vdots \\ M : \sigma^k\end{matrix} \qquad \begin{matrix}\Delta \\ \vdots \\ N : \tau^k\end{matrix}}{\langle M, N\rangle^k : \sigma \wedge \tau^k}\ \wedge\mathcal{I} \qquad \frac{\begin{matrix}\Gamma \\ \vdots \\ M : \sigma \wedge \tau^k\end{matrix}}{(\mathsf{fst}\ M)^k : \sigma^k}\ \wedge\mathcal{E}_l \qquad \frac{\begin{matrix}\Gamma \\ \vdots \\ M : \sigma \wedge \tau^k\end{matrix}}{(\mathsf{snd}\ M)^k : \tau^k}\ \wedge\mathcal{E}_r$$

$$\frac{\begin{matrix}\Gamma\ \ [x^k : \sigma^k] \\ \vdots \\ M : \tau^k\end{matrix}}{(\lambda x^k{:}\sigma^k.M)^k : \sigma \supset \tau^k}\ \supset\mathcal{I} \qquad \frac{\begin{matrix}\Gamma \\ \vdots \\ M : \sigma \supset \tau^k\end{matrix} \qquad \begin{matrix}\Delta \\ \vdots \\ N : \sigma^k\end{matrix}}{(MN)^k : \tau^k}\ \supset\mathcal{E}$$

While we will always attach a level to a formula, in the exposition we will be more liberal for terms, writing both $M : \sigma^k$ and $M^k : \sigma^k$, since no confusion may arise in this way.

If S is one of the systems introduced below, we write $\Gamma \vdash_S M^i : \sigma^i$ when in S there is a derivation:

$$\begin{matrix}\Gamma \\ \vdots \\ M^i : \sigma^i\end{matrix}$$

$\Gamma \vdash_S \sigma^i$ holds when there is a term M^i such that $\Gamma \vdash_S M^i : \sigma^i$.

2.2 System $\lambda^\ell K$

The typed λ-calculus $\lambda^\ell K$ is obtained by adding to λ^ℓ the two rules we have already discussed:

$$\dfrac{\begin{matrix}\Gamma\\ \vdots\\ M^j : \sigma^j\end{matrix}}{\mathsf{gen}(M^j)^{j-1} : \Box\sigma^{j-1}}\ \Box\mathcal{I}\ \ {\scriptstyle j>\#\Gamma} \qquad\qquad \dfrac{\begin{matrix}\Gamma\\ \vdots\\ M^j : \Box\sigma^j\end{matrix}}{\mathsf{ungen}(M^j)^{j+1} : \sigma^{j+1}}\ \Box\mathcal{E}$$

The informal meaning of these rules is thus the following.

- $\Box\mathcal{I}$ If, from the hypotheses in Γ, we derive that σ holds at knowledge level j, and nothing else holds at any level greater than or equal to j, then $\Box\sigma$ has to hold at level $j-1$.
- $\Box\mathcal{E}$ If, from the hypothesis in Γ, we derive that $\Box\sigma$ holds at knowledge level j, then σ has to be true at the next level.

This interpretation corresponds to the semantical property of the minimal normal modal logic K: K is complete with respect to the set of Kripke frames whose accessibility relation is 'irreflexive', 'asymmetric' and 'intransitive'.

The characteristic axiom of K, $\Box(\alpha \supset \beta) \supset (\Box\alpha \supset \Box\beta)$, is derived in $\lambda^\ell K$ in the following way.

$$\dfrac{\dfrac{\dfrac{\dfrac{\dfrac{x^1 : \Box\alpha^1}{\mathsf{ungen}(x^1)^2 : \alpha^2}\ \Box\mathcal{E} \qquad \dfrac{y^1 : \Box(\alpha\supset\beta)^1}{\mathsf{ungen}(y^1)^2 : \alpha\supset\beta^2}\ \Box\mathcal{E}}{(\mathsf{ungen}(y^1)^2\mathsf{ungen}(x^1)^2)^2 : \beta^2}\ \supset\mathcal{E}}{\mathsf{gen}((\mathsf{ungen}(y^1)^2\mathsf{ungen}(x^1)^2)^2)^1 : \Box\beta^1}\ \Box\mathcal{I}}{\lambda x^1{:}\Box\alpha^1.\mathsf{gen}((\mathsf{ungen}(y^1)^2\mathsf{ungen}(x^1)^2)^2)^1 : \Box\alpha\supset\Box\beta^1}\ \supset\mathcal{I}}{\lambda y^1{:}\Box(\alpha\supset\beta)^1.\lambda x^1{:}\Box\alpha^1.\mathsf{gen}((\mathsf{ungen}(y^1)^2\mathsf{ungen}(x^1)^2)^2)^1 : \Box(\alpha\supset\beta)\supset(\Box\alpha\supset\Box\beta)^1}\ \supset\mathcal{I}$$

2.3 System $\lambda^\ell KT$

The typed λ-calculus $\lambda^\ell KT$ is obtained from $\lambda^\ell K$ by extending the elimination rule for $\Box$.

$$\dfrac{\begin{matrix}\Gamma\\ \vdots\\ M^j : \Box\sigma^j\end{matrix}}{\mathsf{ungen}(M^j)^k : \sigma^k}\ \Box\mathcal{E}\ \ {\scriptstyle k\in\{j,j+1\}}$$

The informal meaning of such a rule is thus:

> If, from the hypothesis in Γ, we derive that $\Box\sigma$ holds at knowledge level j, then σ has to hold at the next $(j+1)$ and at the current (j) level.

We have a correspondence with the semantical property of KT: KT is complete with respect to the set of Kripke frames whose accessibility relation is 'reflexive', 'asymmetric' and 'intransitive'.

By using the above rule we can prove the characteristic axiom of KT, $\Box\alpha \supset \alpha$.

$$\dfrac{\dfrac{x^1 : \Box\alpha^1}{\mathsf{ungen}(x^1)^1 : \alpha^1}\,\Box\mathcal{E}}{(\lambda x^1 : \Box\alpha^1.\mathsf{ungen}(x^1)^1)^1 : \Box\alpha \supset \alpha^1}\,\supset\mathcal{I}$$

2.4 System $\lambda^\ell K4$

The typed λ-calculus $\lambda^\ell K4$ is obtained from $\lambda^\ell K$ by extending the elimination rule for $\Box$.

$$\dfrac{\begin{array}{c}\Gamma\\ \vdots\\ M^j : \Box\sigma^j\end{array}}{\mathsf{ungen}(M^j)^k : \sigma^k}\,\Box\mathcal{E}\quad {\scriptstyle k>j}$$

The informal meaning of such a rule is:

> If, from the hypotheses in Γ, we derive that $\Box\sigma$ holds at knowledge level j, then σ has to hold at any level greater than j.

We have a correspondence with the semantical property of K4: K4 is complete with respect to the set of frames whose accessibility relation 'irreflexive', 'asymmetric', and 'transitive'.

The characteristic K4 axiom, $\Box\alpha \supset \Box\Box\alpha$, is proved in the following way.

$$\dfrac{\dfrac{\dfrac{\dfrac{x^1 : \Box\alpha^1}{\mathsf{ungen}(x^1)^3 : \alpha^3}\,\Box\mathcal{E}}{\mathsf{gen}(\mathsf{ungen}(x^1)^3)^2 : \Box\alpha^2}\,\Box\mathcal{I}}{\mathsf{gen}(\mathsf{gen}(\mathsf{ungen}(x^1)^3)^2)^1 : \Box\Box\alpha^1}\,\Box\mathcal{I}}{(\lambda x^1 : \Box\alpha^1.\mathsf{gen}(\mathsf{gen}(\mathsf{ungen}(x^1)^3)^2)^1)^1 : \Box\alpha \supset \Box\Box\alpha^1}\,\supset\mathcal{I}$$

2.5 System $\lambda^\ell S4$

The typed λ-calculus $\lambda^\ell S4$ is obtained from $\lambda^\ell K$ by allowing both the $\lambda^\ell KT$ and the $\lambda^\ell K4$ $\Box$ elimination rules.

$$\frac{\begin{array}{c}\Gamma\\ \vdots\\ M^j : \Box\sigma^j\end{array}}{\mathsf{ungen}(M^j)^k : \sigma^k}\,\Box\mathcal{E}\quad {}_{k\geq j}$$

The informal meaning of such a rule is:

> If, from the hypotheses in Γ, we derive that $\Box\sigma$ holds at knowledge level j, then σ has to hold at each level greater than or equal to j.

We have a correspondence with the semantical property of S4: S4 is complete with respect to the set of Kripke frames whose accessibility relation is 'reflexive', 'asymmetric', and 'transitive'.

Since the characteristic axioms of S4 are those of K, KT, and K4, they can be proved in $\lambda^\ell S4$.

2.6 Implicit systems

From the systems we have introduced, we may obtain 'implicit' calculi, forgetting some of the (redundant) information. First, the implicitly typed λ-calculi $\lambda^* K$, $\lambda^* KT$, $\lambda^* K4$, and $\lambda^* S4$ can be defined, by erasing from the terms the level and type decorations. We loose a 1-to-1 correspondence between proofs and terms, but we gain in readability. The proof terms corresponding to the characteristic axioms of the various logic in the implicit form are thus the following.

$\vdash_{\lambda^* K} \lambda y.\lambda x.\mathsf{gen}(\mathsf{ungen}(y)\mathsf{ungen}(x)) : \Box(\alpha \supset \beta) \supset (\Box\alpha \supset \Box\beta)^1$

$\vdash_{\lambda^* KT} \lambda x.\mathsf{ungen}(x) : \Box\alpha \supset \alpha^1$

$\vdash_{\lambda^* K4} \lambda x.\mathsf{gen}(\mathsf{gen}(\mathsf{ungen}(x))) : \Box\alpha \supset \Box\Box\alpha^1$

Second, we may forget λ-terms altogether; the resulting systems are less verbose and we will often resort to them in the sequel. Figure 1 summarizes these systems.

3 HILBERT-STYLE SYSTEMS, A COMPARISON

In this section we show that provability in the calculi $\lambda^\ell K$, $\lambda^\ell KT$, $\lambda^\ell K4$, and $\lambda^\ell S4$ is equivalent to provability in the positive fragments of the Hilbert-style formulation of the corresponding logics. Let L be the positive fragment of one of the logics in $\{\mathsf{K}, \mathsf{KT}, \mathsf{K4}, \mathsf{S4}\}$.

For the purpose of this section, we omit the term labelling for the derivations in the calculus; moreover, if $\Gamma = \{\sigma_1^i, \ldots, \sigma_n^i\}$, then $\Gamma^\flat = \{\sigma_1, \ldots, \sigma_n\}$.

$$\sigma^k$$

$$\frac{\begin{matrix}\Gamma & \Delta \\ \vdots & \vdots \\ \sigma^k & \tau^k\end{matrix}}{\sigma \wedge \tau^k}\wedge\mathcal{I} \qquad \frac{\begin{matrix}\Gamma \\ \vdots \\ \sigma \wedge \tau^k\end{matrix}}{\sigma^k}\wedge\mathcal{E}_l \qquad \frac{\begin{matrix}\Gamma \\ \vdots \\ \sigma \wedge \tau^k\end{matrix}}{\tau^k}\wedge\mathcal{E}_r$$

$$\frac{\begin{matrix}\Gamma \quad [\sigma^k] \\ \vdots \\ \tau^k\end{matrix}}{\sigma \supset \tau^k}\supset\mathcal{I} \qquad \frac{\begin{matrix}\Gamma & \Delta \\ \vdots & \vdots \\ \sigma \supset \tau^k & \sigma^k\end{matrix}}{\tau^k}\supset\mathcal{E}$$

$$\frac{\begin{matrix}\Gamma \\ \vdots \\ \sigma^k\end{matrix}}{\Box\sigma^{k-1}}\Box\mathcal{I} \quad k>\#\Gamma \qquad \frac{\begin{matrix}\Gamma \\ \vdots \\ \Box\sigma^j\end{matrix}}{\sigma^k}\Box\mathcal{E}^\dagger$$

$\dagger$ $k = j + 1$ in $\lambda^\ell K$;
$k \in \{j, j+1\}$ in $\lambda^\ell KT$;
$k \geq j + 1$ in $\lambda^\ell K4$; $k \geq j$ in $\lambda^\ell S4$.

FIGURE 1. Term-free systems.

THEOREM 2

$$\vdash_L \sigma \iff \vdash_{\lambda^\ell L} \sigma^1$$

The proof of this fact breaks down into the following lemmas.

LEMMA 3 $\quad \vdash_L \sigma \implies \vdash_{\lambda^\ell L} \sigma^1.$

Proof. By induction on the lenght of the derivation. Propositional axioms and modus ponens are trivial. The proofs in $\lambda^\ell L$ of the modal axioms of L have been given in the previous section. For the inference rule (Gen), suppose we proved $\vdash_L \sigma$. By induction hypothesis we have $\vdash_{\lambda^\ell L} \sigma^1$. Replacing now any index level k in this deduction with $k + 1$, we obtain a deduction of $\vdash_{\lambda^\ell L} \sigma^2$. An application of $\Box\mathcal{I}$ gives $\vdash_{\lambda^\ell L} \Box\sigma^1$. ■

The following 'translation function' corresponds to the intended meaning of our judgments, as informally stated in the Introduction.

DEFINITION 4

- $\mathcal{T}[\Gamma \vdash \alpha^1] = \bigwedge \Gamma^\flat \supset \alpha$
- $\mathcal{T}[\Gamma \vdash \alpha^{j+1}] = \bigwedge (\Gamma^{<2})^\flat \supset \Box(\mathcal{T}[\Gamma^{\geq 2}(-1) \vdash \alpha^j])$

where, $\bigwedge \varnothing = \top$ *(any tautology).*

LEMMA 5 $\quad \Gamma \vdash_{\lambda^\ell L} \alpha^j \implies \vdash_L \mathcal{T}[\Gamma \vdash \alpha^j].$

Proof. By trivial, and long, induction on the deduction $\Gamma \vdash_{\lambda^\ell L} \alpha^j$. ■

4 INTERMEZZO

4.1 On the geometry of proofs

One of the most interesting properties of the systems we have proposed (or better of the level machinery), is the geometry they enforce on the deductions. In particular, we show in this section that all (not necessary normal) deductions involving a $\Box\mathcal{I}$ have a very remarkable global structure.

Before showing the technical fact we have in mind, however, we recall a different natural deduction system for S4, namely the 'third version' of the calculus discussed in Prawitz's classical monograph [14, Chapter VI, pag. 79]. The propositional rules of that system can be obtained from those of Definition 1 by simply erasing the terms and the level indexes (thus obtaining the usual rules for intuitionistic propositional logic). As for the rules for $\Box$, the elimination is the same as ours with the omission of terms and levels, while the introduction rule (which we will call $\Box\mathcal{I}_\mathcal{P}$) is rather elaborate, in order to ensure normalization and the subformula property. An application of $\Box\mathcal{I}_\mathcal{P}$ has the form:

$$\begin{array}{ccc} \mathcal{D}_1 & & \mathcal{D}_n \\ [\Box\tau_1 & \dots & \Box\tau_n] \\ & \mathcal{D} & \\ & \dfrac{\sigma}{\Box\sigma}\ \Box\mathcal{I}_{\mathcal{P}} & \end{array}$$

where $\Box\tau_1, \dots, \Box\tau_n$ are all the open assumptions of $\mathcal{D}$, and no open assumption in the deductions $\mathcal{D}_i$ (of conclusion $\Box\tau_i$) is bound in $\mathcal{D}$.[2] In other words, an application of $\Box\mathcal{I}_{\mathcal{P}}$ is obtained by taking a deduction

$$\begin{array}{ccc} \Box\tau_1 & \dots & \Box\tau_n \\ & \mathcal{D} & \\ & \sigma & \end{array}$$

where *all* open assumptions (if any) are boxed, and plugging into these assumptions arbitrary derivations with the right conclusion. Rule $\Box\mathcal{I}_{\mathcal{P}}$ is far from being 'natural': it allows normalization, but at the price of a strong globally stated constraint on its application. [3]

Coming back to our system, we can show that all applications of $\Box\mathcal{I}$, in fact, comply with Prawitz's requirement. In order to give a clearer picture of the involved deductions we will dispense from terms.

THEOREM 6 *Let $\mathcal{E}$ be a deduction of $\Gamma \vdash_{\lambda^\ell S4} \sigma^k$, with $k > \#\Gamma$. Then $\mathcal{E}$ has the form*

$$\begin{array}{ccc} \Gamma_1 & & \Gamma_n \\ \mathcal{D}_1 & & \mathcal{D}_n \\ [\Box\tau_1^{h_1} & \dots & \Box\tau_n^{h_n}] \\ & \mathcal{D} & \\ & \sigma^k & \end{array}$$

where: (i) any Γ_j is non empty; (ii) the union of the Γ_j's is Γ; (iii) any $\mathcal{D}_j$ is a deduction of $\Gamma_j \vdash_{\lambda^\ell S4} \Box\tau_j^{h_j}$; (iv) $\mathcal{D}$ is a deduction of $\Box\tau_1^{h_1}, \dots, \Box\tau_n^{h_n} \vdash_{\lambda^\ell S4} \sigma^k$; (v) for any j, $h_j < k$.

Proof. If Γ is empty, then the statement is vacuous ($n = 0$ and $\mathcal{D} = \mathcal{E}$). Otherwise, the following informal algorithm will produce the required data.

Let r be any path in $\mathcal{E}$ from the conclusion σ^k to an occurrence of an open assumption $\rho \in \Gamma$. In r there must be an application of $\Box\mathcal{E}$, since $k > \#\Gamma$ and $\Box\mathcal{E}$

[2] Prawitz's original definition is formulated via the concept of *essentially modal formula.* It is not difficult to see that our formulation of $\Box\mathcal{I}_{\mathcal{P}}$, though not equivalent (it allows less deductions), it is still complete for the $\wedge, \supset, \Box$ fragment.

[3] In order to introduce $\Box$ we must look *into* the proof tree.

is the only rule allowing its conclusion to be at a higher level of its premise. Pick the first application (starting from the conclusion up) of $\Box\mathcal{E}$ in which the level h of the premise (say, $\Box\tau^h$) is strictly less than k. We choose this $\Box\tau^h$ as $\Box\tau_1^{h_1}$; $\mathcal{D}_1$ is the subtree of $\mathcal{E}$ rooted at $\Box\tau^h$; Γ_1 is the set of all open assumptions of the deduction $\mathcal{D}_1$ with conclusion $\Box\tau^h$. Clearly $\rho \in \Gamma_1$, thus establishing (i) and (iii) for $j = 1$; in order to eventually establish (iv), we have to guarantee that no assumption in Γ_1 can be discharged below $\Box\tau^h$. However, this is obvious, because any formula appearing in the path r below $\Box\tau^h$ have a level $l > h$ (by construction), while $\#\Gamma_1 \leq h$, (see Proposition 9), thus forbidding the application of a $\supset \mathcal{E}$ at level l below $\Box\tau^h$.

The previous procedure can be repeated until there are no more occurrences of assumptions in Γ not already allocated to some Γ_i.

When the algorithm terminates, (ii) is obvious, and (iv) follows by construction. ■

Writing $\vdash_{\mathcal{P}}$ for provability in Prawitz's system and denoting with $(\quad)^\flat$ the obvious function stripping the level indexes out of a formula (deduction, etc.), the following is straightforward.

COROLLARY 7 *If $\mathcal{D}$ is a deduction of $\Gamma \vdash_{\lambda^\ell S4} \sigma^k$, then $\mathcal{D}^\flat$ is a deduction of $\Gamma^\flat \vdash_{\mathcal{P}} \sigma$.*

Not any proof in $\vdash_{\mathcal{P}}$ can be decorated with levels to become a proof in our system, as the following deduction shows:

$$\frac{\Box\sigma}{\Box\Box\sigma}\ \Box\mathcal{I}_{\mathcal{P}}$$

Proofs in our system, however, are still enough for completeness and, moreover, seem to have a more direct interpretation as travelling in a Kripke model (cfr. [3]).

Many consequences can be drawn from Corollary 7. Since any reduction step (see Section 6) in our system becomes also a reduction step in $\vdash_{\mathcal{P}}$ under the $(\quad)^\flat$ translation, we immediately obtain normalization for $\lambda^\ell S4$ from the normalization theorem for $\vdash_{\mathcal{P}}$. We will give a different proof of *strong normalization* for $\lambda^\ell S4$ in Section 6.2

Moreover, by adding to the usual clauses that, for any k, σ^k has to be considered a subformula of σ^h, we also obtain the following *subformula property*.

THEOREM 8 (Subformula principle) *Any formula occurring in a normal deduction of $\Gamma \vdash_{\lambda^\ell S4} \sigma^k$ is a subformula of σ^k or of some formula in Γ.*

Proof.[Sketch] It is enough to prove the theorem for $\vdash_{\mathcal{P}}$, which can be obtained in the same way as the corresponding result for intuitionistic logic [14, Theorem 2, p. 53; cf. p. 80]. ■

Starting from a purely structural extension of the syntax (the levels) we have thus recovered the global, elaborated constraint on the *geometry* of the deductions that other systems are forced to require from the beginning.

4.2 A proof theoretical request... fulfilled

One of the principal features of a good modal proof theory is its capability to suit a class of different modal logics without changing the way modal connectives are manipulated (e.g. introduced and eliminated). The differentiation between logics is delegated instead to the way (general) formulas are manipulated.

A clear discussion on this topics (in the case of modal systems) may be found in recent work of Wansing [17] and Došen [6]. This point of view (that [17] calls *Došen principle*, but that should be traced back to Gentzen, for the differentiation of intuitionistic from classical logic) may be stated as:

> The rules for the logical operations are never changed: all changes are made in the structural rules. [6]

At first look our calculi violate this principle: different systems are obtained with different elimination rules.

One could attempt a 'weak' defense, arguing that all the proposed elimination rules are equal in 'spirit', since they differ only in level decorations. But this would miss the point. Indeed, we totally agree with Došen principle and we claim that our approach does not violate it... when observed in the right context. The distinction between logical and structural rules, in fact, is only relevant in *sequent* calculi, and it is thus there that the challenge posed by Došen principle has to be attacked. The natural deduction system $\lambda^{\ell}K$ has an associated *2-sequent* calculus (see [12]), whose left and right $\Box$ introduction rules are the following:

$$\dfrac{\begin{array}{ccc} \mathfrak{S}_1 & & \varepsilon \\ \mathfrak{S}_2 & & \varepsilon \\ \vdots & \vdash & \vdots \\ \mathfrak{S}_{k-1} & & \varepsilon \\ \mathfrak{S}_k, \sigma & & \tau \end{array}}{\begin{array}{ccc} \mathfrak{S}_1 & & \varepsilon \\ \mathfrak{S}_2 & & \varepsilon \\ \vdots & \vdash & \vdots \\ \mathfrak{S}_{k-1}, \Box\sigma & & \varepsilon \\ \mathfrak{S}_k & & \tau \end{array}}\ \Box\vdash \qquad \dfrac{\begin{array}{ccc} \mathfrak{S}_1 & & \varepsilon \\ \mathfrak{S}_2 & & \varepsilon \\ \vdots & \vdash & \vdots \\ \mathfrak{S}_k & & \varepsilon \\ \varepsilon & & \sigma \end{array}}{\begin{array}{ccc} \mathfrak{S}_1 & & \varepsilon \\ \mathfrak{S}_2 & & \varepsilon \\ \vdots & \vdash & \vdots \\ \mathfrak{S}_k & & \Box\sigma \end{array}}\ \vdash\Box$$

Sequent calculi for KT and K4 are now obtained by adding the *structural rules* (on levels) $\vdash\!\uparrow(KT)$ and $\vdash\!\downarrow(T4)$, respectively:

$$
\dfrac{\begin{array}{ccc}
\mathfrak{S}_1 & & \varepsilon \\
\mathfrak{S}_2 & & \varepsilon \\
\vdots & \vdash & \vdots \\
\mathfrak{S}_k & & \varepsilon \\
\varepsilon & & \sigma
\end{array}}{\begin{array}{ccc}
\mathfrak{S}_1 & & \varepsilon \\
\mathfrak{S}_2 & & \varepsilon \\
\vdots & \vdash & \vdots \\
\mathfrak{S}_k & & \sigma
\end{array}} \vdash\!\uparrow (KT)
\qquad\qquad
\dfrac{\begin{array}{ccc}
\mathfrak{S}_1 & & \varepsilon \\
\mathfrak{S}_2 & & \varepsilon \\
\vdots & \vdash & \vdots \\
\mathfrak{S}_k & & \varepsilon \\
\varepsilon & & \sigma
\end{array}}{\begin{array}{ccc}
\mathfrak{S}_1 & & \varepsilon \\
\mathfrak{S}_2 & & \varepsilon \\
\vdots & \vdash & \vdots \\
\mathfrak{S}_k & & \varepsilon \\
\varepsilon^n & & \varepsilon^n \\
\varepsilon & & \sigma
\end{array}} \vdash\!\downarrow (K4)
$$

where $n \geq 1$ and

$$
\varepsilon^n = \left.\begin{array}{c} \varepsilon \\ \vdots \\ \varepsilon \end{array}\right\} n \text{ times}
$$

The calculus for S4 is obtained by adding to the basic logical rules both $\vdash\!\uparrow (KT)$ and $\vdash\!\downarrow (T4)$. Thus, in a sequent setting, the four systems *are* obtained by means of structural rules.

The structural rules of sequent calculi, on the other hand, should not be explicit in the corresponding natural deduction formulation. Intuitionistic contractions and weakenings, for instance, are realized by the convention on the discharging (meta-) operation. This is why our calculi do not contain rules like

$$
\dfrac{\begin{array}{c}\Gamma \\ \vdots \\ \sigma^r\end{array}}{\sigma^j}
$$

Following the traditional natural deduction approach, such rules have been embedded into the (logical) $\Box\mathcal{E}$ rules.

5 WORKING WITH $\lambda^{\ell}S4$ PROOFS

In this section $\vdash$ will stand for $\vdash_{\lambda^{\ell}S4}$, whose rules are recalled in Figure 2.

PROPOSITION 9 *Let* $\Gamma \vdash M^i : \sigma^i$. *Then* $i \geq \#\Gamma$.

Proof. Inspection of the rules. ∎

DEFINITION 10 *A derivation* $\Gamma \vdash M^i : \sigma^i$ *is* concluded *if* $i = 1$ *and* $\#\Gamma \leq 1$.

$$x : \sigma^k$$

$$\frac{\begin{matrix}\Gamma & & \Delta\\ \vdots & & \vdots \\ M : \sigma^k & & N : \tau^k\end{matrix}}{\langle M, N\rangle : \sigma \wedge \tau^k}\wedge\mathcal{I} \qquad \frac{\begin{matrix}\Gamma\\ \vdots \\ M : \sigma \wedge \tau^k\end{matrix}}{(\mathsf{fst}\, M) : \sigma^k}\wedge\mathcal{E}_l \qquad \frac{\begin{matrix}\Gamma\\ \vdots \\ M : \sigma \wedge \tau^k\end{matrix}}{(\mathsf{snd}\, M) : \tau^k}\wedge\mathcal{E}_r$$

$$\frac{\begin{matrix}\Gamma \quad [x : \sigma^k]\\ \vdots \\ M : \tau^k\end{matrix}}{(\lambda x^k : \sigma^k . M)^k : \sigma \supset \tau^k}\supset\mathcal{I} \qquad \frac{\begin{matrix}\Gamma & & \Delta\\ \vdots & & \vdots \\ M : \sigma \supset \tau^k & & N : \sigma^k\end{matrix}}{(MN)^k : \tau^k}\supset\mathcal{E}$$

$$\frac{\begin{matrix}\Gamma\\ \vdots \\ M : \sigma^j\end{matrix}}{\mathsf{gen}(M)^{j-1} : \Box\sigma^{j-1}}\Box\mathcal{I}\ {\scriptstyle j > \#\Gamma} \qquad \frac{\begin{matrix}\Gamma\\ \vdots \\ M : \Box\sigma^j\end{matrix}}{\mathsf{ungen}(M)^k : \sigma^k}\Box\mathcal{E}\ {\scriptstyle k \geq j}$$

FIGURE 2. Rules for $\lambda^\ell S4$

THEOREM 11 *Any non concluded derivation can be extended to a concluded one.*

Proof. Given a non concluded $\Gamma \vdash \sigma^k$, first replace any $\sigma^h \in \Gamma$ with:

$$\frac{\Box\sigma^1}{\sigma^h}\Box\mathcal{E}$$

At this point apply $k-1$ times the rule $\Box\mathcal{I}$ to the conclusion, obtaining a formula of level 1. ■

The following technical lemmas aim to show, given a deduction for $\Gamma \vdash \sigma^k$, how to obtain a deduction for the same conclusion at a different level, possibly changing also some of the levels in Γ. We will see in Section 6 that this operation is essential for the computational interpretation of proofs.

At first sight the operations on proof trees given in this section may appear heavy and difficult to grasp. Notice, however, that the problem we tackle is essentially the same as the one arising in the first order natural deduction system NJ, when, given a deduction for $\Gamma \vdash \sigma(x)$ and a term t, we want to obtain a deduction for $\Gamma' \vdash \sigma(t)$. Many texts say, in this case, that all it is needed is the uniform substitution of t for x in the proof tree. A careful analysis of the process involved, however, shows that care must be taken in performing this operation; in particular, it is needed a strong discipline in the use of variables. A very clear account of the issues involved is given in [16, Volume II, Ch. 10]. Levels in $\lambda^{\ell}S4$ play a very similar role to variables of NJ; the following lemmas can then be seen as expressing concepts like 'renaming', 'term substitution' and their properties.

LEMMA 12 *Let* $k > 0$. $\Gamma \vdash \alpha^j \quad \Longrightarrow \quad \forall i \leq j \quad \Gamma^{<i}, \Gamma^{\geq i}(+k) \vdash \alpha^{j+k}$

Proof.

- $\alpha^j \vdash \alpha^j \Rightarrow \alpha^{j+k} \in \Gamma^{\geq i}(+k)$; then $\alpha^{j+k} \vdash \alpha^{j+k}$;

- if $\dfrac{\begin{matrix}\Gamma, \alpha^j \\ \vdots \\ \beta^j\end{matrix}}{\alpha \supset \beta^j}$ then $I.H.\left\{\begin{matrix}\Gamma^{<i}, \Gamma^{\geq i}(+k), \alpha^{j+k} \\ \vdots \\ \beta^{j+k}\end{matrix}\right.$ $\dfrac{}{\alpha \supset \beta^{j+k}}$

- if $\dfrac{\begin{matrix}\Gamma \\ \vdots \\ \alpha^j\end{matrix}}{\Box\alpha^{j-1}}$ then $I.H.\left\{\begin{matrix}\Gamma^{<i}, \Gamma^{\geq i}(+k) \\ \vdots \\ \alpha^{j+k}\end{matrix}\right.$ $\dfrac{}{\Box\alpha^{j-1+k}}$

- if $\dfrac{\begin{matrix}\Gamma \\ \vdots \\ \Box\alpha^j\end{matrix}}{\alpha^r}$ we have two subcases:

  $(i \leq j)$ in this case we have:

  $I.H.\left\{\begin{matrix}\Gamma^{<i}, \Gamma^{\geq i}(+k) \\ \vdots \\ \Box\alpha^{j+k}\end{matrix}\right.$ $\dfrac{}{\alpha^{r+k}}$

  $(j < i \leq r)$ in this case, without applying the induction hypothesis and noting that $\Gamma^{\geq i} = \varnothing$ by Proposition 9, we have:

  $$\dfrac{\begin{matrix}\Gamma^{<i}, \Gamma^{\geq i}(+k) \\ \vdots \\ \Box\alpha^j\end{matrix}}{\alpha^{r+k}}$$

∎

LEMMA 13 *Let* $j \geq 2$. $\Gamma \vdash \alpha^j \quad \Longrightarrow \quad \forall i\,(2 \leq i \leq j)\ \Gamma^{<i}, \Gamma^{\geq i}(-1) \vdash \alpha^{j-1}$

Proof.

- $\alpha^j \vdash \alpha^j \Rightarrow \alpha^{j-1} \in \Gamma^{\geq i}(-1)$ then $\alpha^{j-1} \vdash \alpha^{j-1}$

- if $\dfrac{\begin{array}{c}\Gamma, \alpha^j \\ \vdots \\ \beta^j\end{array}}{\alpha \supset \beta^j}$ then $I.H. \left\{ \dfrac{\begin{array}{c}\Gamma^{<i}, \Gamma^{\geq i}(-1), \alpha^{j-1} \\ \vdots \\ \beta^{j-1}\end{array}}{\alpha \supset \beta^{j-1}} \right.$

- if $\dfrac{\begin{array}{c}\Gamma \\ \vdots \\ \alpha^j\end{array}}{\Box\alpha^{j-1}}$ then $I.H. \left\{ \dfrac{\begin{array}{c}\Gamma^{<i}, \Gamma^{\geq i}(-1) \\ \vdots \\ \alpha^{j-1}\end{array}}{\Box\alpha^{j-2}} \right.$

- if $\dfrac{\begin{array}{c}\Gamma \\ \vdots \\ \Box\alpha^j\end{array}}{\alpha^r}$ we have two subcases:

  $(j \geq 2, i \leq j)$ in this case we have:

  $$I.H. \left\{ \dfrac{\begin{array}{c}\Gamma^{<i}, \Gamma^{\geq i}(-1) \\ \vdots \\ \Box\alpha^{j-1}\end{array}}{\alpha^{r-1}} \right.$$

  $(j < i \leq r)$ in this case, without applying the induction hypothesis and noting that $\Gamma^{\geq i} = \varnothing$ by Proposition 9, we have:

  $$\dfrac{\begin{array}{c}\Gamma^{<i}, \Gamma^{\geq i}(+k) \\ \vdots \\ \Box\alpha^j\end{array}}{\alpha^{r-1}}$$

■

The operations on deductions given in the proof of the two previous lemmas is explicitated on proof terms by the following definition.

DEFINITION 14 (Level substitution) *Let* $\Gamma \vdash M : \sigma^v$; *let* $n \in \{-1\} \cup \mathbb{N}$, $i \geq 2$ *if* $v \geq 2$, *and* $n \in \mathbb{N}, i \geq 1$ *otherwise. We inductively define the* level substitution $[n]_i M$ *(read: increment by* n *any level greater than or equal to* i*).*
If $v < i$:

$$[n]_i M^v = M^v$$

If $v \geq i$:

$$[n]_i x^v = x^{v+n}$$

$$[n]_i \langle M, N\rangle^v = \langle [n]_i M, [n]_i N\rangle^{v+n}$$

$$[n]_i (\mathsf{fst}\, M)^v = (\mathsf{fst}\, [n]_i M)^{v+n}$$

$$[n]_i (\mathsf{snd}\, M)^v = (\mathsf{snd}\, [n]_i M)^{v+n}$$

$$[n]_i (\lambda x^v : \sigma^v . M)^v = (\lambda x^{v+n} : \sigma^{v+n} . [n]_i M)^{v+n}$$

$$[n]_i (M N)^v = ([n]_i M\, [n]_i N)^{v+n}$$

$$[n]_i \mathsf{gen}(M)^v = \mathsf{gen}([n]_i M)^{v+n}$$

$$[n]_i \mathsf{ungen}(M)^v = \mathsf{ungen}([n]_i M)^{v+n}$$

Whenever we write a level substitution $[n]_i M$, we assume that all the constraints of the definition are satisfied (in particular those on n). Note that $[0]_i M^v = M^v$.

Lemmas 12 and 13 can then be reformulated as follows.

LEMMA 15 $\Gamma \vdash M : \alpha^j \implies \forall i \leq j, \quad \Gamma^{<i}, \Gamma^{\geq i}(+k) \vdash [+k]_i M : \alpha^{j+k}$

LEMMA 16 *Let* $j \geq 2$, *then:* $\Gamma \vdash M : \alpha^j \implies \forall i, 2 \leq i \leq j, \quad \Gamma^{<i}, \Gamma^{\geq i}(-1) \vdash [-1]_i M : \alpha^{j-1}$

We can picture the effect of the level substitution on proof trees as follows, observing first that Theorem 6 can be generalized.

THEOREM 17 *Let $\mathcal{E}$ be a deduction of $\Gamma \vdash \sigma^k$, and let $i \leq k$. Then $\mathcal{E}$ has the form*

$$\begin{array}{ccc} \Gamma_1^{<i} & & \Gamma_n^{<i} \\ \mathcal{D}_1 & & \mathcal{D}_n \\ [\Box\tau_1^{h_1} & \ldots & \Box\tau_n^{h_n}] \quad \Gamma^{\geq i} \\ & \mathcal{D} & \\ & \sigma^k & \end{array}$$

where: (i) any $\Gamma_j^{<i}$ is non empty; (ii) the union of the $\Gamma_j^{<i}$'s is $\Gamma^{<i}$; (iii) any $\mathcal{D}_j$ is a deduction of $\Gamma_j^{<i} \vdash \Box\tau_j^{h_j}$; (iv) $\mathcal{D}$ is a deduction of $\Box\tau_1^{h_1}, \ldots, \Box\tau_n^{h_n}, \Gamma^{\geq i} \vdash \sigma^k$; (v) for any j, $h_j < i$.

Proof. A simple variant of the proof of Theorem 6, or, for the reader preferring recursion to iteration, by a routine induction on the lenght of $\mathcal{E}$, similarly to Lemmas 12 and 13. ■

In the same hypotheses of the theorem, now, the deduction $[m]_i\mathcal{E}$ is

$$\dfrac{\begin{matrix}\Gamma_1^{<i} & & \Gamma_n^{<i} & \\ \mathcal{D}_1 & & \mathcal{D}_n & \\ [\Box\tau_1^{h_1} & \ldots & \Box\tau_n^{h_n}] & \Gamma^{\geq i}(m)\end{matrix}}{\begin{matrix}[m]_i\mathcal{D} \\ \sigma^k\end{matrix}}$$

where *all* the formulas appearing in $\mathcal{D}$ are affected by the level modification, while all the $\mathcal{D}_j$'s deductions are left unchanged.

Next lemma shows how level substitution interacts with term substitution.

LEMMA 18 $([n]_iM)[[n]_iN/x^{k+n}] = [n]_i(M[N/x^k])$

Proof. Just note that if $x^k \in FV(M)$ then any subterm of M containing x^k has a level $h \geq k$. If $x^k \notin FV(M)$ then the variable-level-convention ensures the thesis. ■

We state, next, a property of nested level substitutions.

LEMMA 19 (Level substitution lemma) *Let* $k \geq 1$, $n \geq -1$, *and* $j \geq 0$.

(i) If $i \leq k-1$, *then for any* $l \geq k$, $[n]_i[j-1]_kM^l = [j-1]_{k+n}[n]_iM^l$.

(ii) If $k-1 < i \leq k-1+j$, *then for any* $l \geq k$, $[n]_i[j-1]_kM^l = [j-1+n]_kM^l$.

6 COMPUTATIONS

In the previous section a class of modal typed λ-calculi has been introduced. It is now necessary to study their computational behavior, by introducing suitable notions of reduction extending the usual typed β-reduction and corresponding to the process of proof normalization. We will be able to show that this reduction is *confluent* (or Church-Rosser) and *strongly normalizing* (or nötherian).

DEFINITION 20 (Contraction) $((\lambda x^j : \sigma^j.M^j)^j N^j)^j \triangleright M^j[N^j/x^j]$

$(\mathsf{fst}\,\langle M^j, N^j\rangle^j)^j \triangleright M^j$

$(\mathsf{snd}\,\langle M^j, N^j\rangle^j)^j \triangleright N^j$

$\mathsf{ungen}(\mathsf{gen}(M^k)^{k-1})^{k-1+j} \triangleright [j-1]_kM^k$ *(for* $j \in \mathbb{N}$*).*

The compatible closure of $\triangleright$ *(or* one-step reduction*) is denoted with* $\rightarrow$*; its transitive closure is* $\xrightarrow{+}$*, while* $\twoheadrightarrow$ *(the* reduction relation*) is the transitive and reflexive closure.*

The following theorem expresses the correctness of Definition 20 with respect to types.

THEOREM 21 *If* $\Gamma \vdash M^j : \sigma^j$ *and* $M^j \twoheadrightarrow N^j$*, then* $\Gamma \vdash N^j : \sigma^j$.

Proof. The only non standard case is when the reduction is a modal contraction. From $M = \mathsf{ungen}(\mathsf{gen}(P^k)^{k-1})^{k-1+j}$ we have $\Gamma \vdash P^k : \Box\sigma^k$ and $k > \#\Gamma$. The level substitution in $[j-1]_k P^k$, then, does not affect the free variables of P and Lemmas 15 and 16 give the thesis. ■

Reduction and level substitution match well: the former is preserved by the latter, as the following results will show.

LEMMA 22 *If* $M^h \rhd N^h$*, then for any* $i \leq h$ *and any* n, $[n]_i M^h \rhd [n]_i N^h$.

LEMMA 23 *If* $M^h \to N^h$ *then for any* $i \leq h$ *and any* n, $[n]_i M^h \to [n]_i N^h$.

THEOREM 24 *If* $M^h \twoheadrightarrow N^h$*, then for any* $i \leq h$ *and any* n, $[n]_i M^h \twoheadrightarrow [n]_i N^h$.

Proof. By induction on the lenght of the reduction, using Lemma 23. ■

6.1 Confluence

We prove the Church-Rosser property for $\twoheadrightarrow$, using Tait's technique as formulated in [9] or [1], here adapted to modal terms. We define first a new auxiliary notion of reduction ($\rightrightarrows$), corresponding to the (possible) parallel contraction of several non overlapping redexes; note that $\rightrightarrows$ is not transitive. Then, $\rightrightarrows$ is shown to respect both term and level substitution; from this we prove that ' $\rightrightarrows$ satisfy the diamond property', Lemma 28. Since $\twoheadrightarrow$ is obviously the transitive and reflexive closure of $\rightrightarrows$, we have a standard proof that $\twoheadrightarrow$ is Church-Rosser.

In the following proofs we will give only the main (modal) cases. For the others, which are standard, the reader is referred to the literature.

DEFINITION 25 *The* one-step parallel reduction *is defined by the following clauses.*

i. $M : \alpha^k \rightrightarrows M : \alpha^k$;

ii. *If* $M : \alpha^k \rightrightarrows M' : \alpha^k$ *and* $N : \beta^k \rightrightarrows N' : \beta^k$*, then* $\langle M, N\rangle^k : \alpha \wedge \beta^k \rightrightarrows \langle M', N'\rangle^k : \alpha \wedge \beta^k$;

iii. *If* $M : \beta^k \rightrightarrows M' : \beta^k$*, then* $(\lambda x^k{:}\alpha^k.M)^k : \alpha \supset \beta^k \rightrightarrows (\lambda x^k{:}\alpha^k.M')^k : \alpha \supset \beta^k$;

iv. *If* $M : \alpha^k \rightrightarrows M' : \alpha^k$*, then* $\mathsf{gen}(M)^{k-1} : \Box\alpha^{k-1} \rightrightarrows \mathsf{gen}(M')^{k-1} : \Box\alpha^{k-1}$;

v. *If* $M:\alpha\wedge\beta^k \rightrightarrows M':\alpha\wedge\beta^k$, *then* $(\mathsf{fst}\,M)^k:\alpha^k \rightrightarrows (\mathsf{fst}\,M')^k:\alpha^k$ *and* $(\mathsf{snd}\,M)^k:\beta^k \rightrightarrows (\mathsf{snd}\,M')^k:\beta^k$;

vi. *If* $M:\alpha^k \rightrightarrows M':\alpha^k$ *and* $N:\alpha\supset\beta^k \rightrightarrows N':\alpha\supset\beta^k$, *then* $(NM)^k:\beta^k \rightrightarrows (N'M')^k:\beta^k$;

vii. *If* $M:\Box\alpha^k \rightrightarrows M':\Box\alpha^k$, *then* $\mathsf{ungen}(M)^w:\alpha^w \rightrightarrows \mathsf{ungen}(M')^w:\alpha^w$;

viii. *If* $M:\alpha^k \rightrightarrows M':\alpha^k$ *and* $N:\beta^k \rightrightarrows N':\beta^k$, *then* $(\mathsf{fst}\,\langle M,N\rangle^k)^k:\alpha^k \rightrightarrows M':\alpha^k$ *and* $(\mathsf{snd}\,\langle M,N\rangle^k)^k:\beta^k \rightrightarrows N':\beta^k$;

ix. *If* $M:\beta^k \rightrightarrows M':\beta^k$ *and* $N:\alpha^k \rightrightarrows N':\alpha^k$, *then* $((\lambda x^k{:}\alpha^k.M)^k N^k)^k:\beta^k \rightrightarrows M'[N'/x^k]:\beta^k$;

x. *If* $M:\alpha^k \rightrightarrows M':\alpha^k$, *then* $\mathsf{ungen}(\mathsf{gen}(M)^{k-1})^{k-1+j}:\alpha^{k-1+j} \rightrightarrows [j-1]_k M':\alpha^{k-1+j}$.

As usual, a *derivation* of $M:\beta^h \rightrightarrows M':\beta^h$ is a sequence of reductions $M_0:\beta^h \rightrightarrows M_1:\beta^h,\ldots,M_{n-1}:\beta^h \rightrightarrows M_n:\beta^h, M:\beta^h \rightrightarrows M':\beta^h$, such that any reduction of the sequence is either an axiom (clause i) or it is obtained by a clause from some previous reduction(s).

LEMMA 26 *If* $N:\alpha^k \rightrightarrows N':\alpha^k$, $M:\beta^h \rightrightarrows M':\beta^h$, *and* $x^k:\alpha^k$, *then*

$$M[N/x^k]:\beta^h \rightrightarrows M'[N'/x^k]:\beta^h.$$

Proof. By induction on the lenght of the derivation of $M:\beta^h \rightrightarrows M':\beta^h$.

Base case: $M=M'$. This case is established by a structural induction on M.

Induction step: by cases on the last clause used to derive $M:\beta^h \rightrightarrows M':\beta^h$. We give some cases, labelled by the clause used.

(iv) $\mathsf{gen}(M)^h:\Box\alpha^h \rightrightarrows \mathsf{gen}(M')^h:\Box\alpha^h$, where $M:\alpha^{h+1} \rightrightarrows M':\alpha^{h+1}$. By induction hypothesis, $M[N/x^k]:\alpha^{h+1} \rightrightarrows M'[N'/x^k]:\alpha^{h+1}$, from which the thesis, since $\mathsf{gen}(M)^h[N/x^k]=\mathsf{gen}(M[N/x^k])^h:\Box\alpha^h$ and similarly for M'.

(vii) similarly to the previous case.

(x) $\mathsf{ungen}(\mathsf{gen}(M)^{h-1})^{h-1+j}:\alpha^{h-1+j} \rightrightarrows [j-1]_h M':\alpha^{h-1+j}$, where $M:\alpha^h \rightrightarrows M':\alpha^h$.

The thesis is now given by the following diagram:

$$\begin{array}{ccc}
\mathsf{ungen}(\mathsf{gen}(M)^{h-1})^{h-1+j}[N/x^k]:\alpha^{h-1+j} & \overset{?}{\rightrightarrows} & ([j-1]_h M')[N'/x^k]:\alpha^{h-1+j} \\
\| & & \|\ (*) \\
\mathsf{ungen}(\mathsf{gen}(M)^{h-1}[N/x^k])^{h-1+j}:\alpha^{h-1+j} & \rightrightarrows & [j-1]_h(M'[N'/x^k]):\alpha^{h-1+j}
\end{array}$$

where the $\rightrightarrows$ in the bottom row is given by induction hypothesis and clause (x), while equality (*) holds as either x^k does not occur in M, or $k < h$.

■

LEMMA 27 *If* $M : \alpha^k \rightrightarrows M' : \alpha^k$, *then for any* i *and any* n

$$[n]_i M : \alpha^{k+n} \rightrightarrows [n]_i M' : \alpha^{k+n}$$

Proof. If $i > k$ the statement is trivial; otherwise we proceed by induction on the length of the derivation of $M : \alpha^k \rightrightarrows M' : \alpha^k$.

Basis: If $M = M'$ the thesis is trivial.

Induction step: we give only the modal cases, labelled by the last clause used.

(vii) $M = \mathsf{ungen}(N^{k-j})^k : \alpha^k$ and $M' = \mathsf{ungen}(N'^{k-j})^k : \alpha^k$, with $N^{k-j} : \beta^{k-j} \rightrightarrows N'^{k-j} : \beta^{k-j}$.
The thesis is given by the following diagram:

$$\begin{array}{ccc}
[n]_i\mathsf{ungen}(N^{k-j})^k : \alpha^{k+n} & \overset{?}{\rightrightarrows} & [n]_i\mathsf{ungen}(N'^{k-j})^k : \alpha^{k+n} \\
\| & & \| \\
\mathsf{ungen}([n]_i N^{k-j})^{k+n} : \alpha^{k+n} & \overset{\text{by IH}}{\rightrightarrows} & \mathsf{ungen}([n]_i N'^{k-j})^{k+n} : \alpha^{k+n}
\end{array}$$

(x) $M = \mathsf{ungen}(\mathsf{gen}(N^{k-j+1})^{k-j})^k : \alpha^k$ and $M' = [j-1]_{k-j+1} N'^{k-j+1} : \alpha^k$,
with $N^{k-j+1} : \alpha^{k-j+1} \rightrightarrows N'^{k-j+1} : \alpha^{k-j+1}$, which by IH implies that

(2) $[n]_i N^{k-j+1} : [n]_i \alpha^{k-j+1} \rightrightarrows [n]_i N'^{k-j+1} : [n]_i \alpha^{k-j+1}$

We have two cases. If $i \leq k-j$ the thesis is given by the following diagram:

$$\begin{array}{ccc}
[n]_i\mathsf{ungen}(\mathsf{gen}(N^{k-j+1})^{k-j})^k : \alpha^{k+n} & \overset{?}{\rightrightarrows} & [n]_i[j-1]_{k-j+1} N'^{k-j+1} : \alpha^{k+n} \\
\| & & \| \ (*) \\
\mathsf{ungen}(\mathsf{gen}([n]_i N^{k-j+1})^{k-j+n})^{k+n} : \alpha^{k+n} & \overset{(**)}{\rightrightarrows} & [j-1]_{k-j+1+n}[n]_i N'^{k-j+1} : \alpha^{k+n}
\end{array}$$

where (*) is Lemma 19(i), and (**) follows by (6.1) and clause x. If $i > k-j$ the following diagram establishes the thesis:

$$\begin{array}{ccc}
[n]_i\mathsf{ungen}(\mathsf{gen}(N^{k-j+1})^{k-j})^k : \alpha^{k+n} & \overset{?}{\rightrightarrows} & [n]_i[j-1]_{k-j+1} N'^{k-j+1} : \alpha^{k+n} \\
\| & & \| \ (*) \\
\mathsf{ungen}(\mathsf{gen}(N^{k-j+1})^{k-j})^{k+n} : \alpha^{k+n} & \overset{(**)}{\rightrightarrows} & [j-1+n]_{k-j+1} N'^{k-j+1} : \alpha^{k+n}
\end{array}$$

where (*) is Lemma 19(ii), and (**) follows by 6.1 and clause x.

■

LEMMA 28 (Diamond Property) *Suppose* $M : \alpha^k \rightrightarrows M' : \alpha^k$ *and* $M : \alpha^k \rightrightarrows M'' : \alpha^k$. *Then we can find a term* $M''' : \alpha^k$ *such that*

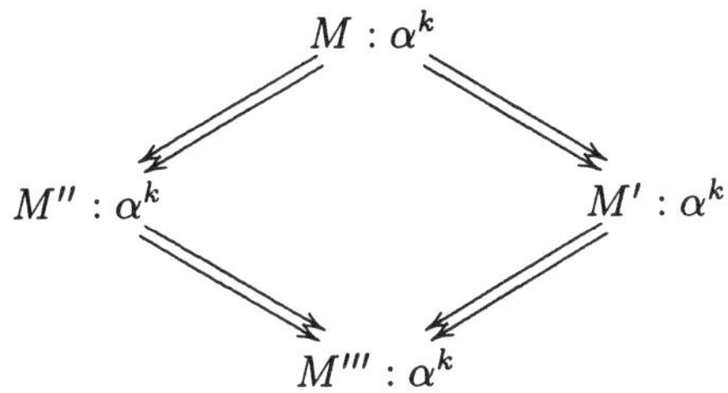

Proof. Let n' and n'' be the lenght of the derivations r' of $M : \alpha^k \rightrightarrows M' : \alpha^k$ and r'' of $M : \alpha^k \rightrightarrows M'' : \alpha^k$. The thesis is extablished by induction on $n' + n''$. We give only the modal cases. We write $M : \alpha^k \overset{n}{\rightrightarrows} M' : \alpha^k$ to denote that $M : \alpha^k \rightrightarrows M' : \alpha^k$ is established with a deduction of lenght n.

- Both r' and r'' end with clause (iv).

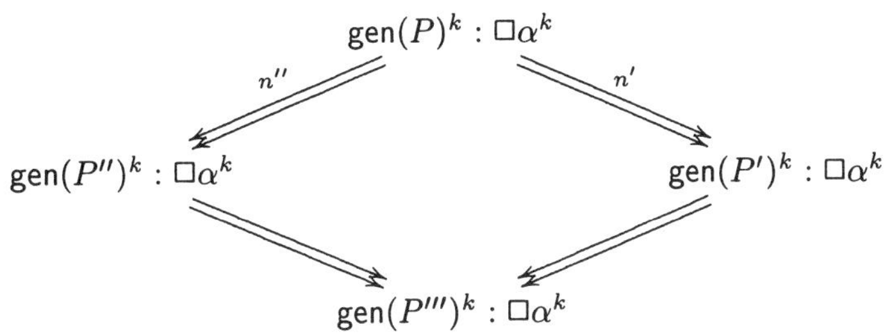

where

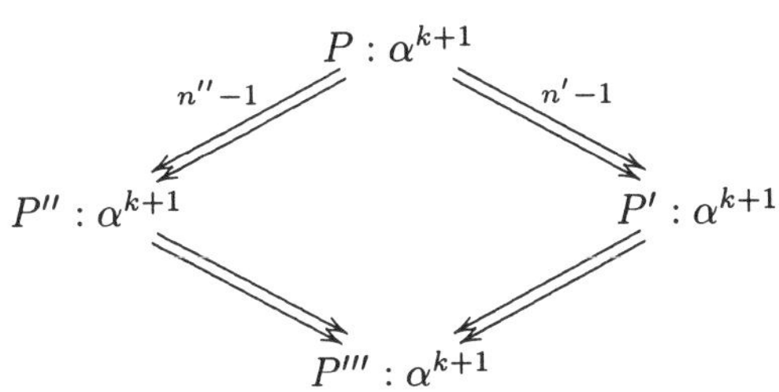

is obtained by I.H.

- Both r' and r'' end with clause (vii). As the previous case.
- Both r' and r'' end with clause (x).

$$\begin{array}{ccccc} & & \mathsf{ungen}(\mathsf{gen}(P)^{k-j})^k : \alpha^k & & \\ & \swarrow\!\!\swarrow^{n''} & & \searrow\!\!\searrow^{n'} & \\ [j-1]_{k-j+1}P'' : \alpha^k & & & & [j-1]_{k-j+1}P' : \alpha^k \\ & \searrow\!\!\searrow & & \swarrow\!\!\swarrow & \\ & & [j-1]_{k-j+1}P''' : \alpha^k & & \end{array}$$

Where

$$\begin{array}{ccccc} & & P : \alpha^{k-1} & & \\ & \swarrow\!\!\swarrow^{n''-1} & & \searrow\!\!\searrow^{n'-1} & \\ P'' : \alpha^{k-1} & & & & P' : \alpha^{k-1} \\ & \searrow\!\!\searrow & & \swarrow\!\!\swarrow & \\ & & P''' : \alpha^{k-1} & & \end{array}$$

is obtained by I.H. and

$$[j-1]_{k-j+1}P' : \alpha^k \rightrightarrows [j-1]_{k-j+1}P''' : \alpha^k \leftleftarrows [j-1]_{k-j+1}P'' : \alpha^k$$

holds by Lemma 27.

- Last clause of r' is (x) and last clause of r'' is (vii).

$$\begin{array}{ccccc} & & \mathsf{ungen}(\mathsf{gen}(P)^{k-j})^k : \alpha^k & & \\ & \swarrow\!\!\swarrow^{n''} & & \searrow\!\!\searrow^{n'} & \\ \mathsf{ungen}(Q^{k-j})^k : \alpha^k & & & & [j-1]_{k-j+1}P' : \alpha^k \\ & \searrow\!\!\searrow & & \swarrow\!\!\swarrow & \\ & & [j-1]_{k-j+1}P''' : \alpha^k & & \end{array}$$

where

$$\mathsf{gen}(P)^{k-j} : \Box\alpha^{k-j} \overset{n''-1}{\rightrightarrows} Q^{k-j} : \Box\alpha^{k-j}$$

Observe now that $Q^{k-j} = \mathsf{gen}(P'')^{k-j}$, with

$$P : \alpha^{k-j+1} \overset{<n''}{\rightrightarrows} P'' : \alpha^{k-j+1}$$

since no clause for $\rightrightarrows$ may remove a toplevel gen. By induction hypothesis we obtain

$$\begin{array}{ccccc} & & P : \alpha^{k-j+1} & & \\ & \overset{n''-1}{\swarrow\!\!\swarrow} & & \overset{n'-1}{\searrow\!\!\searrow} & \\ P'' : \alpha^{k-j+1} & & & & P' : \alpha^{k-j+1} \\ & \searrow\!\!\searrow & & \swarrow\!\!\swarrow & \\ & & P''' : \alpha^{k-j+1} & & \end{array}$$

Lemma 27 allows to conclude.

■

THEOREM 29 (Church-Rosser) *The $\twoheadrightarrow$ relation is confluent.*

Proof. $M \twoheadrightarrow N$ iff $M = M_0 \rightrightarrows \ldots \rightrightarrows M_n = N$ with $n \geq 0$. The usual Church-Rosser diamond can now be closed by using the previous lemma. ■

6.2 Strong normalization

We reduce the problem of strong normalization for $\lambda^{\ell}S4$ to the strong normalization for the system of next definition.

DEFINITION 30 (Calculus $\lambda S4$) Terms *and* derivations *are inductively defined by the following rules.*

$$x : \sigma^k$$

$$\dfrac{\begin{matrix}\Gamma \\ \vdots \\ M : \sigma^k\end{matrix} \qquad \begin{matrix}\Delta \\ \vdots \\ N : \tau^k\end{matrix}}{\langle M, N\rangle : \sigma \wedge \tau^k}\wedge\mathcal{I} \qquad\qquad \dfrac{\begin{matrix}\Gamma \\ \vdots \\ M : \sigma \wedge \tau^k\end{matrix}}{(\mathsf{fst}\,M) : \sigma^k}\wedge\mathcal{E}_l \qquad \dfrac{\begin{matrix}\Gamma \\ \vdots \\ M : \sigma \wedge \tau^k\end{matrix}}{(\mathsf{snd}\,M) : \tau^k}\wedge\mathcal{E}_r$$

$$\frac{\begin{array}{c}\Gamma \quad [x:\sigma^k] \\ \vdots \\ M:\tau^k\end{array}}{\lambda x.M:\sigma\supset\tau^k}\supset\mathcal{I} \qquad \frac{\begin{array}{cc}\Gamma & \Delta \\ \vdots & \vdots \\ M:\sigma\supset\tau^k & N:\sigma^k\end{array}}{MN:\tau^k}\supset\mathcal{E}$$

$$\frac{\begin{array}{c}\Gamma \\ \vdots \\ M:\sigma^j\end{array}}{M:\Box\sigma^{j-1}}\Box\mathcal{I}\ {\scriptstyle j>\#\Gamma} \qquad \frac{\begin{array}{c}\Gamma \\ \vdots \\ M:\Box\sigma^j\end{array}}{M:\sigma^k}\Box\mathcal{E}\ {\scriptstyle k\geq j}$$

Contractions (and reduction) for $\lambda S4$ are defined is the obvious way (note that modal contractions do not affect the terms, but only the levels in the formulas).

Derivations (i.e. terms) in $\lambda^\ell S4$ can be obviously translated into $\lambda S4$, via a straightforward forgetting translation $(\ )^\circ$, establishing easily the following propositions.

PROPOSITION 31 $\Gamma \vdash_{\lambda^\ell S4} M : \sigma^k \iff \Gamma^\circ \vdash_{\lambda S4} M^\circ : \sigma^k$.

PROPOSITION 32 *Let* $M : \sigma^k \rhd M' : \sigma^k$ *be a propositional reduction in* $\lambda^\ell S4$; *then* $M^\circ : \sigma^k \rhd M'^\circ : \sigma^k$ *in* $\lambda S4$, *using the same contraction.*

Let λ be the intuitionistic lambda calculus with implication and conjunction. The types of $\lambda S4$ can be interpreted into the types of λ by means of the following translation.

$$\begin{array}{rcl}(p^k)^* & = & p \\ (\alpha\supset\beta^k)^* & = & (\alpha^k)^*\supset(\beta^k)^* \\ (\alpha\wedge\beta^k)^* & = & (\alpha^k)^*\wedge(\beta^k)^* \\ (\Box\alpha^k)^* & = & (\alpha^{k+1})^*\end{array}$$

PROPOSITION 33 $\Gamma \vdash_{\lambda S4} M : \sigma^k \Longrightarrow \Gamma^* \vdash_\lambda M : (\sigma^k)^*$

Proof. Induction on the derivation, the modal rules becoming vacuous steps in λ. ■

It is well known (e.g. [2]) that λ enjoys strong normalization. It is now easy to obtain as a corollary the same result for $\lambda^\ell S4$.

THEOREM 34 $\lambda^\ell S4$ *enjoys strong normalization.*

Proof. Let $\Gamma \vdash_{\lambda^\ell S4} M : \sigma^k$ and suppose there is an infinite reduction sequence starting from M. Observe now that this sequence cannot be composed (from some point on) only of modal contractions, since a modal contraction strictly decreases the number of nodes of the proof tree. Hence the reduction sequence contains an infinite number of propositional reductions (that is, contractions of non modal redexes). But this is impossibile, because these propositional reductions would also be reductions of M° (by Proposition 32), which is typeable in λ, by the previous propositions, while λ enjoys strong normalization. ■

6.3 Computations in subsystems of $\lambda^\ell S4$

While all the results of the previous sections are stated for the stronger system $\lambda^\ell S4$, a careful inspection of the proofs shows that all the statements which apply to a given system still hold.

THEOREM 35 *Let $\lambda^\ell L$ one of $\lambda^\ell K$, $\lambda^\ell KT$, $\lambda^\ell K4$, or $\lambda^\ell S4$.(i) Reduction in $\lambda^\ell L$ is confluent.(ii) Reduction in $\lambda^\ell L$ enjoys strong normalization.*

7 RELATED WORK

In this brief section we relate our proposal to some of the recent work on natural deduction for modal logics. This is not the place for a detailed proof theoretical analysis of the different systems and, for this reason, we will discuss only the calculi allowing a computational interpretation of proofs, the exception being the work presented next, which we take into account for its similarity with our system. For more detailed analysis of the proof theory of modal logic, we refer to [13] and [17].

7.1 The 'constructive' system of Benevides and Maibaum

Benevides and Maibaum [3], starting from a semantical intuition on Hintikka systems, propose a set of natural deduction rules for the $\Box$-based systems essentially analogous to our proposal. However, their semantical insight seems to hinder a full mastering of the level machinery. They introduce in their system redundant rules for the interaction between $\Box$ and the propositional connectives (see the rules R_2 of their Section 3.2.2), which are derivable in the rest of the system. From a proof theoretical point of view these rules are hardly justified, since they do not obey to the tenet: "For each connective, (only) one introduction and (only) one elimination". For this reason, it seems impossible to relate these systems to any clean sequent calculus for the same logics (see, *contra*, our Section 4.2)

They prove the equivalence of their systems to the classical Hilbert style formulations. However, no notion of normalization of proofs is introduced. It is our belief that any study of the constructive content of a logical system cannot dispense from the study of proof normalization. If one does not tackle normalization, why not using the simple, naive approaches outlined in the Introduction?

7.2 *The intuitionistic calculus of Bierman, Meré and de Paiva*

Bierman et al. [4] present an approach to the $\Box$ modality essentially based on a variation of Prawitz proposal. The problematic $\Box\mathcal{I}$ rule has the form

$$\dfrac{\begin{matrix}\Gamma_1 \\ \vdots \\ M_1 : \Box\sigma_1\end{matrix} \quad \ldots \quad \begin{matrix}\Gamma_n \\ \vdots \\ M_n : \Box\sigma_n\end{matrix} \qquad \begin{matrix}[x_1 : \Box\sigma_1, \ldots, x_n : \Box\sigma_n] \\ \vdots \\ N : \tau\end{matrix}}{\mathsf{box}\ N\ \mathsf{with}\ M_1, \ldots, M_n\ \mathsf{for}\ x_1, \ldots, x_n : \Box\tau}\ \Box\mathcal{I}$$

For this system, the authors introduce several reduction rules corresponding to a categorical model and show its equivalence with the standard sequent calculus for S4.

It is clear that this system has some advantages over Prawitz's, since the rule $\Box\mathcal{I}$ is not as global as $\Box\mathcal{I}_{\mathcal{P}}$ (see Section 4.1) and allows a clearer reduction step. However, the approach seems specifically designed for S4, since it is not clear how the rule would scale to the normal subsystems of S4. Moreover, many of the critiques of the Introduction still apply to the proposed rule. In particular, a good introduction rule should construct its conclusion from *all* its premises and possibly some assumptions, a requirement clearly violated by the proposed $\Box\mathcal{I}$.

Finally, [15] shows a normal deduction in the positive fragment of the calculus in [4] which does not satisfy the subformula property. It is then necessary to add suitable permutative reductions, which are not needed in our (and Prawitz's) approach.

7.3 *The labelled natural deduction of Gabbay and de Queiroz*

A completely different approach is the one proposed by Gabbay and de Queiroz in a sequel of papers, e.g. [7, 8], as an attempt to give a general theory of natural deduction extending the Curry–Howard isomorphism to a large class of logics (classical, linear, relevant, modal). As for modalities,

> it offers a *deductive* (as opposed to *model theoretic*) account of the connections between modal logic and it propositional counterpart when world-variables are introduced in the functional calculus of the labels (i.e. when a little of the semantics is *brought to the syntax,* so to speak). [8]

The calculus is a kind of second order system, where second order objects, which syntactically are only variables, correspond to possible worlds. The basic rules for introduction and elimination of $\Box$ are:

$$\dfrac{\begin{matrix}\Gamma\,[\mathbb{W} : \mathcal{U}] \\ \vdots \\ M : \sigma\end{matrix}}{\Lambda\mathbb{W}.M : \Box\sigma}\ \Box\mathcal{I} \qquad\qquad \dfrac{\begin{matrix}\Gamma \\ \vdots \\ M : \Box\sigma\end{matrix} \qquad \begin{matrix}\Delta \\ \vdots \\ \mathbb{T} : \mathcal{U}\end{matrix}}{\mathcal{EXTR}(M, \mathbb{T}) : \sigma(\mathbb{T})}\ \Box\mathcal{E}$$

where $\mathcal{U}$ stands for an intended collection of 'worlds'.

There is an certain analogy between our levels and the world-variables, which is especially clear in the case of the logic K. Moving to stronger logics, however, obscures this analogy. The characteristic axiom of KT, for instance, is not provable *as is* in their system; they can only prove its modal closure, $\Box(\Box\sigma \supset \sigma)$ (it seems they would need a kind of second order constants to prove $\Box\sigma \supset \sigma$). In the case of K4, the proofs of the characteristic axioms are remarkably different in the two systems; while their proof consists of just a $\Box\mathcal{I}$ rule, in our case we have a $\Box\mathcal{E}$ followed by two $\Box\mathcal{I}$.

Simone Martini
University of Udine, Italy.

Andrea Masini
University of Pisa, Italy.

REFERENCES

[1] H. Barendregt. *The Lambda Calculus: its Syntax and Semantics*. Studies in Logic and the Foundations of Mathematics. North-Holland, 1984. Revised Edition.

[2] H. Barendregt. Lambda calculus with types. In: S. Abramsky, D. M. Gabbay, T. S. E. Maibaum, (eds.), *Handbook of Logic in Computer Science*, Vol. II Background: Computational Structures, pp. 118–310. Oxford University Press, 1992.

[3] M. R. F. Benevides, T. S. E. Maibaum. A constructive presentation for the modal connective of necessity (□). *Journal of Logic and Computation*, **2**, 31–50, 1992.

[4] G. Bierman, C. Meré, V. de Paiva. Intuitionistic necessity revisited. In: *Logic at Work: Applied Logic Conference*, 1992.

[5] R. Bull, K. Segerberg. Basic modal logic. In: D. Gabbay, F. Guenthner, (eds.), *Handbook of Philosophical Logic*, Vol. II, pp. 1–88. Reidel, 1984.

[6] K. Došen. Sequent-systems for modal logic. *Journal of Symbolic Logic*, **50**, 149–168, 1985.

[7] D. M. Gabbay, R. J.G.B. de Queiroz. Extending the Curry-Howard interpretation to linear, relevant and other resource logics. *Journal of Symbolic Logic*, **57**, 1319–1365, 1992.

[8] D. M. Gabbay, R. J.G.B. de Queiroz. An introduction to labelled natural deduction. *Proceedings of Third Advanced Summer School in Artificial Intelligence*, 1992.

[9] J.-Y. Girard. *Proof Theory and Logical Complexity*. Bibliopolis, 1987.

[10] S. Kripke. Semantical analysis of modal logic I. *Zeitschr. f. math. Logik und Grund. d. Mathematik*, **9**, 67–96, 1963.

[11] A. Masini. 2-sequent calculus: A proof theory of modalities. *Annals of Pure and Applied Logic*, **58**, 229–246, 1992.

[12] A. Masini. 2-sequent calculus: Intuitionism and natural deduction. *Journal of Logic and Computation*, **3**, 533–562, 1993.

[13] G. Mints. *Selected Papers in Proof Theory*. Bibliopolis, 1992.

[14] D. Prawitz. *Natural Deduction*. Acta Universitatis Stockholmiensis, Stockholm Studies in Philosophy 3, Almqvist & Wiksell, Stockholm, 1965.

[15] A. S. Troelstra. Natural deduction for intuitionistic linear logic. *Annals of Pure and Applied Logic*, **73**, 79–108, 1995.

[16] A. S. Troelstra and Dirk van Dalen. *Constructivism in Mathematics*, Vol. II. North-Holland, 1988.

[17] H. Wansing. Sequent calculi for normal modal propositional logic. *Journal of Logic and Computation*, **4**, 125–142, 1994.

SZABOLCS MIKULÁS

GABBAY-STYLE CALCULI

It is well known that there are logics, e.g. classical first-order logic with n variables, $\mathcal{L}_n$, that do not have strongly sound and complete Hilbert-style inference systems. However, some of the logics of the above kind have weakly sound and complete Gabbay-style inference systems, cf. Gabbay [6], Venema [13], [14], and Simon [12].

In this paper, we will consider $\mathcal{L}_n$ as a multi-modal logic and give weakly sound and complete Gabbay-style inference systems for two versions. These completeness results follow from representation theorems for the corresponding classes of algebras.

1 INTRODUCTION

Usually logics are defined in a semantical way. That is, one gives a logic by defining its language (the set of formulae), the class of models, and the validity relation (between models and formulae), or the semantical consequence relation (between sets of formulae and formulae). In addition, sometimes we have an interpretation (or meaning) function as well.

The question of whether a logic has a sound and complete inference system, i.e. whether we can grasp the above semantical relations by pure syntactical means, naturally arises.

There is a class of logics that have *strongly* sound and complete inference systems, i.e. where the semantical consequence relation can be simulated by a syntactical relation. For instance, classical propositional logic and the modal logic $S5$ have strongly sound and complete Hilbert-style inference systems, cf., e.g. Andréka, Németi & Sain [4]. Moreover, the calculi referred to above have the following property: the rules of inference can be applied without any restriction (such a restriction could be done, e.g. on the shape of the formulae occurring in the rule). These calculi will be called *Hilbert-style* in this paper, cf. Definition 2.

However, there are very interesting logics that do not have strongly sound and complete Hilbert-style inference systems, e.g. classical first-order logic with n variables, $\mathcal{L}_n$, does not have one, cf., e.g. Andréka [1], Németi [9], and [4]. But they may have *weakly* sound and complete inference systems, i.e. validity may be imitated. (In passing we note that it is an open problem whether $\mathcal{L}_n$ has a weakly sound

H. Wansing (ed.), Proof Theory of Modal Logic, 243–252.

and complete Hilbert-style calculus; we conjecture, it does not have one.) Beside Hilbert-style calculi we may consider inference systems in which some inference rules can be applied only if certain ("reasonable") condition is met. These calculi will be called *Gabbay-style*, cf. Definition 2. The reason for this name is that in Gabbay [6] there is a complete inference system of this kind for a previously incomplete logic, and with this Gabbay initiated a series of new completeness theorems for various logics. In passing we note that Martin Abadi's inference system for first-order temporal logic is also Gabbay-style, cf. Sain [10] for latest results and for references.

In this paper, we will prove that various versions of first-order logic with finitely many variables have sound and complete Gabbay-style inference systems. In particular, we will give calculi for the so-called ordinary and restricted versions of first-order logic with n variables. For these logics Yde Venema gave Gabbay-style (or unorthodox) calculus in Venema [13], [14] and [15]. However, he uses the difference operator D that is defined by equality. Our complete calculi below are defined without using D, partially answering question (1) on p.111 in [13]. For a similar completeness result for first-order logic *without* equality see [8].

2 DEFINITIONS AND MAIN RESULTS

First of all, we give the precise definitions of the logics we are dealing with, cf. Henkin, Monk & Tarski [7].

DEFINITION 1 **Ordinary first-order logic with n variables with equality** *is defined as the ordered triple*

$$\mathcal{L}_n^= \stackrel{\text{def}}{=} \langle Fml_{\mathcal{L}_n^=}, Mod_{\mathcal{L}_n^=}, \models_{\mathcal{L}_n^=} \rangle$$

for which the following conditions hold.

1. *Let* $V \stackrel{\text{def}}{=} \{v_0, \ldots, v_{n-1}\}$ *be the set of* **variables**. *Let* P *denote the set of* **atomic formulae**, *i.e.*

$$P \stackrel{\text{def}}{=} \{r_i(v_{j_0}, \ldots, v_{j_{n-1}}) : i \in I \& j_0, \ldots, j_{n-1} < n\}$$

for some set I*; the symbols* r_i $(i \in I)$ *are called* **relation symbols**. *Then the set* $Fml_{\mathcal{L}_n^=}$ *of* **formulae** *is the smallest set* H *satisfying:*

- $P \subseteq H$
- $v_i = v_j \in H$ *for every* $i, j \in n$
- $\varphi, \psi \in H, i \in n \Rightarrow (\varphi \wedge \psi), \neg\varphi, \exists v_i \varphi \in H.$

Sometimes we will use the notation $\exists_i\varphi$ *instead of* $\exists v_i\varphi$ *and* E_{ij} *instead of* $v_i = v_j$. *By* $\exists_{(\{i_0,\dots,i_m\})}\varphi$, $\{i_0,\dots,i_m\} \subseteq n$, *we abbreviate* $\exists_{i_0}\dots\exists_{i_m}\varphi$.[1] *By the set of* **connectives** *of* $\mathcal{L}_n^=$, $Cn(\mathcal{L}_n^=)$, *we mean the set* $\{\wedge, \neg, \exists_i, E_{ij} : i,j \in n\}$.

2. *The class* $Mod_{\mathcal{L}_n^=}$ *of* **models** *is defined by*

$$Mod_{\mathcal{L}_n^=} \stackrel{\text{def}}{=} \{\langle M, R_i\rangle_{i\in I} : M \neq \varnothing, R_i \subseteq {}^nM \quad (i \in I)\}.$$

If $\mathcal{M} = \langle M, R_i\rangle_{i\in I} \in Mod_{\mathcal{L}_n^=}$, *then* M *is called the* **universe** *of* $\mathcal{M}$.

3. *Let* $\mathcal{M} = \langle M, R_i\rangle_{i\in I} \in Mod_{\mathcal{L}_n^=}$, $k \in {}^nM$ *and* $\varphi \in Fml_{\mathcal{L}_n^=}$. *We define the ternary relation* $\langle \mathcal{M}, k\rangle \models \varphi$ *(sometimes also denoted by* $\mathcal{M} \models \varphi[k]$*) by induction on the complexity of* φ:

   - $\langle \mathcal{M}, k\rangle \models r_i(v_{j_0},\dots,v_{j_{n-1}}) \stackrel{\text{def}}{\iff} \langle k_{j_0},\dots,k_{j_{n-1}}\rangle \in R_i \quad (i \in I)$
   - $\langle \mathcal{M}, k\rangle \models v_i = v_j \stackrel{\text{def}}{\iff} k_i = k_j \quad (i,j \in n)$
   - *if* $\psi_1, \psi_2 \in Fml_{\mathcal{L}_n^=}$ *and* $i \in n$, *then*

$$\begin{array}{rcl} \langle \mathcal{M}, k\rangle \models \neg\psi_1 & \stackrel{\text{def}}{\iff} & \textit{not } \langle \mathcal{M}, k\rangle \models \psi_1 \\ \langle \mathcal{M}, k\rangle \models \psi_1 \wedge \psi_2 & \stackrel{\text{def}}{\iff} & \langle \mathcal{M}, k\rangle \models \psi_1 \ \&\ \langle \mathcal{M}, k\rangle \models \psi_2 \\ \langle \mathcal{M}, k\rangle \models \exists v_i\psi_1 & \stackrel{\text{def}}{\iff} & (\exists k' \in {}^nM)(\forall j \neq i) \\ & & (k'_j = k_j \ \&\ \langle \mathcal{M}, k'\rangle \models \psi_1). \end{array}$$

If $\langle \mathcal{M}, k\rangle \models \varphi$, *then we say that the evaluation* k **satisfies** *the formula* φ *in the model* $\mathcal{M}$. *We say that* $\mathcal{M}$ **satisfies** φ, *in symbols* $\mathcal{M} \models_{\mathcal{L}_n^=} \varphi$, *iff for every* $k \in {}^nM$, $\langle \mathcal{M}, k\rangle \models \varphi$. *The* **interpretation**, *or* **meaning**, *of a formula* φ *in a model* $\mathcal{M}$ *is defined as*

$$mean_{\mathcal{M}}\varphi \stackrel{\text{def}}{=} \{k \in {}^nM : \langle \mathcal{M}, k\rangle \models \varphi\}.$$

Instead of $mean_{\mathcal{M}}\varphi$ *sometimes we will use the notation* $\varphi^{\mathcal{M}}$.

Restricted first-order logic with n **variables with equality**, ${}^r\mathcal{L}_n^=$, *differs from the corresponding ordinary logic in the following: in restricted logic the order of the variables in atomic formulae* $r(v_0,\dots,v_{n-1})$ *is fixed.*

If no confusion is likely, we will omit the subscript $\mathcal{L}_n^=$ in $\models_{\mathcal{L}_n^=}$, etc.

---

[1] This definition makes sense, since the quantifiers commute, cf. the semantics below.

DEFINITION 2 *By an* **inference system** $\vdash$ *we mean a finite set of axiom schemata and rules of inference. Generally, a rule has the form*

$$\frac{\vdash A_1 \ldots \vdash A_n}{\vdash A_0} \quad \textit{provided } C$$

where $A_0, \ldots, A_n$ *are formula schemata and* C *is some "reasonable" condition. A rule is called* **Hilbert-style** *if the condition* C *is empty, otherwise* **Gabbay-style**. $\vdash$ *is a Hilbert-style inference system if all of its rules are Hilbert-style, and Gabbay-style otherwise.*

The inference system $\vdash$ *is* **strongly complete** *with respect to the logic* $\mathcal{L}$ *if for every set* $\Gamma \cup \{\varphi\}$ *of formulae,*

$$\Gamma \vdash \varphi \Leftarrow \Gamma \models_{\mathcal{L}} \varphi$$

and **strongly sound** *if*

$$\Gamma \vdash \varphi \Rightarrow \Gamma \models_{\mathcal{L}} \varphi.$$

In both cases, if $\Gamma = \varnothing$, *then we use the adverb* **weakly** *instead of strongly.*

REMARK 3 *In the above definition of inference rules the requirement that the condition* C *must be 'reasonable' is rather vague. We made this restriction to exclude some useless rules (e.g.* C *might be the condition '*A_0 *is valid'). In this paper,* C *is always the condition that 'some atomic formula does not occur in the conclusion* A_0 *of the rule', an easily decidable syntactic property.*

Let us formulate our main theorem.

THEOREM 4 *Let* $\mathcal{L}$ *be one of the following logics:*

1. *ordinary first-order logic with* n *variables with equality,* $\mathcal{L}_n^=$,
2. *restricted first-order logic with* n *variables with equality,* ${}^r\mathcal{L}_n^=$.

Let us assume that there are infinitely many relation symbols in the language of $\mathcal{L}$. *Then there is a Gabbay-style inference system that is weakly sound and complete with respect to* $\mathcal{L}$.

3 MODAL FORMALISM AND CONNECTION WITH ALGEBRAS

We will prove Theorem 4 applying an algebraic representation theorem. Representation theorems for algebras state that certain axiomatically given algebras are isomorphic to set algebras (e.g. algebras whose elements are relations, and the operations are natural operations on relations). Algebraic representation theorems correspond to logical completeness results, cf. [7] and [4]. We have to give an alternative definition of our logics, so that the algebraic counterparts of these logics have the appropriate similarity type. First we define a new logical connective.

DEFINITION 5 *Let n be fixed and $\sigma \in {}^n n$, i.e. a function from n into n, then we interpret the connective S_σ, called* **substitution**, *as*

$$\langle \mathcal{M}, k\rangle \models S_\sigma \varphi \overset{\text{def}}{\iff} \langle \mathcal{M}, k\sigma\rangle \models \varphi.$$

For example, if σ sends 1 to 0 and leaves everything else fixed,

$$\langle \mathcal{M}, \langle k_0, k_1, k_2\rangle\rangle \models S_\sigma \varphi \overset{\text{def}}{\iff} \langle \mathcal{M}, \langle k_0, k_0, k_2\rangle\rangle \models \varphi.$$

Thus the intuitive meaning of $S_\sigma\varphi(v_0, v_1, v_2)$ is $\varphi(v_0, v_0, v_1)$.

We will need the following fact, the proof of which is simple and omitted. For more details see [8].

COROLLARY 6 *If we define ordinary first-order logic with n variables by adding the connectives S_σ ($\sigma \in {}^n n$) to the set of connectives of the corresponding restricted logic, then the new definition is equivalent with the old one in the following sense: there is a one-one translation function between the sets of formulae such that for every formula and every model, the model satisfies the formula iff it satisfies the translation of the formula.*

By the above corollary we can simply omit the variables in any formula. Thus, first-order logic can be considered as a multi-modal logic, cf. Venema [13] and [15]. The existential quantifiers $\exists_i$ ($i \in n$) are diamond type modalities, and the equations E_{ij} ($i, j \in n$) are modal constants. Then the possible worlds are the evaluations k of the variables, and two worlds k and k' are related by the ith accessibility relation if $(\forall j \neq i) k_j = k'_j$. E_{ij} holds at k if $k_i = k_j$. Substitutions can be treated in a similar way.

Now we define some algebraic notions concerning logics. We will use the notation of Henkin, Monk & Tarski [7] and Sain & Thompson [11]. Thus **I**, **H**, **S**, and **P** denote the operators taking isomorphic copies, homomorphic copies, subalgebras, and isomorphic copies of direct products, respectively. We denote the classes of n-dimensional cylindric, representable cylindric, and quasi-polyadic equality algebras by CA_n, RCA_n, and QPEA_n, respectively. We will use the notions of variety and of discriminator variety, cf. Burris & Sankappanavar [5]. Given a variety V, we will denote the class of its simple elements by $\mathrm{Sim}\mathsf{V}$.

From now on, by a logic we will mean any of the logics in Definition 1.

DEFINITION 7 *The* **formula algebra** *of a logic $\mathcal{L}$ is defined as*

$$\mathcal{F} \overset{\text{def}}{=} \langle Fml_\mathcal{L}, c\rangle_{c \in Cn(\mathcal{L})}.$$

The corresponding class of algebras for a logic $\mathcal{L}$ is

$$\mathsf{Alg}\mathcal{L} \overset{\text{def}}{=} \{mean_\mathcal{M}''\mathcal{F} : \mathcal{M} \in Mod_\mathcal{L}\},$$

where $mean_\mathcal{M}''\mathcal{F}$ is the (homomorphic) image of the formula algebra $\mathcal{F}$ along the interpretation $mean_\mathcal{M}$ of the formulae.

We mention, without proof, the following consequence of the above alternative definition of our logics. For more detail about the correspondence of logics and algebras, how to translate algebraic equations to logical formulae and vice versa, etc. we refer to [4], [7] and [9].

COROLLARY 8 *$\mathsf{Alg}^r\mathcal{L}_n^=$ and $\mathsf{Alg}\mathcal{L}_n^=$ generate the varieties of representable CA_n's, and representable QPEA_n's, respectively.*

DEFINITION 9 *Let α be any ordinal and V_α be one of CA_α, or QPEA_α. Let $\mathcal{A} \in \mathsf{V}_\alpha$ and $a \in A$. We say that a is* **rectangular** *iff*

$$\mathsf{c}_{(\Gamma)}a \cdot \mathsf{c}_{(\Delta)}a = \mathsf{c}_{(\Gamma\cap\Delta)}a$$

for all finite subsets Γ and Δ of α.

We say that $\mathcal{A}$ is **rectangularly dense** *iff*

$$(\forall 0 \neq a \in A)(\exists 0 \neq b \in A)(b \leq a \;\&\; b \text{ is rectangular}).$$

Let us formulate the representation theorem referred to above, which is proved in Andréka et al. [2].

THEOREM 10 *Let α be any ordinal and $\mathsf{V}_\alpha \in \{\mathsf{CA}_\alpha, \mathsf{QPEA}_\alpha\}$. Let RV_α denote the class of representable elements of V_α and VR_α denote the class of rectangularly dense elements of V_α. Then*

$$\mathsf{RV}_\alpha = \mathbf{SP}\mathsf{VR}_\alpha.$$

4 PROOF

Now we are ready to prove Theorem 4. In the following proof, we will define a Gabbay-style calculus given by the translation of the axioms for CA_n, and a rule corresponding to rectangular density. The intuitive meaning of this rule is the following: a given formula φ is satisfiable iff there is another formula ψ such that

1. ψ is satisfiable,
2. ψ has a special form (viz., it is rectangular),
3. $\psi \to \varphi$ is satisfiable.

Proof. [Theorem 4] We give the precise proof for ${}^r\mathcal{L}_n^=$ (i.e. for 4.2), and indicate the differences for the other logic.

First we note that for $\mathcal{A} \in \mathsf{CA}_n$ and $a \in \mathcal{A}$,

$$a \text{ is rectangular} \iff \mathcal{A} \models \tau(a) = 1$$

where

$$\tau(a) = -\mathsf{c}_{(n)} - (-\prod\{\mathsf{c}_{(n\smallsetminus\{i\})}a : i \in n\} + a),$$

cf. [2]. Further, by setting $\mathsf{V}_\alpha = \mathsf{CA}_n$ in the representation Theorem 10, we get that the rectangularly dense elements of CA_n are in RCA_n.

Let Ax be a finite set of equations axiomatizing CA_n. Let the inference system $\vdash$ be defined as follows. Its set of axiom schemata is the translations[2] of the elements of Ax into the language of ${}^r\mathcal{L}_n^=$: for every $i, j, k\langle n$,

1. enough propositional tautologies,
2. $\exists_i(\varphi \vee \psi) \leftrightarrow \exists_i\varphi \vee \exists_i\psi$,
3. $\varphi \rightarrow \exists_i\varphi$,
4. $\exists_i\neg\exists_i\varphi \rightarrow \neg\exists_i\varphi$,
5. $\exists_i\exists_j\varphi \leftrightarrow \exists_j\exists_i\varphi$,
6. E_{ii},
7. $E_{ik} \leftrightarrow \exists_j(E_{ij} \wedge E_{jk})$ if $j \notin \{i, k\}$,
8. $E_{ij} \wedge \exists_i(E_{ij} \wedge \varphi) \rightarrow \varphi$ if $i \neq j$,

its inference rules are Modus Ponens, Universal Generalization and the following rule:

$$(N\text{-rule}) \qquad \frac{\vdash (p \wedge \bar{\tau}(p \wedge \neg\varphi)) \rightarrow \varphi}{\vdash \varphi} \qquad \text{provided } p \notin \varphi$$

where $\bar{\tau}(p \wedge \neg\varphi)$ is the translation of $\tau(p \wedge \neg\varphi)$:

$$\neg\exists_{(n)}\neg(\neg\bigwedge_{i\in n}\exists_{(n\smallsetminus\{i\})}(p \wedge \neg\varphi) \vee (p \wedge \neg\varphi)),$$

and $p \notin \varphi$ denotes that p is an atomic proposition not occurring in the formula φ. SOUNDNESS: We check the validity of the N-rule. Assume that φ is not valid, i.e. that there are $\mathcal{M}$ and k such that

$$\langle\mathcal{M}, k\rangle \not\models \varphi.$$

Then $\neg\varphi^{\mathcal{M}} \neq \varnothing$, so we can choose $\langle a_0, \ldots, a_{n-1}\rangle \in \neg\varphi^{\mathcal{M}}$. Let p be a relation symbol not occurring in φ, and let $p^{\mathcal{M}} = \{\langle a_0, \ldots, a_{n-1}\rangle\}$. Let the evaluation of variables $\langle a_0, \ldots, a_{n-1}\rangle$ be denoted by a. Then

$$\langle\mathcal{M}, a\rangle \models p \wedge \neg\varphi.$$

[2]This translation is defined by substituting formula schemata for algebraic variables, and replacing algebraic operations by the corresponding logical connectives.

We claim that for every evaluation k,

$$\langle \mathcal{M}, k \rangle \models \bar{\tau}(p \wedge \neg\varphi).$$

This amounts to prove that for an arbitrary evaluation k',

$$\langle \mathcal{M}, k' \rangle \models \neg \bigwedge_{i \in n} \exists_{(n \smallsetminus \{i\})}(p \wedge \neg\varphi) \vee (p \wedge \neg\varphi).$$

To prove this, first let $k' = a$. Then

$$\langle \mathcal{M}, k' \rangle \models p \wedge \neg\varphi.$$

If $k' \neq a$, then for some $j \in n$, $a_j \neq k'_j$, whence

$$\langle \mathcal{M}, k' \rangle \not\models \exists_{(n \smallsetminus \{j\})}(p \wedge \neg\varphi).$$

Thus in both cases,

$$\langle \mathcal{M}, k' \rangle \models \neg \bigwedge_{i \in n} \exists_{(n \smallsetminus \{i\})}(p \wedge \neg\varphi) \vee (p \wedge \neg\varphi)$$

whence for any evaluation k,

$$\langle \mathcal{M}, k \rangle \models \neg\exists_{(n)}\neg(\neg \bigwedge_{i \in n} \exists_{(n \smallsetminus \{i\})}(p \wedge \neg\varphi) \vee (p \wedge \neg\varphi)),$$

i.e.

$$\langle \mathcal{M}, k \rangle \models \bar{\tau}(p \wedge \neg\varphi)$$

as desired. Then we have

$$\langle \mathcal{M}, a \rangle \models p \wedge \neg\varphi \ \ \& \ \ \langle \mathcal{M}, a \rangle \models \bar{\tau}(p \wedge \neg\varphi)$$

whence

$$\langle \mathcal{M}, a \rangle \not\models (p \wedge \bar{\tau}(p \wedge \neg\varphi)) \to \varphi.$$

So the rule is valid. The other rules and the (instances of the) axiom schemata are clearly valid, since $\mathsf{Alg}^r \mathcal{L}_n^{=} = \mathsf{RCA}_n \subseteq \mathsf{CA}_n$.

COMPLETENESS: First we prove that the syntactical Lindenbaum–Tarski algebra $\mathcal{A}$ is a rectangularly dense element of CA_n. Clearly $\mathcal{A} \in \mathsf{CA}_n$. Now let $0 \neq a \in A$. Then a is the equivalence class of a formula ψ such that $\not\vdash \neg\psi$. By the N-rule above, $\not\vdash (p \wedge \bar{\tau}(p \wedge \psi)) \to \neg\psi$ whenever $p \notin \psi$. Let us fix such a p, and let b be the equivalence class of the formula $(p \wedge \psi) \wedge \bar{\tau}(p \wedge \psi)$. Then $b \neq 0$ and $b \leq a$. It remains to show that b is rectangular, i.e. $\tau(b)$ holds in $\mathcal{A}$. In fact we will show that

$$\mathsf{CA}_n \models \tau(x \cdot \tau(x)) = 1.$$

Let $\mathcal{B}$ be a simple element of CA_n. Since τ is closed under the cylindrifications, for every $x \in B$,

$$\tau(x) \in \{0,1\}.$$

Now assume that $\tau(x) = 1$ holds in $\mathcal{B}$. Then

$$\tau(x \cdot \tau(x)) = \tau(x) = 1.$$

If $\tau(x) = 0$, then

$$\tau(x \cdot \tau(x)) = \tau(0) = 1,$$

since 0 is rectangular. Then $\mathbf{Sim}\mathsf{CA}_n \models \tau(x \cdot \tau(x)) = 1$ whence

$$\mathsf{CA}_n \models \tau(x \cdot \tau(x)) = 1,$$

since CA_n is a discriminator variety. In particular, $\tau(b) = 1$ for the b above. Thus, for arbitrary non-zero a, we found a non-zero $b \leq a$ that is rectangular. This means that $\mathcal{A}$ is rectangularly dense, so it is in RCA_n by the representation Theorem 10 above.

Now assume that $\not\vdash \varphi$. Then $\mathcal{A} \not\models \varphi' = 1$ where φ' denotes the equivalence class of φ. Then, since by the above argument $\mathcal{A} \in \mathsf{RCA}_n = \mathbf{HSP}\mathsf{Alg}^r\mathcal{L}_n^=$, we have

$$\mathsf{Alg}^r\mathcal{L}_n^= \not\models \varphi' = 1,$$

i.e. there is a model $\mathcal{M}$ such that $\varphi^{\mathcal{M}} \neq {}^nM$, whence

$$\mathcal{M} \not\models \varphi.$$

Thus φ is not valid whenever it is not derivable.

The proof is essentially the same: just use QPEA_n instead of CA_n, and its representable subclass instead of RCA_n.

So Theorem 4 has been proved. ■

REMARK 11 *The N-rule above is only weakly sound. That is, there are formulae φ and ψ such that $\psi \models (p \wedge \bar{\tau}(p \wedge \neg\varphi)) \rightarrow \varphi$ but $\psi \not\models \varphi$. Indeed, let φ be a non-valid formula, and let p and $\mathcal{M}$ be such that $\neg p^{\mathcal{M}} \subseteq \neg\varphi^{\mathcal{M}}$. Then $\neg p \models (p \wedge \bar{\tau}(p \wedge \neg\varphi)) \rightarrow \varphi$. On the other hand, $\neg p \not\models \varphi$.*

ACKNOWLEDGEMENTS

The author is grateful to Hajnal Andréka, István Németi and Ildikó Sain. Thanks are also due to Yde Venema. Research is supported by the Hungarian National Foundation for Scientific Research grant Nos. F17452 and T16448.

University of Amsterdam, The Netherlands

REFERENCES

[1] H. Andréka. *Complexity of the Equations Valid in Algebras of Relations*, Thesis for DSc, Hungarian Academy of Sciences, 1991.

[2] H. Andréka, S. Givant, Sz. Mikulás, I. Németi. *Rectangular Density Implies Representability*, Mathematical Institute of the Hungarian Academy of Sciences, 1993.

[3] H. Andréka, J. D. Monk, I. Németi, (eds.). *Algebraic Logic*. North-Holland, Amsterdam, 1991.

[4] H. Andréka, I. Németi, I. Sain. *Methodology of Applying Algebraic Logic to "Plain" Logic*, Mathematical Institute of the Hungarian Academy of Sciences, 1993.

[5] S. Burris, H. P. Sankappanavar. *A Course in Universal Algebra*. Springer-Verlag, Berlin, 1981.

[6] D. M. Gabbay. *An Irreflexivity Lemma with Applications to Axiomatizations of Conditions on Tense Frames*. In: U. Mönnich, (ed.), *Aspects of Philosophical Logic*, pp. 67–89. Reidel, Dordrecht, 1981.

[7] L. Henkin, J. D. Monk, A. Tarski. *Cylindric Algebras, Parts I., II.* North-Holland, Amsterdam, 1971, 1985.

[8] Sz. Mikulás. *Taming Logics*. PhD thesis, University of Amsterdam, ILLC dissertation series 1995-12.

[9] I. Németi. *An Algebraization of Quantifier Logics, an Introductory Overview, Version 12*. Mathematical Institute of the Hungarian Academy of Sciences, 1994. A shortened version is available in: *Studia Logica*, **50**, 485–570, 1991.

[10] I. Sain. *Temporal Logics Need Their Clocks*. Theoretical Computer Science, **95**, 1992.

[11] I. Sain, R. J. Thompson. *Strictly Finite Schema Axiomatization of Quasi-polyadic Algebras*. In: [3], pp. 539–571.

[12] A. Simon. *Finite Schema Completeness for Typeless Logic and Representable Cylindric Algebras*. In: [3], pp. 665–670.

[13] Y. Venema. *Many-dimensional Modal Logic*. PhD thesis, University of Amsterdam, 1991.

[14] Y. Venema. *Cylindric Modal Logic*. *Journal of Symbolic Logic*, **60**, 591–623, 1995.

[15] Y. Venema. A Modal Logic of Quantification and Substitution. *Bulletin of the IGPL*, **2**, 31–45, 1994.

PART III

TRANSLATION-BASED PROOF SYSTEMS

H. J. OHLBACH, R. SCHMIDT AND U. HUSTADT

TRANSLATING GRADED MODALITIES INTO PREDICATE LOGICS

1 INTRODUCTION

From Minsky's early frame systems, which were defined purely operationally, and Brachman's KL-ONE knowledge representation system [4, 35] to the language $\mathcal{ALC}$ of Schmidt-Schauß and Smolka's [28] paper there has been a continuous trend in designing knowledge representation systems more and more according to logical principles with clear syntax and semantics and logical inferences as basic operations. $\mathcal{ALC}$ in particular is a language with the usual logical connectives $\sqcap$, $\sqcup$, $\neg$ and the additional constructs (all R C) and (some R C). For example, the following is an $\mathcal{ALC}$ definition which defines a 'concept' proud-father as a father all of whose children are successful persons.

$$\text{proud-father} = \text{father} \sqcap (\text{all has-child successful-person}),$$

The fragment of $\mathcal{ALC}$ that includes the operations $\sqcap$, $\sqcup$, $\neg$, all, some is just a variant of the multi-modal logic $\mathbf{K}_{(m)}$ [27]. The concept (all R C) corresponds to $[R]C$ where the relational term R (a 'role' in KL-ONE jargon) is the parameter of the modal operator, and is interpreted as a binary accessibility relation. $\mathcal{ALC}$ is still limited in its expressiveness. In pure $\mathcal{ALC}$ it is not possible to define concepts like, for example, a city as a place with more than, say, 100 000 inhabitants. There are extensions of $\mathcal{ALC}$, like $\mathcal{ALCN}$, with additional operators, called 'number restrictions'.

(1) city = place $\sqcap$ (atleast 100 001 inhabited-by people)

is a suitable $\mathcal{ALCN}$ definition. (atleast n R C) and (atmost n R C) restrict the number of so-called 'role fillers', i.e. they restrict the number of elements in the range of the relation R to at least n and at most n, respectively. The corresponding modal logic of $\mathcal{ALCN}$ is the multi-modal version of the system of 'graded modalities', which was introduced by Goble [17] and Fine [13, 14] and which is investigated in Fattorosi-Barnaba and de Caro [12, 11], and van der Hoek [34, 33].

Graded modalities are modal operators indexed with cardinals which fix the number of worlds in which a formula is true. The formula $\Diamond_n\varphi$ is true in a world iff there are more than n accessible worlds in which φ is also true. The dual formula $\Box_n\varphi$, given by $\neg\Diamond_n\neg\varphi$, is then true in a world iff there are at most n accessible

H. Wansing (ed.), Proof Theory of Modal Logic, 253–291.

worlds in which $\neg\varphi$ is true. More formally, the semantics is defined in terms of one accessibility relation, say R, by

$$\begin{array}{lll} \mathcal{M}, x \models_{\overline{\mathrm{K}}} \Diamond_n \varphi & \text{iff} & |\{y \mid R(x,y) \;\&\; \mathcal{M}, y \models_{\overline{\mathrm{K}}} \varphi\}| \rangle n \\ \mathcal{M}, x \models_{\overline{\mathrm{K}}} \Box_n \varphi & \text{iff} & |\{y \mid R(x,y) \;\&\; \mathcal{M}, y \models_{\overline{\mathrm{K}}} \neg\varphi\}| \leq n, \end{array}$$

where $\mathcal{M}$ denotes a model and x, y denote possible worlds. For any set A, $|A|$ denotes the cardinality of A. This semantics is very natural and intuitive, but it has one disadvantage. All inference systems based on this semantics, in particular, tableaux systems, deal with these $\Diamond_n$-operators by generating a corresponding number of terms explicitly. For example, the formula $\Diamond_{100\,000}$*people* triggers the generation of 100 001 constant symbols as representatives for the individuals denoting *people*. Except for counting these constant symbols and comparing the length of lists, known tableaux systems do not provide for arithmetical computation. In particular, reasoning with symbolic arithmetic terms is impossible. For example, in tableaux systems the formula $\Diamond_{n+1}p \rightarrow \Diamond_n p$ which is true for all n can only be verified for concrete values of n, but in general it cannot be verified for arbitrary values of n.

This is not the case for the Hilbert system axiomatizing the graded modalities. It is formulated with arithmetical terms, and in principle, this allows for invoking arithmetical computations. However, Hilbert systems have other disadvantages that makes them unsuitable to form the basis for automated reasoning. For example they do propositional reasoning just with modus ponens and the instantiation rule. Even for trivial theorems one gets large proofs and the search space is very unstructured and enormously big.

A direct translation of formulae with graded modalities into predicate logic requires the axiomatization of finite domains. This is feasible only for small cardinalities. We may translate sentence (1) as follows:

$$\begin{array}{lll} \forall x\; \mathit{city}(x) & \leftrightarrow & \mathit{place}(x) \wedge \\ & & \exists y_1 \ldots y_{100\,001} \quad y_1 \neq y_2 \wedge y_1 \neq y_3 \wedge \ldots \wedge y_1 \neq y_{100\,001} \wedge \\ & & \qquad\qquad\qquad y_2 \neq y_3 \wedge \ldots \wedge y_2 \neq y_{100\,001} \wedge \\ & & \qquad\qquad\qquad\qquad\qquad \vdots \\ & & \qquad\qquad\qquad\qquad y_{100\,000} \neq y_{100\,001} \wedge \\ & & \mathit{inhabited\text{-}by}(x, y_1) \wedge \ldots \wedge \mathit{inhabited\text{-}by}(x, y_{100\,001}) \wedge \\ & & \mathit{people}(y_1) \wedge \ldots \wedge \mathit{people}(y_{100\,001}). \end{array}$$

The translation of $\Diamond_n$-expressions requires $(n+1)n/2$ equations. Even for small n this is more than current theorem provers can cope with. One immediate alternative is introducing set variables and a cardinality function. For sentence (1) an alternative formulation is:

$$\forall x\; \mathit{city}(x) \quad \leftrightarrow \quad \mathit{place}(x) \wedge \exists Y\; (|Y| \rangle 100\,000 \wedge$$

$$\forall y\ (y \in Y \rightarrow (\textit{inhabited-by}(x, y) \wedge \textit{people}(y)))).$$

This is not really a feasible alternative, for the axiomatization of the cardinality function then requires the above $(n+1)n/2$ equations, and this for every n:

$$\begin{array}{lclr} \forall Y\ |Y|\rangle n & \leftrightarrow & \exists y_1 \ldots y_{n+1} & y_1 \in Y \wedge \ldots \wedge y_{n+1} \in Y \wedge \\ & & & y_1 \neq y_2 \wedge y_1 \neq y_3 \wedge \ldots \wedge y_1 \neq y_{n+1} \wedge \\ & & & y_2 \neq y_3 \wedge \ldots \wedge y_2 \neq y_{n+1} \wedge \\ & & & \vdots \\ & & & y_n \neq y_{n+1}. \end{array}$$

In this paper we present a two step translation of graded modal logics into predicate logic. In the first step, we transform graded modal logics into another multi-modal logic with standard interpretation. In particular, we accommodate modal logics with graded modalities in a multi-modal logic with two kinds of modalities:

(i) $\langle n \rangle$, $[n]$ characterized by a relational structure (over a universe U) of infinitely but countably many different relations R_n ($n \in \mathbb{N}_0$), and
(ii) $\Diamond$, $\Box$ characterized by a designated relation E.

We translate formulae of the form $\Diamond_n \varphi$ into $\langle n \rangle \Box \varphi$ and the intuitive idea underlying this translation is this: If φ is true in a set Y of worlds with more than n elements then we introduce an accessibility relation R_n that connects the actual world and a world w_Y which we can think of as being a representative for the set Y. This defines the $\langle n \rangle$-operator. $\Box \varphi$ and its associated accessibility relation E expresses that φ is true in all the worlds of the set Y. E connects the world w_Y with all the worlds in Y and can be thought of as the membership relation. Thus, $\langle n \rangle \Box \varphi$ encodes 'there is a set with more than n elements (encoded by $\langle n \rangle$) and φ is true for all the elements of this set (encoded by $\Box$)'. Our first problem now is to find a sound and adequate axiomatization of the modalities $\langle n \rangle$, $[n]$, $\Diamond$ and $\Box$ as to capture the graded modalities $\Diamond_n$ and $\Box_n$. It turns out that this is not entirely possible. The axiomatization we present in this paper has some non-standard models which do not reflect our intuition. But this does no harm, as we will see. We show: A formula φ is a theorem of a graded modal logic iff the translation of φ is a theorem in the new logic. This translation is only an intermediary step in a translation to predicate logic.

In the second step, we translate the multi-modal logic into a predicate logic using the functional translation of [23, 24, 10, 18, 2, 36]. The reason for using the functional translation instead of the usual relation translation is this: The multi-modal logic of graded modalities can have frame properties that are not first-order definable in terms of R_n relations. However, the frame properties can be formulated in a weak fragment of second-order logic and it is possible to formulate them in an alternative adapted language as first-order expressions. The alternative language is a *functional language* in which binary relations are encoded as sets AF of functions. The set AF of accessibility functions defining the accessibility relation R is given by:

$$R(x,y) \leftrightarrow \exists f \in AF\, y = f(x).$$

This sequence of translations of a system of numerical modalities first into another multi-modal logic and then into a many-sorted predicate logic (using the functional translation) yields an axiomatization, in particular, an axiomatization of properties of finite sets. Instead of counting symbols this system uses arithmetical reasoning.

This paper is structured as follows. In Section 2 we give a short overview of modal logics with graded modalities. In Section 3 we introduce the normal multi-modal logic for accommodating graded modal logics. We define a translation from logics of graded modalities into the multi-modal logic that we exhibit to be sound and complete. In Section 5 we present the functional translation of the multi-modal logic into predicate logic. We conclude with Section 6 in which we apply the new techniques to the knowledge representation language $\mathcal{ALCN}$.

2 GRADED MODALITIES, THE SYSTEM $\overline{\mathbf{K}}$

Normal modal logics like **K**, **T**, **S4** and **S5** have one modal operator, the *necessity* (or *box*) operator $\Box$. The *possibility* (or *diamond*) operator $\Diamond$ is defined as its dual. By definition,

$$\Diamond\varphi \leftrightarrow \neg\Box\neg\varphi.$$

In [17] Goble investigates modal logics with more than one modality. His logics have a fixed and finite number of modalities. Each modality represents a different grade of necessity. For example, the formula

$$N_m\varphi \wedge N_n\psi$$

for positive integers $m\langle n$, is read to mean ψ is more necessary than φ. Kit Fine [13, 14] generalizes this idea and introduces modal logics with *numerical modalities*. These are now commonly referred to as modal logics with *graded modalities*. In a series of papers Fattorosi-Barnaba, de Caro and Cerrato [12, 7, 11, 6] rediscover and analyze various modal logics of graded modalities.

Recent investigations of graded modal logics are by van der Hoek in [34] and [33]. Together with de Rijke he applies graded modalities to linguistics and artificial intelligence. In [31] they show that generalized quantifiers can be modelled with graded modalities. In [32] they also show that certain numerical quantifier operations available in KL-ONE-based knowledge representation languages can be modelled with graded modalities.

In this paper we adopt the definition of the graded modal logic $\overline{\mathbf{K}}$ of van der Hoek [34]. $\overline{\mathbf{K}}$ is an extension of the normal modal logic **K** with graded modalities. Formally, the vocabulary of $\overline{\mathbf{K}}$ consists of the set of propositional symbols

$p, p_1, p_2, \ldots, q, q_1, q_2, \ldots$, the constant $\perp$ (falsity), the logical symbol $\rightarrow$ (implication) and the modal operator symbols $\Diamond_n$ for $n \in \mathbb{N}_0$, and the usual punctuation symbols. Formulae of $\overline{\mathbf{K}}$ have the following forms:

$$p, q, \ldots, \quad \perp, \quad \varphi \rightarrow \psi, \quad \Diamond_n \varphi.$$

As usual we define $\neg\varphi$ (negation), $\top$ (truth), $\varphi \vee \psi$ (disjunction), $\varphi \wedge \psi$ (conjunction) and $\varphi \leftrightarrow \psi$ (double-implication) as abbreviations for $\varphi \rightarrow \perp$, $\neg\perp$, $\neg\varphi \rightarrow \psi$ and $\neg\varphi \vee \neg\psi$, respectively. Furthermore, $\Box_n\varphi$ abbreviates $\neg\Diamond_n\neg\varphi$ for $n \geq 0$, $\Diamond!_0\varphi$ abbreviates $\Box_0\neg\varphi$, and $\Diamond!_n\varphi$ is the abbreviation for $\Diamond_{n-1}\varphi \wedge \neg\Diamond_n\varphi$ with $n \rangle 0$.

$\Diamond_n\varphi$	is read to mean	φ is true in more than n accessible worlds,
$\Box_n\varphi$	is read to mean	$\neg\varphi$ is true in at most n accessible worlds, and
$\Diamond!_n\varphi$	is read to mean	φ is true in exactly n accessible worlds.

DEFINITION 1 *The system* $\overline{\mathbf{K}}$ *of graded modalities is defined by the following axioms*

A1 *the axioms of propositional logic*

A2 $\vdash_{\overline{\mathrm{K}}} \Diamond_{n+1}\varphi \rightarrow \Diamond_n\varphi$

A3 $\vdash_{\overline{\mathrm{K}}} \Box_0(\varphi \rightarrow \psi) \rightarrow (\Diamond_n\varphi \rightarrow \Diamond_n\psi)$

A4 $\vdash_{\overline{\mathrm{K}}} \Box_0\neg(\varphi \wedge \psi) \rightarrow ((\Diamond!_n\varphi \wedge \Diamond!_m\psi) \rightarrow \Diamond!_{n+m}(\varphi \vee \psi))$

together with the uniform substitution rule, Modus Ponens, and the necessitation rule for $\Box_0$*:*

US *if* φ *is a theorem so is every substitution instance of* φ*,*

MP *if* $\vdash_{\overline{\mathrm{K}}} \varphi$ *and* $\vdash_{\overline{\mathrm{K}}} \varphi \rightarrow \psi$ *then* $\vdash_{\overline{\mathrm{K}}} \psi$

N *if* $\vdash_{\overline{\mathrm{K}}} \varphi$ *then* $\vdash_{\overline{\mathrm{K}}} \Box_0\varphi$.

Observe that $\Diamond_0$ and $\Box_0$ coincide with the standard modal operators $\Diamond$ and $\Box$. **K** is therefore a subsystem of $\overline{\mathbf{K}}$.

THEOREM 2 *[34] The following are theorems of* $\overline{\mathbf{K}}$.

A5 $\vdash_{\overline{\mathrm{K}}} \Box_0(\varphi \rightarrow \psi) \rightarrow (\Box_n\varphi \rightarrow \Box_n\psi)$

A6 $\vdash_{\overline{\mathrm{K}}} \Diamond_n(\varphi \wedge \psi) \rightarrow (\Diamond_n\varphi \wedge \Diamond_n\psi)$

A7 $\vdash_{\overline{\mathrm{K}}} \Diamond!_n\varphi \wedge \Diamond!_m\varphi \rightarrow \perp$ *for* $n \neq m$

A8 $\vdash_{\overline{\mathrm{K}}} \Box_n\neg\varphi \leftrightarrow (\Diamond!_0\varphi \nabla \Diamond!_1\varphi \nabla \ldots \nabla \Diamond!_n\varphi)$

A9 $\vdash_{\overline{\mathbf{K}}} \neg\Diamond_n(\varphi \vee \psi) \rightarrow \neg\Diamond_n\varphi$

A10 $\vdash_{\overline{\mathbf{K}}} \Diamond_{n+m}(\varphi \vee \psi) \rightarrow (\Diamond_n\varphi \vee \Diamond_m\psi)$

A11 $\vdash_{\overline{\mathbf{K}}} \Diamond!_n\varphi \wedge \Diamond_m\varphi \rightarrow \bot$ *for* $m \geq n$

A12 $\vdash_{\overline{\mathbf{K}}} \Diamond_n(\varphi \wedge \psi) \wedge \Diamond_m(\varphi \wedge \neg\psi) \rightarrow \Diamond_{n+m+1}\varphi$

(∇ denotes exclusive-or.)

The semantics of $\overline{\mathbf{K}}$ is given by a *frame* $\mathcal{F} = (W, R)$ consisting of a non-empty set W, called the set of *worlds*, and a binary relation R over W, called the *accessibility relation*. We define a *model* (based on a frame $\mathcal{F}$) of $\overline{\mathbf{K}}$ as a triple $\mathcal{M} = (W, R, V)$. V denotes a *valuation* function mapping propositional variables to subsets of W. *Truth* in a model $\mathcal{M}$ for formulae of $\overline{\mathbf{K}}$ at any world x is defined (in terms of the *satisfiability relation* $\models_{\overline{\mathbf{K}}}$) as follows:

$$(2)\quad \begin{array}{lll} \mathcal{M}, x \models_{\overline{\mathbf{K}}} p & \text{iff} & x \in V(p) \\ \mathcal{M}, x \models_{\overline{\mathbf{K}}} \Diamond_n\varphi & \text{iff} & |\{y \in W \mid R(x,y) \wedge \mathcal{M}, y \models_{\overline{\mathbf{K}}} \varphi\}| \rangle n \\ \mathcal{M}, x \models_{\overline{\mathbf{K}}} \Box_n\varphi & \text{iff} & |\{y \in W \mid R(x,y) \wedge \mathcal{M}, y \models_{\overline{\mathbf{K}}} \neg\varphi\}| \leq n \end{array}$$

and as usual for the other connectives. For any binary relation R, let $R(x)$ denote the set of images of x under R. That is, define $R(x) = \{y \mid R(x, y)\}$. Then, satisfiability of $\Diamond_n\varphi$ and $\Box_n\varphi$ can be reformulated as follows:

$$(3)\quad \begin{array}{l} \mathcal{M}, x \models_{\overline{\mathbf{K}}} \Diamond_n\varphi \\ \quad \text{iff} \quad \exists Y \subseteq R(x) \text{ with } |Y|\rangle n \text{ and } \forall y \in Y : \mathcal{M}, y \models_{\overline{\mathbf{K}}} \varphi \\ \mathcal{M}, x \models_{\overline{\mathbf{K}}} \Box_n\varphi \\ \quad \text{iff} \quad \forall Y \subseteq R(x) \text{ with } |Y|\rangle n, \exists y \in Y : \mathcal{M}, y \models_{\overline{\mathbf{K}}} \varphi. \end{array}$$

(The proof is routine.) It is now easy to see that the usual duality for box and diamond also hold for $\Diamond_n$ and $\Box_n$, i.e. we have $\Diamond_n\varphi \leftrightarrow \neg\Box_n\neg\varphi$.

A formula φ is a $\overline{\mathbf{K}}$ thoerem, i.e. $\models_{\overline{\mathbf{K}}} \varphi$, iff φ holds in all worlds of all $\overline{\mathbf{K}}$ frames.

THEOREM 3 *The axiomatization of $\overline{\mathbf{K}}$ is sound and complete, i.e. for all formulae φ, we have*

$$\vdash_{\overline{\mathbf{K}}} \varphi \quad \textit{iff} \quad \models_{\overline{\mathbf{K}}} \varphi.$$

A proof can be found in [12].

In the remainder of this paper we assume the formulae of $\overline{\mathbf{K}}$ to be in *negation normal form* which can be obtained by systematically applying the following equivalences from left to right.

$$\begin{array}{rcl} \neg(\varphi \vee \psi) & \leftrightarrow & \neg\varphi \wedge \neg\psi \\ \neg(\varphi \wedge \psi) & \leftrightarrow & \neg\varphi \vee \neg\psi \\ \varphi \rightarrow \psi & \leftrightarrow & \neg\varphi \vee \psi \\ \varphi \leftrightarrow \psi & \leftrightarrow & (\varphi \rightarrow \psi) \wedge (\psi \rightarrow \psi) \\ \neg\Box_n\varphi & \leftrightarrow & \Diamond_n\neg\varphi \\ \neg\Diamond_n\varphi & \leftrightarrow & \Box_n\neg\varphi \end{array}$$

3 FROM GRADED MODALITIES TO MULTI-MODAL LOGIC

In this section we present a new interpretation for $\overline{\mathbf{K}}$ and investigate its logical counterpart.

A formula $\Diamond_n\varphi$ of $\overline{\mathbf{K}}$ is interpreted as an expression in which a subset of the accessible worlds with more than n worlds is selected. More concretely, the formula $\Diamond_n\varphi$ is true in a world x iff there is a set of worlds (a subset of W) accessible by R from x containing more than n worlds in which φ holds. Equivalently, $\Diamond_n\varphi$ is true in x iff there is subset Y of the range of R from x with cardinality strictly greater than n such that φ is true in every world in this subset, see (2). In our alternative interpretation of $\overline{\mathbf{K}}$ we introduce a new class of worlds W_Y, each world representing subsets of accessible worlds of W. That is, we represent the set Y by a designated world in this new class of worlds. Furthermore, instead of having just one accessibility relation R, here, we have for each $n \in \mathbb{N}_0$ a different accessibility relation R_n. Their domain is W and their range is restricted to W_Y. More precisely, for any n, R_n relates worlds in W to those elements in W_Y which represent sets (of worlds in W) of cardinality greater than n. And finally, there is an additional designated accessibility relation, denoted E for 'element of', which relates the new kind of worlds W_Y to the worlds in W again. The relation E represents the element-of relation (strictly speaking the converse of the 'element of' relation) between a subset Y of W and its elements. For example, consider the formula $\Diamond_3\varphi$. According to the definition of the previous section $\Diamond_3\varphi$ is true in a world x iff there are at least 4 worlds to which x is R-related. This definition is depicted in the first picture below. The second picture depicts our new alternative view.

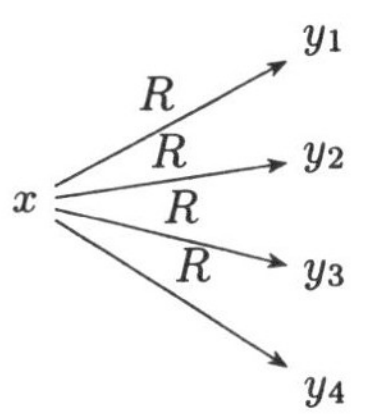

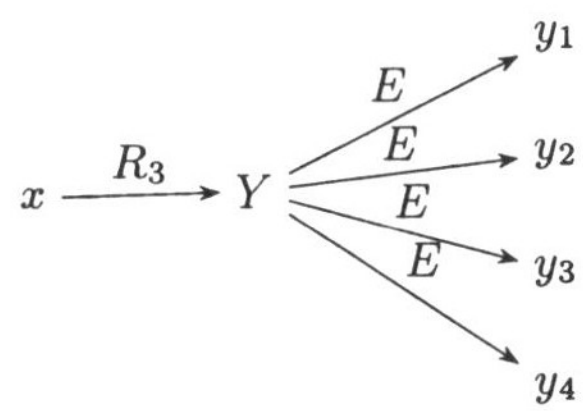

The relation R is replaced by the relational composition of the two new relations R_3 and E. In the process we have introduced a new world which we labelled Y as it is meant to represent the set of worlds y_1, y_2, y_3 and y_4.

The alternative semantics for $\overline{\mathbf{K}}$ sketched above characterizes a new graded modal logic which we describe now. We call the system $\overline{\mathbf{K}}_E$. It is a normal multi-modal logic system with graded modalities. In this logic the operators $\Diamond_n$ are replaced by a combination of two operators. $\overline{\mathbf{K}}_E$ is more expressive than the system $\overline{\mathbf{K}}$. Nevertheless, it has similar properties as $\overline{\mathbf{K}}$ as we will show below.

Our system $\overline{\mathbf{K}}_E$ differs from an alternative translation into a multi-modal logic 'Lcount', developed by [1]. In their system there are n-place operators $\langle n\rangle$ with semantics

$$\mathcal{M}, x \models \langle n \rangle \varphi_1, \ldots, \varphi_n \quad \text{iff} \quad \begin{array}{l}\text{there are distinct } y_1, \ldots, y_n \text{ such that} \\ \mathcal{M}, y_1 \models \varphi_1 \text{ and } \ldots \text{ and } \mathcal{M}, y_n \models \varphi_n\end{array}$$

Calculi based on this semantics, however, seem not to be much different to the calculi based on the original semantics for graded modalities. In a corresponding tableaux system one has to introduce witnesses for the worlds again, but this is what we want to avoid.

3.1 *The system* $\overline{\mathbf{K}}_E$

The language of $\overline{\mathbf{K}}_E$ is that of $\overline{\mathbf{K}}$ with the graded modalities $\Diamond_n$ and $\Box_n$ replaced by the symbols $\langle n \rangle$, $[n]$, $\Diamond$ and $\Box$. Formulae of $\overline{\mathbf{K}}_E$ have the following forms:

$$p, q, \ldots, \quad \bot, \quad \varphi \to \psi, \quad \langle n \rangle \varphi, \quad \Diamond \varphi.$$

As in Section 2 we define the classical connectives in the usual way. The duals of $\langle n \rangle$ and $\Diamond$ are abbreviated as follows: For $n \in \mathbb{N}_0$, $[n]\varphi$ abbreviates $\neg\langle n \rangle\neg\varphi$ and $\Box\varphi$ abbreviates $\neg\Diamond\neg\varphi$. The intended meaning of $\langle n \rangle \varphi$ is,

φ is true in some world accessible by the binary relation R_n.

The intended meaning of $\Diamond\varphi$ is

φ is true in some world accessible by the binary relation E.

We call $\langle n \rangle$ and $[n]$ the *numerical operators* and $\Diamond$ and $\Box$ the *membership operators*. The relations R_n and E are defined as sketched above. Namely, W forms the domain of the R_n and the co-domain of E and the new class of worlds W_Y forms the co-domain of the R_n and the domain of E. Dually, the intended meaning of $[n]\varphi$ is

φ is true in all worlds accessible by R_n.

And the intended meaning of $\Box\varphi$ is

φ is true in all worlds accessible by E.

So, the syntax $\langle n \rangle \varphi$ (respectively $[n]\varphi$) is the shorthand for $\langle R_n \rangle \varphi$ (respectively $[R_n]\varphi$), $\Diamond\varphi$ (respectively $\Box\varphi$) is the shorthand for $\langle E \rangle \varphi$ (respectively $[E]\varphi$).

$\overline{\mathbf{K}}$-formulae of the form $\Diamond_n\varphi$ and $\Box_n\varphi$ can be formulated as $\overline{\mathbf{K}}_E$-expressions of the form

$$\langle n \rangle \Box \varphi \quad \text{and} \quad [n]\Diamond\varphi,$$

respectively. The logic $\overline{\mathbf{K}}_E$ is more expressive than $\overline{\mathbf{K}}$. It permits arbitrary combinations of modal operators, not only alternate combinations of necessity and possibility operators. For example, $[3][4]\Box\Box\varphi$ is a well-formed formula of $\overline{\mathbf{K}}_E$, although it may not make much sense in our intended semantics. However, there are combinations of modal operators which have no equivalent formulation in $\overline{\mathbf{K}}$, but which have interesting applications. Here is an example of such a formula:

$$[10](\textit{football-team} \rightarrow \Box \textit{football-player}).$$

It says that, if there is a set with more than 10 elements for which the proposition *football-team* holds then the proposition *football-player* must be true for all its elements. In this way we can distinguish between notions like teams which we interpret as sets and notions like players which we interpret as elements.

We now give a Hilbert axiomatization for $\overline{\mathbf{K}}_E$ and investigate its characteristic frames.

DEFINITION 4 *The axioms and rules of the system* $\overline{\mathbf{K}}_E$ *are:*

N1 *the axioms of propositional logic and Modus Ponens*

N2 *the K-axioms for* $[n]$ *and* $\Box$*:*

$\vdash_{\overline{\mathbf{K}}_E} [n](\varphi \rightarrow \psi) \rightarrow ([n]\varphi \rightarrow [n]\psi)$ $\qquad$ $\vdash_{\overline{\mathbf{K}}_E} \Box(\varphi \rightarrow \psi) \rightarrow (\Box\varphi \rightarrow \Box\psi)$

N3 *the necessitation rules for* $[n]$ *and* $\Box$*:*

If $\vdash_{\overline{\mathbf{K}}_E} \varphi$ *then* $\vdash_{\overline{\mathbf{K}}_E} [n]\varphi$ $\qquad$ *if* $\vdash_{\overline{\mathbf{K}}_E} \varphi$ *then* $\vdash_{\overline{\mathbf{K}}_E} \Box\varphi$

N4 $\vdash_{\overline{\mathbf{K}}_E} [0]\Diamond\varphi \rightarrow [n]\Box\varphi$

N5 $\vdash_{\overline{\mathbf{K}}_E} \langle n\rangle\Box\varphi \rightarrow \langle n\rangle\Diamond\varphi$

N6 $\vdash_{\overline{\mathbf{K}}_E} [n]\varphi \rightarrow [n+1]\varphi$

N7 $\vdash_{\overline{\mathbf{K}}_E} \langle n+m\rangle\Box(\varphi \vee \psi) \rightarrow (\langle n\rangle\Box\varphi \vee \langle m\rangle\Box\psi)$

N8 $\vdash_{\overline{\mathbf{K}}_E} (\langle n\rangle\Box(\varphi \wedge \psi) \wedge \langle m\rangle\Box(\varphi \wedge \neg\psi)) \rightarrow \langle n+m+1\rangle\Box\varphi$

Although the box operators can be treated as abbreviations in terms of diamond operators, or vice versa, we use both operators and allow for arbitrary conversions back and forth with

$$\begin{array}{ccc} [n]\varphi & \leftrightarrow & \neg\langle n\rangle\neg\varphi \\ \Box\varphi & \leftrightarrow & \neg\Diamond\neg\varphi. \end{array}$$

N1–N3 are the basic axioms for every normal modal logic. N4 captures that, if something holds for every set of worlds with more than zero elements, that is, if it holds for every non-empty set of worlds, then it holds also for all sets with more than n elements. This means, the composition R_n ; E for n arbitrary is a subrelation of the composition R_0 ; E. N5 ensures that no set with more than n elements is empty. A contrapositive version of N6 is $\langle n+1\rangle\Box\varphi \rightarrow \langle n\rangle\Box\varphi$. It captures that sets with more than $n+1$ elements are sets with more than n elements.

N7 corresponds to A10 and is a bit more complicated to explain. As an example suppose $n = 2$ and $m = 4$. For these values N7 is

$$\langle 6\rangle\Box(\varphi \vee \psi) \rightarrow (\langle 2\rangle\Box\varphi \vee \langle 4\rangle\Box\psi)$$

which is equivalent to

$$(\langle 6\rangle\Box(\varphi \vee \psi) \wedge \neg\langle 2\rangle\Box\varphi) \rightarrow \langle 4\rangle\Box\psi.$$

In words, if there are more than 6, say 7, worlds in which the formula $\varphi \vee \psi$ is true, but it is not the case that φ holds in more than two worlds (i.e. $\neg\varphi$ is true in all but possibly two worlds) then in the remaining 5 of 7 worlds ψ is true. The axiom says that every $(n+m)$-element set can be decomposed into an n-element set and an m-element set, but note, the axiom is slightly stronger.

The intuition underlying N8 is the following: Suppose there is a set Y_1 with at least $n+1$ elements where $\varphi \wedge \psi$ holds and there is another set Y_2 with at least $m+1$ elements where $\varphi \wedge \neg\psi$ holds. Since ψ and $\neg\psi$ cannot hold simultaneously in one world, Y_1 and Y_2 must be disjoint. Thus, φ holds in $Y_1 \cup Y_2$ which is of cardinality at least $n+m+2$. Therefore, $\langle n+m+1\rangle\Box\varphi$ is true.

Now we turn to the semantics of $\overline{\mathbf{K}}_E$. The K-axioms and necessitation rules allow us to use the standard Kripke semantics. We choose the Kripke semantics for the multi-modal logic $\mathbf{K}_{(m)}$ where m is the number of modal operators. Note that $\overline{\mathbf{K}}_E$ has infinitely but countably many modal operators. A $\overline{\mathbf{K}}_E$*-frame* is a relational structure

$$\mathcal{F} = (W, \{R_n\}_{n\in\mathbf{N}_0}, E).$$

W is a non-empty set of worlds. The R_n are binary relations over W (each is associated with a modality $\langle n\rangle$) and E is a designated binary relation over W (associated with the modality $\Diamond$). The relations satisfy the properties N4–N8 given below. A *model* of $\overline{\mathbf{K}}_E$ *based on a frame* $\mathcal{F}$ is a pair $\mathcal{M} = (\mathcal{F}, V)$ where V is a function mapping propositional variables to subsets of W. Truth and satisfaction for the propositional fragment of $\overline{\mathbf{K}}_E$ is defined as for the propositional fragment of $\overline{\mathbf{K}}$. See (2) in the previous section. A modal formula is *satisfied* (is *true* or *holds*) in a world x iff depending on its form the following holds:

$\mathcal{M}, x \models_{\overline{\mathbf{K}}_E}$	$\langle n\rangle\varphi$	iff	there is a y such that $R_n(x,y)$ and $\mathcal{M}, y \models_{\overline{\mathbf{K}}} \varphi$
$\mathcal{M}, x \models_{\overline{\mathbf{K}}_E}$	$[n]\varphi$	iff	for all y such that $R_n(x,y)$, $\mathcal{M}, y \models_{\overline{\mathbf{K}}} \varphi$
$\mathcal{M}, x \models_{\overline{\mathbf{K}}_E}$	$\Diamond\varphi$	iff	there is a y such that $E(x,y)$ and $\mathcal{M}, y \models_{\overline{\mathbf{K}}} \varphi$
$\mathcal{M}, x \models_{\overline{\mathbf{K}}_E}$	$\Box\varphi$	iff	for all y such that $E(x,y)$, $\mathcal{M}, y \models_{\overline{\mathbf{K}}} \varphi$.

A formula is a *tautology* if for any frame $\mathcal{F}$ the formula is satisfied in all $\mathcal{F}$-based models.

The following are the characteristic properties of $\overline{\mathbf{K}}_E$-frames that correspond to the axioms N4–N8:

N4′ $\forall xyz\, ((R_n(x,y) \wedge E(y,z)) \rightarrow \exists u\, (R_0(x,u) \wedge \forall v\, (E(u,v) \rightarrow v = z)))$

N5′ $\forall xy\, (R_n(x,y) \rightarrow \exists z\, E(y,z))$

N6′ $\forall xy\, (R_{n+1}(x,y) \rightarrow R_n(x,y))$

N7′ $$\begin{aligned}&\forall xy\, R_{n+m}(x,y) \rightarrow\\ &\quad \forall fg\, \exists uv\, (R_n(x,u) \rightarrow E(u,f(u)) \wedge\\ &\qquad R_m(x,v) \rightarrow E(v,g(v))) \rightarrow\\ &\qquad (R_n(x,u) \wedge R_m(x,v) \wedge E(y,f(u)) \wedge f(u) = g(v))\end{aligned}$$

N8′ $$\begin{aligned}&\forall xyz\, (R_n(x,y) \wedge R_m(x,z) \wedge \forall u\, (E(y,u) \rightarrow \neg E(z,u)) \rightarrow\\ &\quad \exists v\, (R_{n+m+1}(x,v) \wedge \forall w\, (E(v,w) \rightarrow E(y,w) \vee E(z,w))))\end{aligned}$$

for any $n, m \in \mathbb{N}_0$.

We computed these properties with a tool, called SCAN[1], which is an implementation of the quantifier elimination algorithm of Gabbay and Ohlbach [15]. To this end we used the standard Kripke semantics as translation rules for translating the axioms into predicate logic. In this *relational translation* (this is the *standard translation* ST of van Benthem [29, 30]), the formula variables become universally quantified predicate variables. The quantifier elimination algorithm produces for these second-order formulae equivalent formulae without predicate quantifiers. That means ST(N4) $\leftrightarrow$ N4′ etc. This procedure guarantees soundness of the semantics with respect to the axiom system, i.e.

(4) if $\vdash_{\overline{\mathbf{K}}_E} \varphi$ then $\models_{\overline{\mathbf{K}}_E} \varphi$.

If all these ‘frame axioms’ (N4–N8) were first-order then the Sahlqvist Theorem [26, 29, 30] would ensure completeness of this frame class relative to the axioms. Unfortunately, N7′ is again second-order. Therefore we have to prove completeness explicitly. We do this indirectly *for translated* $\overline{\mathbf{K}}$ *formulae* by using the completeness of $\overline{\mathbf{K}}$ and the soundness and completeness of the translation into $\overline{\mathbf{K}}_E$ which is proven below. General completeness for arbitrary formulae is still open, but for the purpose of our translation, this is fortunately not necessary.

The correspondence property N4′ states that all singleton subsets of the set of worlds accessible by R_n are uniquely represented by a world accessible by R_0. N5′ asserts that every world accessible by R_n leads via E to another world. We say E is *weakly serial*. (Recall, a relation R is said to be *serial* (or *total*) iff $\forall x\, \exists y\, R(x,y)$). By N6′ the set $\{R_n\}_{n \in \mathbb{N}_0}$ of R_n relations forms a linear order with R_0 being the largest element, since for any $m \rangle n$, R_m is a subrelation of R_n.

[1] SCAN is accessible via World Wide Web at

`http://www.mpi-sb.mpg.de/guide/staff/ohlbach/scan/scan.html.`

This is a WWW interface for activating the program remotely. We invite the reader to use the tool and verify the above correspondence properties for N4–N8.

The correspondence property N7′ of N7 expresses intuitively that every set y with more than $n+m$ elements can be decomposed into a set u with more than n elements and a set v with more than m elements, and if y happens to have exactly $n+m+1$ elements then u and v overlap in at least one element.

N8′ expresses, as already mentioned, that for disjoint sets the cardinality of their union is the sum of the cardinalities of the sets.

For a better understanding of the frame properties it is helpful to think of the variable y in $R_n(x,y)$ and $E(y,z)$ as representing a set Y, $R_n(x,y)$ as representing that the cardinality of Y is greater than n, and $E(y,z)$ as representing that z is an element of Y. Then N4′–N8′ represent:

N4″ $\forall Yz\,((|Y|\rangle n \wedge z \in Y) \rightarrow \{z\} \subseteq Y$

N5″ $\forall y\,(|Y|\rangle n \rightarrow Y \neq \varnothing)$

N6″ $\forall Y\,(|Y|\rangle n+1 \rightarrow |Y|\rangle n)$

N7″ $\forall Y\,(|Y|\rangle n+m \rightarrow$
$\forall fg\,\exists UV(|U|\rangle m \wedge |V|\rangle m) \rightarrow$
if f selects from U and g from V then $f(U) \in Y \wedge f(U) = g(V))$

N8″ $\forall YZ\,((|Y|\rangle n \wedge |Z|\rangle m \wedge Y \cap Z = \varnothing) \rightarrow$
$\exists V\,(|V|\rangle n+m+1 \wedge V \subseteq Y \cup Z))$.

We can show that the standard class of frames associated with $\overline{\mathbf{K}}_E$ have the expected structure, namely that all worlds accessible by R_n have more than n E-successors. However, non-standard $\overline{\mathbf{K}}_E$-frames exist which do not have this intended structure. The problem is, we cannot enforce in a Hilbert system that R_1-accessible worlds have *more than* one E-successor. This may be captured by an axiom like

$$[1](\exists p\,(\Diamond p \wedge \Diamond \neg p)),$$

or a rule similar to Gabbay's irreflexivity rule (but this gives no new theory) [16]. See also [25]. The modal language of $\overline{\mathbf{K}}$ and $\overline{\mathbf{K}}_E$ is not expressive enough to characterize this class of frames. On the other hand, we can show using an inductive argument that whenever an R_1-successor has more than one E-successor, then for any positive integer n every R_n-successor has more than n E-successors. This is to say, the induction step goes through, but unfortunately the base case of the induction cannot be guaranteed. Because the translation of the logic $\overline{\mathbf{K}}$ into the logic $\overline{\mathbf{K}}_E$ is sound and complete (we show this below), we know whenever a translated $\overline{\mathbf{K}}$-formula has a model then it has a model with the expected structure.

We did not investigate the non-standard models further. It may turn out that they are p-morphic images of standard models, in which case they are completely irrelevant because normal modal logics cannot distinguish p-morphic images.

3.2 From $\overline{\mathbf{K}}$ to $\overline{\mathbf{K}}_E$

Next we define a translation function mapping formulae of $\overline{\mathbf{K}}$ into formulae of $\overline{\mathbf{K}}_E$. We show that the translation is sound and complete.

DEFINITION 5 *The translation function* Π *maps* $\overline{\mathbf{K}}$*-formulae into* $\overline{\mathbf{K}}_E$*-formulae according to the following constraints:*

$$\begin{array}{rcl} \Pi(p) & = & p \\ \Pi(\neg\varphi) & = & \neg\Pi(\varphi) \\ \Pi(\varphi\ @\ \psi) & = & \Pi(\varphi)\ @\ \Pi(\psi) \\ \Pi(\Diamond_n\varphi) & = & \langle n\rangle\Box\Pi(\varphi) \\ \Pi(\Box_n\varphi) & = & [n]\Diamond\Pi(\varphi), \end{array}$$

where p denotes any propositional variable and @ denotes any binary logical connective $\wedge$, $\vee$, $\rightarrow$ *or* $\leftrightarrow$.

THEOREM 6 (Soundness of Π) *The translation* Π *from* $\overline{\mathbf{K}}$ *into* $\overline{\mathbf{K}}_E$ *is sound. That is, for any formula* φ *of* $\overline{\mathbf{K}}$

$$\textit{if } \vdash_{\overline{\mathbf{K}}} \varphi \textit{ then } \vdash_{\overline{\mathbf{K}}_E} \Pi(\varphi).$$

Proof. Suppose φ is a theorem in $\overline{\mathbf{K}}$. We proceed by induction on the length of the proof of φ and show that the proof sequence of φ in $\overline{\mathbf{K}}$ determines a proof sequence of $\Pi(\varphi)$ in $\overline{\mathbf{K}}_E$. We are done if we show that the Π-translations of the axioms and the rules of $\overline{\mathbf{K}}$ are $\overline{\mathbf{K}}_E$-theorems.

Π leaves the propositional axioms and Modus Ponens unchanged. The translation of the necessitation rule N is:

$$\vdash_{\overline{\mathbf{K}}_E} \varphi \quad \text{implies} \quad \vdash_{\overline{\mathbf{K}}_E} [0]\Diamond\varphi.$$

If φ holds then, by the necessitation rule for $\Box$ and $[0]$, $[0]\Box\varphi$ holds. Apply modus ponens using the contrapositive instance with $n = 0$ of N5 and get $[0]\Diamond\varphi$[2].

The translation of A2 is a contrapositive version of N6. It remains to prove the translations of A3 and A4 can be derived from the axioms of $\overline{\mathbf{K}}_E$ using the rules of $\overline{\mathbf{K}}_E$.

For A3 we prove

$$\Pi(\text{A3}) = [0]\Diamond(\varphi \rightarrow \psi) \rightarrow (\langle n\rangle\Box\varphi \rightarrow \langle n\rangle\Box\psi)$$

is a theorem in $\overline{\mathbf{K}}_E$. Suppose $[0]\Diamond(\varphi \rightarrow \psi)$ and $\langle n\rangle\Box\varphi$ hold. Suppose further that $\neg\langle n\rangle\Box\psi$, i.e. $[n]\Diamond\neg\psi$, holds. From $[0]\Diamond(\varphi \rightarrow \psi)$ we infer by N4 that $[n]\Box(\varphi \rightarrow \psi)$ holds. By the K-axiom for $\Box$, it follows that $[n](\Box\varphi \rightarrow \Box\psi)$. This is equivalent

[2]Formally, instead of φ one has to consider $\Pi(\varphi)$. But for the proofs this makes no difference.

to $[n](\neg\Box\psi \rightarrow \neg\Box\varphi)$, i.e. $[n](\Diamond\neg\psi \rightarrow \Diamond\neg\varphi)$. Using the **K**-axiom for $[n]$ we infer $[n]\Diamond\neg\psi \rightarrow [n]\Diamond\neg\varphi$. From $\neg\langle n\rangle\Box\psi$, which is equivalent to $[n]\Diamond\neg\psi$, using Modus Ponens we get $[n]\Diamond\neg\varphi$, or equivalently $\neg\langle n\rangle\Box\varphi$. This contradicts $\langle n\rangle\Box\varphi$. Thus $\Pi(\text{A3})$ is derivable in $\overline{\mathbf{K}}_E$.

For A4: Let

$$\phi = [0]\Diamond\neg(\varphi \wedge \psi) \wedge \langle n-1\rangle\Box\varphi \wedge \neg\langle n\rangle\Box\varphi \wedge \langle m-1\rangle\Box\psi \wedge \neg\langle m\rangle\Box\psi.$$

Then $\Pi(\text{A4})$ is equivalent to

$$\phi \rightarrow (\langle n+m-1\rangle\Box(\varphi \vee \psi) \wedge \neg\langle n+m\rangle\Box(\varphi \vee \psi)).$$

We prove this in two steps. First, we prove $\phi \rightarrow \langle n+m-1\rangle\Box(\varphi \vee \psi)$. Suppose ϕ holds. It suffices to show

(5) $\langle n-1\rangle\Box(\varphi \wedge \neg\psi)$.

From $\langle n-1\rangle\Box(\varphi \wedge \neg\psi)$, or equivalently $\langle n-1\rangle\Box((\varphi \vee \psi) \wedge \neg\psi)$, and $\langle m-1\rangle\Box\psi$, or equivalently $\langle m-1\rangle\Box((\varphi \vee \psi) \wedge \psi)$, using axiom N8 we deduce $\langle n+m-1\rangle\Box(\varphi \vee \psi)$.

For proving that (3.2) follows from ϕ we proceed by contradiction. Suppose that $\neg\langle n-1\rangle\Box(\varphi \wedge \neg\psi)$, i.e. $[n-1]\Diamond(\neg\varphi \vee \psi)$ holds. From $[0]\Diamond\neg(\varphi \wedge \psi)$ using N4 we get $[n-1]\Box\neg(\varphi \wedge \psi)$ Since, in general, in any normal modal logic $\Box$ (and, in particular, $[n-1]$) distributes over conjunction, we obtain

(6) $[n-1](\Diamond(\neg\varphi \vee \psi) \wedge \Box(\neg\varphi \vee \neg\psi))$.

The **K**-axiom for $\Box$ is equivalent to $(\Box\varphi \wedge \Diamond\psi) \rightarrow \Diamond(\varphi \wedge \psi)$. Thus (3.2) is equivalent to $[n-1](\Diamond((\neg\varphi \vee \psi) \wedge (\neg\varphi \vee \neg\psi)))$. This in turn is equivalent to $[n-1]\Diamond\neg\varphi$. Thus $\neg\langle n-1\rangle\Box\varphi$ which contradicts $\langle n-1\rangle\Box\varphi$.
Next, we prove $\phi \rightarrow \neg\langle n+m\rangle\Box(\varphi \vee \psi)$. Suppose ϕ holds. Then, in particular, $\neg\langle n\rangle\Box\varphi$ and $\neg\langle m\rangle\Box\psi$ hold, and $\neg\langle n+m\rangle\Box(\varphi \vee \psi)$ is derivable by (the contraposition of) N7. Therefore, $\Pi(\text{A4})$ is a theorem in $\overline{\mathbf{K}}_E$. ■

For proving completeness a semantic proof suffices. We prove (in the next the theorem) that for a translated formula $\Pi(\varphi)$ which is true in all $\overline{\mathbf{K}}_E$-frames, φ is true in all $\overline{\mathbf{K}}$-frames. If $\Pi(\varphi)$ is provable in $\overline{\mathbf{K}}_E$ then the soundness of the $\overline{\mathbf{K}}_E$-semantics guarantees that $\Pi(\varphi)$ is true in all $\overline{\mathbf{K}}_E$-frames, then φ is true in all $\overline{\mathbf{K}}$-frames and then it is provable in $\overline{\mathbf{K}}$ (by the completeness of $\overline{\mathbf{K}}$).

THEOREM 7 *For any formula φ of $\overline{\mathbf{K}}$,*

$$\textit{if } \models_{\overline{\mathrm{K}}_E} \Pi(\varphi) \textit{ then } \models_{\overline{\mathrm{K}}} \varphi.$$

Proof. Our strategy is the following. Suppose $\models_{\overline{\mathrm{K}}_E} \Pi(\varphi)$. (i) For an arbitrary $\overline{\mathbf{K}}$-frame $\mathcal{F}$ we construct a $\overline{\mathbf{K}}_E$-frame $\mathcal{F}'$. $\Pi(\varphi)$ is valid in this particular frame $\mathcal{F}'$. Then we show, (ii) φ is valid in $\mathcal{F}$.

(i): Take any $\overline{\mathbf{K}}$-frame $\mathcal{F} = (W, R)$. We construct a $\overline{\mathbf{K}}_E$-frame $\mathcal{F}'$ as an extension of the frame $\mathcal{F}$ as follows. For any world $x \in W$ let $R(x)$ be the R-image of x. For any finite subset Y of $R(x)$ with $|Y| = n + 1$ for n a non-negative integer, we add Y as a new world to $\mathcal{F}$. We call Y a 'set-world'. Note, every Y is non-empty. We define every relation R_m for $m \leq n$ to contain the pair (x, Y), and, we define the relation E to contain all pairs (Y, z) for $z \in Y$. Furthermore, we assume the relations R_n and E are the smallest relations satisfying these conditions. Now, define $\mathcal{F}'$ to be the relational structure

$$(W', \{R_n\}_{n \in \mathbf{N}_0}, E)$$

with W' being the set of worlds of $\mathcal{F}'$ that includes the set of worlds W of $\mathcal{F}$ and all set-worlds Y. Note, the set-worlds have no R_n-successors and the worlds in W have no E-successors. We show that $\mathcal{F}'$ is a frame for $\overline{\mathbf{K}}_E$ by showing that $\mathcal{F}'$ satisfies the properties N4′–N8′.

N4′: If $R_n(x, y) \wedge E(y, z)$ holds then y must be a set-world with $|y|\rangle n$ and $z \in y$. For $u = \{z\}$ we obtain $R_0(x, u) \wedge \forall v\ (E(u, v) \rightarrow v = z)$.

N5′: If $R_n(x, y)$ then y is a non-empty set-world, i.e. $\exists z\ E(y, z)$ is true.

N6′: If $R_{n+1}(x, y)$ then y is a set-world with $|y|\rangle n + 1\rangle n$, i.e. $R_n(x, y)$ holds as well.

Recall N7′:

$$\begin{aligned}\forall xy\ R_{n+m}(x, y) \rightarrow\ & \forall fg\ \exists uv(R_n(x, u) \rightarrow E(u, f(u)) \wedge \\ & R_m(x, v) \rightarrow E(v, g(v))) \rightarrow \\ & (R_n(x, u) \wedge R_m(x, v) \wedge E(y, f(u)) \wedge f(u) = g(v))\end{aligned}$$

$R_{n+m}(x, y)$ means that y is a set-world with $|y|\rangle n + m$. We distinguish two cases.

<u>Case 1</u>: $|y| = n+m+1$. Let f and g be any functions mapping worlds to worlds. If there is at least one R_n-accessible set-world u with $|u|\rangle n$ and f does not map u to one of its elements ($\neg E(u, f(u))$) or there is at least one R_m-accessible set-world v with $|v|\rangle n$ and g does not map v to one of its elements ($\neg E(v, g(v))$) then we can choose this u or v, respectively. Then the implication

$$\begin{aligned}&(R_n(x, u) \rightarrow E(u, f(u)) \wedge R_m(x, v) \rightarrow E(v, g(v))) \rightarrow \\ &(R_n(x, u) \wedge R_m(x, v) \wedge E(y, f(u)) \wedge f(u) = g(v))\end{aligned} \quad (7)$$

is true because the premiss is false.

Now assume, f chooses for every set-world u with $|u|\rangle n$ some element $f(u) \in u$ and g chooses for every set-world v with $|v|\rangle m$ some element $g(v) \in v$. The key observation for the proof is that for every set with $n + m + 1$ elements every decomposition into a set u with $n + 1$ elements and a set v with $m + 1$ elements overlaps in at least one element. Thus, the situation is as depicted in the following figure.

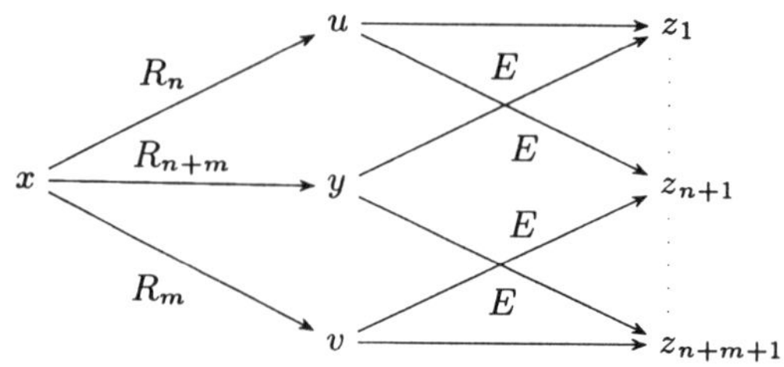

For finding the right u and v we follow this procedure: we start by choosing a subset $u_1 \subseteq y$ with $n+1$ elements. Suppose $f(u_1) = x_1$. If there is a subset $v \subseteq y$ with $|v|\rangle m$ and $g(v) = x_1$ we are done. Suppose for no such subset we have $g(v) = x_1$. x_1 is marked as 'not an image of g'. Now we choose another $n+1$-element subset u_2 of y which *does not contain* x_1. Suppose $f(u_2) = x_2$. Again, if for some subset v with $|v|\rangle m$ we find $g(v) = x_2$ we are done. If not, we mark x_2 as 'not an image of g'. We continue until we have found a suitable u and v, or until exactly $n+1$ worlds remain which are not marked 'not an image of g'. In the latter case we choose this set for u. Suppose $f(u) = x$. Take $v = y \setminus u \cup \{x\}$. $|v| = m+1$ and $g(v) \neq z$ for all $z \in y \setminus u$. Since g must select some element in v, $g(v) = x$ is the only choice. Thus, suitable u and v exist that satisfy (7).

Case 2: $|y|\rangle n+m+1$. Take any subset $y' \subseteq y$ with $|y'| = n+m+1$. By Case 1, we can find for any f and g subsets $u \subseteq y'$ and $v \subseteq y'$ with the property (7). But these are also subsets of y and therefore the property holds as well.

N8$'$: This property expresses that the union of two disjoint sets of cardinality $\rangle n$ and $\rangle m$ is a set with cardinality $\rangle n+m+1$, and this is true in $\mathcal{F}'$.

We have proved $\mathcal{F}'$ is a frame for $\overline{\mathbf{K}}_E$.

(ii): Let $\mathcal{M} = (\mathcal{F}, V)$ be any model based on $\mathcal{F}$ with V an arbitrary valuation. Define $\mathcal{M}'$ to be the model $(\mathcal{F}', V)$. (Observe that $V(p)$ does not, and need not contain set-worlds.) (ii) follows from

$$\mathcal{M}', x \models_{\overline{\mathbf{K}}_E} \Pi(\varphi) \quad \text{iff} \quad \mathcal{M}, x \models_{\overline{\mathbf{K}}} \varphi \tag{8}$$

where x is any world in W. We prove (8) by induction on the structure of φ. The base case in which φ is any propositional variable is trivial. The inductive step for the propositional connectives goes through easily. We consider the case φ is of the form $\Diamond_n\psi$. (The case for φ of the form $\Box_n\psi$ is dual.) The inductive hypothesis is:

$$\mathcal{M}', x \models_{\overline{\mathbf{K}}_E} \Pi(\psi) \quad \text{iff} \quad \mathcal{M}, x \models_{\overline{\mathbf{K}}} \psi.$$

Suppose $\mathcal{M}', x \models_{\overline{\mathbf{K}}_E} \Pi(\Diamond_n\psi)$, i.e. $\langle n\rangle\Box\psi$ is true at x in $\mathcal{M}'$. Then, $R_n(x, Y)$ in $\mathcal{F}'$ for some set $Y \subseteq R(x)$ with $n+1$ elements and for all $z \in Y$ we have $\mathcal{M}', z \models_{\overline{\mathbf{K}}_E} \psi$ and by the inductive hypothesis $\mathcal{M}, z \models_{\overline{\mathbf{K}}} \psi$. There are at least n such z, therefore, $\mathcal{M}, x \models_{\overline{\mathbf{K}}} \Diamond_n\psi$.

Conversely, suppose $\mathcal{M}, x \models_{\overline{\mathbf{K}}} \Diamond_n \psi$. This means the world x has more than n successors by R in all of which ψ is true. Consequently, a set Y with cardinality $n+1$ exists that contains R-successors y of x and in all y, ψ is true. This implies, in $\mathcal{F}'$, x and Y are connected by R_n and Y is connected to all its elements by E. Thus, $\langle n \rangle \Box \psi$ is true in x.

This completes the proof. ■

As consequences we get the following two theorems.

THEOREM 8 (Completeness of Π) *The translation* Π *from* $\overline{\mathbf{K}}$ *into* $\overline{\mathbf{K}}_E$ *is complete. That is, for any formula* φ *of* $\overline{\mathbf{K}}$,

$$\textit{if } \vdash_{\overline{\mathrm{K}}_E} \Pi(\varphi) \textit{ then } \vdash_{\overline{\mathrm{K}}} \varphi.$$

Proof. Suppose $\Pi(\varphi)$ is a theorem in $\overline{\mathbf{K}}_E$, i.e. $\vdash_{\overline{\mathrm{K}}_E} \Pi(\varphi)$. Then, since $\overline{\mathbf{K}}_E$ is sound (4), $\models_{\overline{\mathrm{K}}_E} \Pi(\varphi)$. By the previous theorem $\models_{\overline{\mathrm{K}}} \varphi$. $\overline{\mathbf{K}}$ is sound and complete (Theorem 3). Therefore, it follows that $\vdash_{\overline{\mathrm{K}}} \varphi$. ■

Now, we can show the completeness of the semantics of $\overline{\mathbf{K}}_E$ with respect to its axiomatization *for translated formulae*.

THEOREM 9 (Relative completeness of $\overline{\mathbf{K}}_E$) *For any* $\overline{\mathbf{K}}$ *formula* φ

$$\textit{if } \models_{\overline{\mathrm{K}}_E} \Pi(\varphi) \textit{ then } \vdash_{\overline{\mathrm{K}}_E} \Pi(\varphi).$$

Proof. If $\Pi(\varphi)$ holds in all $\overline{\mathbf{K}}_E$-frames then φ holds in all $\overline{\mathbf{K}}$-frames (by Theorem 7), then φ is provable in $\overline{\mathbf{K}}$ (by the completeness of $\overline{\mathbf{K}}$, Theorem 3), and then $\Pi(\varphi)$ is provable in $\overline{\mathbf{K}}_E$ (by the soundness of the translation, Theorem 6). ■

4 FROM MULTI-MODAL LOGIC TO PREDICATE LOGIC

We aim at making available first-order theorem proving methods for reasoning with graded modal expressions. In the previous section we have embedded the logic $\overline{\mathbf{K}}$ in the multi-modal logic $\overline{\mathbf{K}}_E$. Unfortunately, one of the axioms, namely N7, is not first-order definable in the standard Kripke semantics. Its relational translation N7$'$ is a second-order formula. So, instead of using the standard relational translation we use the functional translation as proposed in Ohlbach and Schmidt [22] for non-first-order axioms like McKinsey's axiom.

The functional translation method was proposed by various authors, for example Ohlbach [23, 24], [10], [18], [2] and [36]. It exploits the fact that every binary relation can be decomposed into a set AF_R of functions, called *accessibility functions*. Any (non-empty) relation R is defined by:

$$R(x, y) \leftrightarrow \exists \gamma \in AF_R \; y = \gamma(x).$$

In the functional translation we quantify over the accessibility functions instead of worlds. For modalities determined by serial (i.e. total) accessibility relations, that is, for modalities satisfying the D-axiom the functional translation rules for modal formulae are:

$$\pi_f([R]\psi, x) = \forall\gamma{:}AF_R\ \pi_f(\psi, \downarrow(\gamma, x))$$
$$\pi_f(\langle R\rangle\psi, x) = \exists\gamma{:}AF_R\ \pi_f(\psi, \downarrow(\gamma, x)).$$

The target logic is a many-sorted predicate logic, in which AF_R is the sort for accessibility functions defining R and the symbol $\downarrow$ is a function symbol for the 'apply' function (this means, the application of a function f to x, i.e. $f(x)$, is encoded by $\downarrow(f, x)$). For modalities determined by accessibility relations that are not serial the set of accessibility functions AF_R contains partial functions. Accordingly, the functional translation π_f of modal formulae must compensate for partiality by an extra condition involving a predicate de_R. For non-serial modalities π_f is defined by:

$$\neg de_R(x) \rightarrow \forall\gamma{:}AF_R\ \pi_f(\psi, \downarrow(\gamma, x)) \quad \text{and} \quad \neg de_R(x) \wedge \exists\gamma{:}AF_R\ \pi_f(\psi, \downarrow(\gamma, x)).$$

The term $\neg de_R(x)$ is meant to capture that x is not a dead-end in the relation R.

We now give the formal definition of the functional translation Π_f for multi-modal logics of which $\mathbf{K}_{(m)}$ is the weakest. Π_f is a function mapping formulae of $\mathbf{K}_{(m)}$ to formulae of a many-sorted predicate logic PL_M (with predicate variables) with a signature specified by:

(i) sort symbols $\perp$ (for the bottom sort), W (for the world sort) and $W^\perp$.
(ii) sort declarations $W \sqsubseteq W^\perp$ and $\perp \sqsubseteq W^\perp$.
(iii) sort symbols AF_R for every modality $\langle R\rangle$
(iv) a binary function symbol $\downarrow$ declared by $\downarrow: AF_R \times W^\perp \rightarrow W^\perp$ for all AF_R.
(v) predicate symbols de_R for every modality $\langle R\rangle$ (*de* is short for dead-end, or in our application, $de_n(x)$ means the set-world x does *not* represent more than n elements) and
(vi) for each propositional variable p there is a unary predicate symbol p (we purposely use the same symbols).

This signature defines a many-sorted predicate logic we refer to as PL_M.

DEFINITION 10 (The functional translation) *Let π_f be a function that takes two arguments: a modal formula and a 'world term' which are mapped to a formula in PL_M. π_f is defined inductively by*

$$\pi_f(p, w) = p(w) \qquad \textit{for } p \textit{ a propositional variable}$$
$$\pi_f([R]\psi, x) = \neg de_R(x) \rightarrow \forall\gamma{:}AF_R\ \pi_f(\psi, \downarrow(\gamma, x))$$
$$\pi_f(\langle R\rangle\psi, x) = \neg de_R(x) \wedge \exists\gamma{:}AF_R\ \pi_f(\psi, \downarrow(\gamma, x))$$

and for the propositional connectives π_f is a homomorphism.

The functional translation *for a multi-modal formula φ with propositional variables $p_1, \ldots, p_n$ is defined by*

$$\Pi_f(\varphi) = \forall p_1, \ldots, p_n \forall w{:}W\ \pi_f(\varphi, w).$$

For a Hilbert rule of the form 'from φ_1 and ... and φ_n infer φ'

$$\Pi_f(\varphi_1) \wedge \ldots \wedge \Pi_f(\varphi_n) \rightarrow \Pi_f(\varphi)$$

is the functional translation. Π_f maps a set Φ of Hilbert axioms and rules to the conjunction of the functional translation of the members:

$$\Pi_f(\Phi) = \bigwedge_{\varphi \in \Phi} \Pi_f(\varphi).$$

Π_f *is called the* functional translation function.

For serial modalities the $\neg de_R(x) \ldots$ part of the definition in π_f can be omitted.

As with the relational translation, the functional translation of the Hilbert axioms yields second-order formulae with universally quantified predicate variables. This translation is sound, i.e. whatever can be proved from the axioms, holds in the models of these second-order formulae. If these second-order formulae are equivalent to a first-order formula then the Sahlqvist theorem together with the completeness of the transition from the relational to the functional representation (which is easy) guarantees completeness. If the second-order formulae are not equivalent to first-order formulae then completeness is still an open problem. Unfortunately this is the case for $\overline{\mathbf{K}}_E$.

Ohlbach and Schmidt [22] prove the following relativized soundness and completeness result for the functional translation Π_f.

THEOREM 11 **(Relative soundness and completeness of the functional translation)** *Let Φ be additional Hilbert axioms in a propositional modal logic $\mathbf{K}_{(m)}$ and φ any modal formula. If the relational second-order translation of Φ is complete then*

φ is a Φ-theorem iff $\Pi_f(\Phi) \rightarrow \Pi_f(\varphi)$ is a predicate logic theorem in PL_M.

The result of the translation by Π_f is in general not a first-order expression. The translation is useful only if the axioms in $\Pi_f(\Phi)$ can be described by a set Φ' of first-order formulae. If $\Pi_f(\Phi)$ is equivalent to such a set Φ' the implication

$$\Pi_f(\Phi) \rightarrow \Pi_f(\varphi)$$

can be proved by refuting the formula

$$\Phi' \wedge \neg\Pi_f(\varphi).$$

$\Pi_f(\varphi)$ is a monadic second-order formula only with second-order universal quantifiers. In the negation normal form of its negation $\neg\Pi_f(\varphi)$ only existentially quantified second-order predicate variables occur and these are treated as ordinary first-order predicates. Therefore, $\Phi' \rightarrow \Pi_f(\varphi)$ can be proved with the standard first-order procedures.

Not every axiomatization Φ has an equivalent first-order formulation. The theorem above can be strengthened for certain axioms without first-order relational characterizations when we use the following quantifier exchange rule.

DEFINITION 12 (Quantifier exchange rule) *Let φ be any modal formula in $\mathbf{K}_{(m)}$. Define an operation Υ on PL_M which transforms the functional translation $\Pi_f(\varphi)$ into its prenex normal form according to the rule*

$$(9) \quad \exists\gamma{:}AF_R\, \forall\delta{:}AF_R\, \psi \quad \leadsto \quad \forall\delta{:}AF_R\, \exists\gamma{:}AF_R\, \psi.$$

The operation Υ moves existential functional quantifiers inwards thus weakening the original formula. $\Upsilon(\Pi_f(\varphi))$ implies $\Pi_f(\varphi)$, but not conversely. The quantifier exchange rule exploits that one relational frame in general corresponds to many 'functional frames', and there is always one which is rich enough to allow for moving existential quantifiers over universal quantifiers. This is investigated in Ohlbach and Schmidt [22] where a stronger theorem than Theorem 11 is proved, namely:

THEOREM 13 **(Relative soundness and completeness of the functional translation with the quantifier exchange rule)** *Let Φ be additional Hilbert axioms in a propositional modal logic $\mathbf{K}_{(m)}$ and φ any modal formula. If the relational second-order translation of Φ is complete then*

$$\varphi \textit{ is a } \Phi\textit{-theorem} \quad \textit{iff} \quad \Upsilon(\Pi_f(\Phi)) \rightarrow \Upsilon(\Pi_f(\varphi)) \textit{ is a theorem in } PL_M,$$

provided in $\Upsilon(\Pi_f(\varphi))$ all existential functional quantifiers are moved inward as far as possible.

This theorem says that the original formula φ can be proved to be a theorem in the system Φ by proving its weakened translation $\Upsilon(\Pi_f(\varphi))$ using the weakened forms $\Upsilon(\Pi_f(\Phi))$ of the translations of the axioms in Φ. In $\Upsilon(\Pi_f(\varphi))$ the existential functional quantifiers are pushed inward as far as possible. Negating $\Upsilon(\Pi_f(\varphi))$ we simultaneously replace universal quantifiers by existential quantifiers and existential quantifiers by universal quantifiers. The quantifier prefix of the prenex normal form of $\neg\Upsilon(\Pi_f(\varphi))$ consists of a sequence of existentially quantified predicate variables p_i (ended with an existentially quantified world variable) followed by a sequence of existentially quantified functional variables, followed by a sequence of universally quantified functional variables. In the Skolemized clause form of $\neg\Upsilon(\Pi_f(\varphi))$ no Skolem functions occur, only Skolem constants. This simplifies the translation considerably. More importantly, Υ allows us to move existential quantifiers inward as far as we like. This weakens an axiom like the McKinsey axiom just enough so that we get a first-order translation for the axiom. This operation works for axiom N7 as well, as we will see in the next section.

The functional translation generates nested $\downarrow$-terms as arguments to predicates. We can avoid these by using the world path notation of Ohlbach [23]. To this end we add a new sort symbol AF^* to PL_M and we let $\circ$ be a new binary function symbol. Furthermore, we include the following axioms and sort declarations:

$$AF_R \sqsubseteq AF^*$$
$$\circ : AF^* \times AF^* \to AF^*$$
$$\forall x{:}W\ \forall \gamma, \delta{:}AF^*\ \downarrow(\gamma \circ \delta, x) = \downarrow(\delta, \downarrow(\gamma, x))$$
$\circ$ is associative.

$\circ$ denotes composition operation of accessibility functions and AF^* denotes the set of all possible compositions of all accessibility functions in the union of AF_R. Instead of nested $\downarrow$-terms, like $\downarrow(\delta, \downarrow(\gamma, x))$, we use a more economic notation and write $\downarrow((\gamma \circ \delta), x)$ or $\downarrow([\gamma\delta], x)$ (omitting $\circ$), instead. The latter uses the world path syntax which we prefer from here on.

Notice that the conditions in the two theorems requiring that the relational translation of the axioms into second-order logic is complete means that all formulae which are valid in the frames characterized by these second-order formulae are provable from the axioms. For $\overline{\mathbf{K}}_E$ we showed this for the translated $\overline{\mathbf{K}}$ formulae only. Since this is sufficient for our purpose, we can assume completeness of these transformations.

5 FROM $\overline{\mathbf{K}}_E$ TO PREDICATE LOGIC

In this section we apply the functional translation method explained in the previous section to the modal logic $\overline{\mathbf{K}}_E$ which we introduced in Section 3.

Recall, $\overline{\mathbf{K}}_E$ is a multi-modal logic with infinitely but countably many numerical modal operators ($\langle n \rangle$ and $[n]$) and two special membership operators $\Diamond$ and $\Box$. In the relational semantics the numerical operators are interpreted by the set $\{R_n\}_{n \in \mathbb{N}_0}$ of binary relations and the membership operators by the special relation E. The functional translation for serial modalities is considerably simpler than for non-serial modalities. The accessibility relations R_n ($n \in \mathbb{N}_0$) and E are not serial. Axiom N5, $\langle n \rangle \Box\varphi \to \langle n \rangle \Diamond\varphi$, specifies a weak form of seriality for E. Every world accessible by some R_n (i.e. every set-world) has a successor by E.

We do not need the full expressiveness of the language of $\overline{\mathbf{K}}_E$. A subset of formulae with characteristic patterns of modal operators $\langle n \rangle$, $[n]$, $\Diamond$ and $\Box$ will do. For example, in the axiomatization defining $\overline{\mathbf{K}}_E$ the operators $\Diamond$ and $\Box$ do not occur in the scope of $\Diamond$ and $\Box$ operators, they always occur in the scope of $\langle n \rangle$ and $[n]$ operators. This is intentional. Only these patterns make sense in our application of $\overline{\mathbf{K}}_E$. We think of the numeric modalities picking only *sets* and the $\Diamond$ and $\Box$ operators picking only elements of these sets. For this we need a special class of formulae in which the E-successors of worlds accessible by E are irrelevant, only E-successors of worlds accessible by the relations R_n count. We are therefore permitted to assume E is serial (which we prove in Theorem 14).

The language of $\overline{\mathbf{K}}_E$ that we will use is restricted to the set of *admissible* formulae. We say a formula φ of $\overline{\mathbf{K}}_E$ is admissible iff all $\Diamond$ and $\Box$ operators appear in the scope of a $\langle n\rangle$ or $[n]$ operator (for some n). Examples of admissible formulae are:

$$\langle n\rangle\Box p, \quad [n]\Diamond p, \quad \langle n\rangle(p \wedge \Diamond q) \quad \text{and} \quad [n](\neg\Box p \rightarrow \Diamond q).$$

The formulae

$$\Diamond p, \quad \text{and} \quad \langle n\rangle\Box\Diamond q$$

on the other hand, are not admissible. We note that the translations of $\overline{\mathbf{K}}$ (presented in Section 3.2) are admissible formulae, since the corresponding modalities for modal operators of $\overline{\mathbf{K}}$ are $\langle n\rangle\Box$ and $[n]\Diamond$. The axioms N4–N8 are also admissible, whenever φ, ψ and ϕ_i are admissible.

For admissible formulae we may assume the relation E is serial.

THEOREM 14 *Let φ be an admissible formula of $\overline{\mathbf{K}}_E$. If φ is valid in a model then φ is valid in a model in which the relation E (associated with the modalities $\Diamond$ and $\Box$) is serial.*

Proof. Let φ be valid in a model $\mathcal{M} = (\mathcal{F}, V)$ based on the frame

$$\mathcal{F} = (W, \{R_n\}_{n\in\mathbf{N}_0}, E).$$

Define $\mathcal{F}^*$ to be the structure $(W, \{R_n\}_{n\in\mathbf{N}_0}, E^*)$ obtained from $\mathcal{F}$ by replacing E with E^*. E^* includes E and all pairs (x, x) of $x \in W$ for which no $y \in W$ exists such that $E(x, y)$. Then, E^* is serial.

We show that φ is valid in $\mathcal{M}^* = (\mathcal{F}^*, V)$. The only critical case is where φ has a formula equivalent to $\psi = \Box\phi \wedge \Box\neg\phi$ as subformula. For ψ is true at any world x in $\mathcal{M}$ iff x has no E-successor in $\mathcal{M}$. In $\mathcal{M}^*$, however, ψ is false at x whenever ψ is true at x in $\mathcal{M}$(otherwise there is an inconsistency). ψ occurs in the scope of either $\langle n\rangle$ or $[n]$ for some n. First, we consider the case that ψ occurs in the scope of $\langle n\rangle$. Let φ' be the $\langle n\rangle$ subformula of φ with ψ in its scope. We may assume φ' is of the form $\langle n\rangle((\psi \vee \alpha) \wedge \beta)$. Now, suppose φ' is true in a world x of $\mathcal{M}$. Then there is a $y \in W$ such that $R_n(x, y)$. Since E is weakly serial there is a $z \in W$ such that $E(y, z)$, which implies ψ is false in y of $\mathcal{M}$. Hence, φ' is true in x of $\mathcal{M}$ iff it is also true in x of $\mathcal{M}^*$. Next, we consider the case that ψ occurs in the scope of $[n]$. Assume φ' is of the form $[n]((\psi \vee \alpha) \wedge \beta)$ and suppose φ' is true in $x \in W$ of $\mathcal{M}$. Then either there are or there are no y's in W such that $R_n(x, y)$. If there are y's then we argue as above. If there are no y's then φ' is trivially true in x of $\mathcal{M}^*$. We conclude, φ is indeed valid in $\mathcal{M}^*$. ∎

This theorem licenses the translation of the $\Diamond$ and $\Box$ operators without the dead-end predicate de_E. For admissible formulae in $\overline{\mathbf{K}}_E$ the translation function π_f is modified as follows:

$$\begin{aligned}
\pi_f([n]\psi, x) &= \neg de_n(x) \to \forall\gamma{:}AF_n\ \pi_f(\psi, \downarrow(\gamma, x))\\
\pi_f(\langle n\rangle\psi, x) &= \neg de_n(x) \wedge \exists\gamma{:}AF_n\ \pi_f(\psi, \downarrow(\gamma, x))\\
\pi_f(\Box\psi, x) &= \forall\gamma{:}AF_E\ \pi_f(\psi, \downarrow(\gamma, x))\\
\pi_f(\Diamond\psi, x) &= \exists\gamma{:}AF_E\ \pi_f(\psi, \downarrow(\gamma, x)).
\end{aligned}$$

We are now set to compute the functional translation of the $\overline{\mathbf{K}}_E$-axioms N4–N8. This is a mechanical and tedious task, which we left to an implementation of the general translation procedure and our tool for eliminating second-order quantifiers.

In our listing of the result we use the following notation. $\gamma{:}n$ is the abbreviation for $\gamma{:}AF_n$ and $\gamma{:}E$ for $\gamma{:}AF_E$. Variables without sort declarations are assumed to be of sort W. In the respective clause forms variables and Skolem functions are indexed by their sort. The value in the subscript of a Skolem function, say f^n_m, associates the function with the m-th modality $\langle m\rangle$. AF_m is the sort of the terms formed with f^n_m. The superscript is part of the name of the Skolem function. It is only used to distinguish the different Skolem functions for the different instances of the clauses. (In an actual implementation, these numbers are the objects of symbolic arithemtical manipulations.) γ_n indicates γ is a variable of sort AF_n.

In the following we list for each axiom N4–N8, (i) the functional translation, (ii) the first-order equivalent formulation and (iii) its clause form.

N4: $\vdash_{\overline{\mathbf{K}}_E}\ [0]\Diamond\varphi \to [n]\Box\varphi$.

The functional translation $\Pi_f(\text{N4})$ is

$$\begin{aligned}
\forall\varphi\forall x\ [(\neg de_0(x) \to \forall\gamma'{:}0\ \exists\delta'{:}E\ \varphi(\downarrow([\gamma'\delta'], x))) \to\\
(\neg de_n(x) \to \forall\gamma{:}n\ \forall\delta{:}E\ \varphi(\downarrow([\gamma\delta], x)))].
\end{aligned}$$

This has a first-order equivalent formulation, namely

$$\begin{aligned}
&\forall x\ de_0(x) \to de_n(x) \wedge\\
(10)\quad &\forall x\ [\neg de_n(x) \to (\forall\gamma{:}n\ \forall\delta{:}E\ \exists\gamma'{:}0\ \forall\delta'{:}E\\
&\downarrow([\gamma\delta], x) = \downarrow([\gamma'\delta'], x))].
\end{aligned}$$

The clause form is:

$$\begin{aligned}
&\neg de_0(x) \vee de_n(x).\\
&de_n(x) \vee \downarrow([\gamma_n\delta_E], x) = \downarrow([f^n_0(x, \gamma_n, \delta_E)\delta'], x).
\end{aligned}$$

N5: This axiom becomes a tautology because we assume E is a serial relation.

N6: $\vdash_{\overline{\mathbf{K}}_E}\ [n]\varphi \to [n+1]\varphi$.

$\Pi_f(\text{N6})$ is given by

$$\begin{aligned}
\forall\varphi\forall x\ [(\neg de_n(x) \to \forall\delta{:}n\ \varphi(\downarrow(\delta, x))) \to\\
(\neg de_{n+1}(x) \to \forall\gamma{:}n{+}1\ \varphi(\downarrow(\gamma, x)))],
\end{aligned}$$

the first-order equivalent by:

$$(11)\quad \begin{array}{l} \forall x\, de_n(x) \rightarrow de_{n+1}(x) \wedge \\ \forall x\, [\neg de_{n+1}(x) \rightarrow (\forall \gamma{:}n{+}1\ \exists \delta{:}n\ \downarrow(\gamma, x) = \downarrow(\delta, x))], \end{array}$$

and the clause form by:

$$\begin{array}{l} \neg de_n(x) \vee de_{n+1}(x). \\ de_{n+1}(x) \vee \downarrow(\gamma_{n+1}, x) = \downarrow(g_n^n(x, \gamma_{n+1}), x). \end{array}$$

N7: $\vdash_{\overline{\mathbf{K}}_E} \langle n+m\rangle\Box(\varphi \vee \psi) \rightarrow (\langle n\rangle\Box\varphi \vee \langle m\rangle\Box\psi)$
is translated to $\Pi_f(\text{N7})$:

$$\begin{array}{l} \forall\varphi\psi\forall x\, [(\neg de_{n+m}(x) \wedge \exists\gamma{:}n{+}m\ \forall\delta{:}E\ (\varphi(\downarrow([\gamma\delta], x)) \vee \psi(\downarrow([\gamma\delta], x)))) \\ \quad \rightarrow\ [(\neg de_n(x) \wedge \exists\gamma{:}n\ \forall\alpha{:}E\ \varphi(\downarrow([\gamma\alpha], x))) \vee \\ \qquad (\neg de_m(x) \wedge \exists\gamma{:}m\ \forall\beta{:}E\ \psi(\downarrow([\gamma\beta], x)))]]. \end{array}$$

This is equivalent to the formula

$$\begin{array}{l} \forall x\, [de_{n+m}(x) \vee [\neg de_n(x) \wedge \neg de_m(x) \wedge \\ \quad \forall\gamma{:}n{+}m\ \forall\alpha, \beta{:}E\ \exists\delta{:}E\ \exists\gamma{:}n\ \exists\gamma{:}m \\ \qquad (\downarrow([\gamma_{n+m}\delta], x) = \downarrow([\gamma_n\alpha(\gamma_n)], x) \wedge \\ \qquad \downarrow([\gamma_{n+m}\delta], x) = \downarrow([\gamma_n\beta(\gamma_m)], x)))]]. \end{array}$$

(Note, here we have used the sorts as indices to distinguish three different variables: γ_n, γ_m and γ_{n+m}.) This formula is still second-order (note the $\alpha(\gamma_n)$ and $\beta(\gamma_m)$ terms). We get a first-order equivalent formula if we apply the quantifier exchange rule Υ to $\Pi_f(\text{N7})$. $\Upsilon(\Pi_f(\text{N7}))$ is

$$\begin{array}{l} \forall\varphi\psi\forall x\, [(\neg de_{n+m}(x) \wedge \\ \qquad \exists\gamma{:}n{+}m\ \forall\delta{:}E\ (\varphi(\downarrow([\gamma\delta], x)) \vee \psi(\downarrow([\gamma\delta], x)))) \rightarrow \\ \qquad [(\neg de_n(x) \wedge \exists\gamma{:}n\ \forall\alpha{:}E\ \varphi(\downarrow([\gamma\alpha], x))) \vee \\ \qquad (\neg de_m(x) \wedge \forall\beta{:}E\ \exists\gamma{:}m\ \psi(\downarrow([\gamma\beta], x)))]] \end{array}$$

which is equivalent to the first-order formula

$$\begin{array}{l} \forall x\, [de_{n+m}(x) \vee [\neg de_n(x) \wedge \neg de_m(x) \wedge \\ \quad \forall\gamma{:}n{+}m\ \forall\alpha{:}E\ \exists\delta{:}E\ \exists\gamma{:}m\ \forall\beta{:}E\ \exists\gamma{:}n \\ \quad (\downarrow([\gamma_{n+m}\delta], x) = \downarrow([\gamma_n\alpha], x) \vee \downarrow([\gamma_{n+m}\delta], x) = \downarrow([\gamma_n\beta], x))]]. \end{array}$$

The clause form is:

$$\begin{array}{l} de_{n+m}(x) \vee \neg de_n(x). \\ de_{n+m}(x) \vee \neg de_m(x). \\ de_{n+m}(x) \vee\ \downarrow([\gamma_{n+m} h1_E^{nm}(x, \gamma_{n+m}, \alpha_E)], x) \\ \qquad\qquad = \downarrow([h2_n^{nm}(x, \gamma_{n+m}, \alpha_E, \beta_E)\alpha_E], x). \\ de_{n+m}(x) \vee\ \downarrow([\gamma_{n+m} h1_E^{nm}(x, \gamma_{n+m}, \alpha_E)], x) \\ \qquad\qquad = \downarrow([h3_n^{nm}(x, \gamma_{n+m}, \alpha_E)\beta_E], x). \end{array}$$

N8: $\vdash_{\overline{\mathbf{K}}_E} (\langle n\rangle\Box(\varphi\wedge\psi)\wedge\langle m\rangle\Box(\varphi\wedge\neg\psi))\rightarrow\langle n+m+1\rangle\Box\varphi$
is translated to $\Pi_f(\text{N8})$:

$$\begin{aligned}&\forall\varphi\psi\forall x\ [[(\neg de_n(x)\wedge\exists\gamma{:}n\ \forall\alpha{:}E\ (\varphi(\downarrow([\gamma\alpha],x))\wedge\psi(\downarrow([\gamma\alpha],x))))\wedge\\&\quad(\neg de_m(x)\wedge\exists\gamma{:}m\ \forall\beta{:}E\ (\varphi(\downarrow([\gamma\beta],x))\wedge\neg\psi(\downarrow([\gamma\beta],x))))]\rightarrow\\&\quad(\neg de_{n+m+1}(x)\wedge\exists\gamma{:}n{+}m{+}1\ \forall\delta{:}E\ \varphi(\downarrow([\gamma\delta],x)))]\end{aligned}$$

This is equivalent to

$$\begin{aligned}&\forall x\ [de_n(x)\vee de_m(x)\vee\\&\quad\forall\gamma{:}n\ \forall\gamma{:}m\ \exists\gamma{:}n{+}m{+}1\ \forall\delta{:}E\\&\qquad(\neg de_{n+m+1}(x)\vee\exists\alpha,\beta{:}E\downarrow([\gamma_n\alpha],x)=\downarrow([\gamma_m\beta],x))\wedge\\&\qquad[\exists\alpha,\beta{:}E\downarrow([\gamma_n\alpha],x)=\downarrow([\gamma_m\beta],x)\vee\\&\qquad\exists\alpha{:}E\downarrow([\gamma_n\alpha],x)=\downarrow([\gamma_{n+m+1}\delta],x)\vee\\&\qquad\exists\beta{:}E\downarrow([\gamma_m\beta],x)=\downarrow([\gamma_{n+m+1}\delta],x)]].\end{aligned}$$

The clause form is

$$\begin{aligned}&de_n(x)\vee de_m(x)\vee\neg de_{n+m+1}(x)\vee\\&\qquad\downarrow([\gamma_n k1_E^{nm}(x,\gamma_n,\gamma_m)],x)=\downarrow([\gamma_m k2_E^{nm}(x,\gamma_n,\gamma_m)],x).\\&de_n(x)\vee de_m(x)\vee\\&\qquad\downarrow([\gamma_n k3_E^{nm}(x,\gamma_n,\gamma_m)],x)=\downarrow([\gamma_m k4_E^{nm}(x,\gamma_n,\gamma_m)],x)\vee\\&\qquad\downarrow([\gamma_n k5_E^{nm}(x,\gamma_n,\delta_E)],x)=\downarrow([k_{n+m+1}^{nm}(x,\gamma_n,\gamma_m)\delta_E],x)\vee\\&\qquad\downarrow([\gamma_m k6_E^{nm}(x,\gamma_m,\delta_E)],x)=\downarrow([k_{n+m+1}^{nm}(x,\gamma_n,\gamma_m)\delta_E],x).\end{aligned}$$

The $\overline{\mathbf{K}}_E$-axioms and their translations are schemas. They represent the conjunction of all instances with the n and m taking on concrete non-negative integer values. This can be exploited in certain generalizations. For example, the subformula $\forall x\ de_n(x)\rightarrow de_{n+1}(x)$ of (5) can be generalized to

$$\forall x\ de_n(x)\rightarrow de_m(x)\qquad\text{for all } m\rangle n.$$

This formula subsumes the subformula $\forall x\ de_0(x)\rightarrow de_n(x)$ of the translation (5) of N4. The remaining part of (5) can also be generalized to:

$$\forall x\ [\neg de_m(x)\rightarrow(\forall\gamma{:}m\ \exists\delta{:}n\downarrow(\gamma,x)=\downarrow(\delta,x))]\qquad\text{for all } m\rangle n.$$

The clause form is

(12) $de_m(x)\vee\downarrow(\gamma_m,x)=\downarrow(g_n^{nm}(x,\gamma_m),x)$ for all $m\rangle n$.

Recall the relation translation of N6. We noted N6′ generalizes to

$$R_m\subseteq R_n\qquad\text{for all } m\geq n.$$

This ordering on the accessibility relations $\{R_n\}_{n\in\mathbf{N}_0}$ induces a linear ordering on the set $\{AF_n\}_{n\in\mathbf{N}_0}$ of sets of accessibility functions. We capture this ordering by the subsort declaration

(13) $AF_m \sqsubseteq AF_n$ for all $m \geq n$.

In a resolution calculus this declaration has the same effect as clause (5). We therefore replace (5) in PL_M by the subsort declaration (5).

The final translation for $\overline{\mathbf{K}}_E$ in PL_M is still to come. We add yet more syntactic sugar and hope this makes the translations more easily readable. Every translated $\overline{\mathbf{K}}_E$-formula φ in negated form, which we aim to *refute*, contains only terms which have the form $\downarrow([s_1 \ldots s_n], x_0)$ (in world-path notation). x_0 is the Skolem constant originating from the $\forall x$ quantifier in $\Pi_f(\varphi)$. Such terms can be replaced by just $[s_1 \ldots s_n]$ or just the empty list [] for formulae not containing modal operators.

In the clause form of the translations of the $\overline{\mathbf{K}}_E$-*axioms* the equations contain terms of the form $\downarrow([t_1 \ldots t_m], x)$ with x a universally quantified variable. We may instantiate x with, say $\downarrow([s_1 \ldots s_n], x_0)$, and obtain $\downarrow([t_1 \ldots t_m], \downarrow([s_1 \ldots s_n], x_0))$ which is the same as $\downarrow([s_1 \ldots s_n t_1 \ldots t_m], x_0))$. We get the same result if we introduce a new variable w_* of sort AF^* and replace $\downarrow([t_1 \ldots t_m], x)$ with $[w_* t_1 \ldots t_m]$. We further require that w_* can be unified with arbitrary strings $[s_1 \ldots s_n]$.

In this notation the axiomatization of $\overline{\mathbf{K}}_E$ reduces to the following set of PL_M formulae which defines our predicate logic theory for $\overline{\mathbf{K}}_E$.

P1 $de_n(w_*),\ [w_* x_n z] = [w_* f_0^n(w_*, x_n, z) y]$

P2 $AF_m \sqsubseteq AF_n$ for all $m \rangle n$

P3 $\neg de_n(w_*),\ de_m(w_*)$ for all $m \rangle n$

P4 $de_{n+m}(w_*)$,
$[w_* x_{n+m} h1^{nm}(w_*, x_{n+m}, y)] = [w_* h2_n^{nm}(w_*, x_{n+m}, y, z) y]$

P5 $de_{n+m}(w_*),\ [w_* x_{n+m} h1^{nm}(w_*, x_{n+m}, y)] = [w_* h3_n^{nm}(w_*, x_{n+m}, y) z]$

P6 $de_{\max(n,m)}(w_*),\ \neg de_{n+m+1}(w_*)$,
$[w_* x_n k1^{nm}(w_*, x_n, y_m)] = [w_* y_m k2^{nm}(w_*, x_n, y_m)]$

P7 $de_{\max(n,m)}(w_*)$,
$[w_* x_n k3^{nm}(w_*, x_n, y_m)] = [w_* y_m k4^{nm}(w_*, x_n, y_m)]$,
$[w_* x_n k5^{nm}(w_*, x_n, z)] = [w_* k_{n+m+1}^{nm}(w_*, x_n, y_m) z]$,
$[w_* y_m k6^{nm}(w_*, y_m, z)] = [w_* k_{n+m+1}^{nm}(w_*, x_n, y_m) z]$.

(The variables y and z and the functions $h1^{nm}$ and $k1^{nm}$–$k6^{nm}$ without index are variables and functions of sort AF_E.)

THEOREM 15 *For any* $\overline{\mathbf{K}}$*-formula* φ,

φ *is a* $\overline{\mathbf{K}}$*-theorem* *iff* (P1–P7) $\rightarrow \Upsilon(\Pi_f(\varphi))$ *is a theorem in* PL_M,

By way of examples we illustrate how P1–P7 can be used during inference with a resolution-based theorem prover. Bernhard Nebel provided the following examples.

EXAMPLE 16 *The set* $A = \{\Diamond_0\Diamond_3\top,\ \Diamond_0\Box_3\bot,\ \Box_1\bot\}$ *is an inconsistent set of* $\overline{\mathbf{K}}$*-formulae. With theory resolution we can show the inconsistency in a single step. The* $\overline{\mathbf{K}}_E$ *and the* PL_M *formulations of* $\Diamond_0\Diamond_3\top$, $\Diamond_0\Box_3\bot$ *and* $\Box_1\bot$ *are respectively*

$$\begin{array}{lll} \langle 0\rangle\Box\langle 3\rangle\Box\top & \textit{and} & \neg de_0([]) \wedge \neg de_3([a_0x]), \\ \langle 0\rangle\Box[3]\Diamond\bot & \textit{and} & \neg de_0([]) \wedge de_3([b_0y]), \\ [1]\Diamond\bot & \textit{and} & de_1([]). \end{array}$$

The set A is represented by the following set of clauses:

$$\begin{array}{llll} C_1 & \neg de_0([]) & C_3 & de_3([b_0y]) \\ C_2 & \neg de_3([a_0x]) & C_4 & de_1([]), \end{array}$$

where x *and* y *are variables and* a_0 *and* b_0 *are Skolem constants. Letting* $n = 0$ *and* $m = 0$ *in* P6 *and using the substitution*

$$\{x_0 \mapsto a_0, y_0 \mapsto b_0, x \mapsto k1^{00}([], a_0, b_0), y \mapsto k2^{00}([], a_0, b_0)\}$$

P6 *simultaneously resolves with* C_1*–*C_4 *yielding the empty clause.*

The next example is more complicated.

EXAMPLE 17 *The set* $B = \{\Diamond_0\Diamond_0\Diamond_3\top,\ \Diamond_0\Diamond_0\Box_3\bot,\ \Diamond_0\Box_1\bot,\ \Box_1\bot\}$ *of* $\overline{\mathbf{K}}$*-formulae is inconsistent. B contains the formulae of A (from Example 16) prefixed with a* $\Diamond_0$ *and the formula* $\Box_1\bot$*. The set of expressions in B is represented by the following clauses (derived as in the previous example via* $\overline{\mathbf{K}}_E$ *and* Π_f *translations, which we omit):*

$$\begin{array}{llll} C_1 & \neg de_0([]) & C_5 & de_3([c_0x'd_0y']) \\ C_2 & \neg de_0([a_0x]) & C_6 & de_1([e_0x'']) \\ C_3 & \neg de_3([a_0xb_0y]) & C_7 & de_1([]) \\ C_4 & \neg de_0([c_0x']) & & \end{array}$$

For the refutation we use P1 *with* $n = 0$ *and* P6 *with* $n = 0$ *and* $m = 0$. P1 *can immediately be simplified with clause* C_1*. The instances are:*

P1′ $\quad [f_0^0([], x_0, z)y] = [x_0z]$

P6′ $\quad de_0(w_*),\ \neg de_1(w_*),\ [w_*x_0k1^{00}(w_*, x_0, y_0)] = [w_*y_0k2^{00}(w_*, x_0, y_0)].$

The result of simultaneously resolving P6′, C_1, *and* C_7 *with unifier* $\{w_* \mapsto []\}$ *is*

$$C_8 \quad [x_0k1^{00}([], x_0, y_0)] = [y_0k2^{00}([], x_0, y_0)].$$

Paramodulating with C_8 and with unifier $\{x_0 \mapsto a_0, x \mapsto k1^{00}([], a_0, y_0)\}$, C_3 becomes (this means we do equality replacement with unification in C_3 using the equation C_8)

$$C_9 \quad \neg de_3([y_0 k2^{00}([], a_0, y_0) b_0 y]).$$

This becomes

$$C_{10} \quad \neg de_3([x_0 z b_0 y])$$

when paramodulating with P1′ *using the unifier*

$$\{y_0 \mapsto f_0^0([], x_0, z), y \mapsto k2^{00}([], a_0, y_0)\}.$$

We resolve P6′ *and C_4 using unifier $\{w_* \mapsto [c_0 x], x' \mapsto x\}$ to get*

$$C_{11} \quad \neg de_1([c_0 x]), \quad [c_0 x x'_0 k1^{00}([c_0 x], x'_0, y'_0)] = [c_0 x y'_0 k2^{00}([c_0 x], x'_0, y'_0)].$$

Now, use the unifier

$$\{x_0 \mapsto c_0, x' \mapsto z \mapsto x, x'_0 \mapsto d_0, y'_0 \mapsto b_0, y' \mapsto k1^{00}([c_0 x], d_0, b_0), y \mapsto k2^{00}([c_0 x], d_0, b_0)\}$$

and apply E-resolution to C_5, C_{10} and C_{11} and get

$$C_{12} \quad \neg de_1([c_0 x]).$$

(This means we resolve between C_5 and C_{10} using an equation in C_{11}.) Resolving this with C_6 using E-resolution with C_8 yields the empty clause. The unifier is

$$\{x_0 \mapsto e_0, x'' \mapsto k1^{00}([], e_0 c_0), y_0 \mapsto c_0, x'' \mapsto k2^{00}([], e_0 c_0)\}.$$

EXAMPLE 18 *In this example we show*

(14) $(\Diamond_n \varphi \wedge \Diamond_m \psi \wedge \Box_0 \neg(\varphi \wedge \psi)) \rightarrow \Diamond_{n+m+1}(\varphi \vee \psi)$

is a theorem in $\overline{\mathbf{K}}$ by showing that the following set of clauses is refutable. The set represents the negation of the theorem.

C_1	$\neg de_n([])$	C_5	$de_0([]),\ \neg\varphi([y_0 c]),\ \neg\psi([y_0 c])$
C_2	$\varphi([a_n x])$	C_6	$de_{n+m+1}([]),\ \neg\varphi([x_{n+m+1} d])$
C_3	$\neg de_m([])$	C_7	$de_{n+m+1}([]),\ \neg\psi([x_{n+m+1} d])$
C_4	$\psi([b_m x'])$		

C_5 can be resolved with P3, *letting $n = 0$ and $m = n$, and C_1 yielding*

$$C'_5 \quad \neg\varphi([y_0 c]),\ \neg\psi([y_0 c]).$$

This can be paramodulated using the equation in P1 *and using the unifier*

$$\{w_* \mapsto [], y_0 \mapsto f_0^n([], x_n, z), y \mapsto c\}.$$

The result is:

$$C_5'' \quad \neg\varphi([x_n z]),\ \neg\psi([x_n z]).$$

Resolve this with C_2 and get

$$C_8 \quad \neg\psi([a_n z]).$$

Now, take C_6 and paramodulate with P7 *using the second equation and the unifier*

$$\{w_* \mapsto [], x_{n+m+1} \mapsto k_{n+m+1}^{nm}([], x_n, y_m), z \mapsto d\}$$

and obtain

$$\begin{array}{ll} C_9 & de_{n+m+1}([]),\ \neg\varphi([x_n k5^{nm}([], x_n, d)]), \\ & de_{\max(n,m)}([]), \\ & [x_n k3^{nm}([], x_n, y_m)] = [y_m k4^{nm}([], x_n, y_m)], \\ & [y_m k6^{nm}([], y_m, d)] = [k_{n+m+1}^{nm}([], x_n, y_m) d]. \end{array}$$

The $de_{\max(n,m)}([])$ literal can be eliminated from C_9 with either C_1 or C_3 in one resolution step. The clause

$$\begin{array}{ll} C_9' & de_{n+m+1}([]),\ \neg\varphi([x_n k5^{nm}([], x_n, d)]), \\ & [x_n k3^{nm}([], x_n, y_m)] = [y_m k4^{nm}([], x_n, y_m)], \\ & [y_m k6^{nm}([], y_m, d)] = [k_{n+m+1}^{nm}([], x_n, y_m) d] \end{array}$$

remains. Take C_7 and paramodulate with C_9' and unifier

$$\{x_{n+m+1} \mapsto k_{n+m+1}^{nm}([], x_n, y_m)\}.$$

We obtain

$$\begin{array}{ll} C_{10} & de_{n+m+1}([]),\ \neg\varphi([x_n k5^{nm}([], x_n, d)]),\ \neg\psi([y_m k6^{nm}([], y_m, d)]) \\ & [x_n k3^{nm}([], x_n, y_m)] = [y_m k4^{nm}([], x_n, y_m)]. \end{array}$$

Use C_2 and C_4 to get rid of the $\neg\varphi$ and the $\neg\psi$ literals. The unifier is

$$\{x_n \mapsto a_n, y_m \mapsto b_m, x \mapsto k5^{nm}([], a_n, d), x' \mapsto k6^{nm}([], b_m, d)\}.$$

C_{10} becomes

$$C_{11} \quad de_{n+m+1}([]),\ [a_n k3^{nm}([], a_n, b_m)] = [b_m k4^{nm}([], a_n, b_m)].$$

This we can use to paramodulate with C_8. The unifier is

$$\{z \mapsto k3^{nm}([], a_n, b_m)\}$$

and the result is:

$C_{12}\quad de_{n+m+1}([]),\ \neg\psi([b_m k4^{nm}([], a_n, b_m)])$.

Resolve this with C_4 which yields

$C_{13}\quad de_{n+m+1}([])$.

Now we use P6 *and get*

$C_{14}\quad de_{\max(n,m)}([]),\ [x_n k1^{nm}([], x_n, y_m)] = [y_m k2^{nm}([], x_n, y_m)]$.

Get rid of the $de_{\max(n,m)}([])$ literal by resolving with either C_1 or C_3. The equation

$C'_{14}\quad [x_n k1^{nm}([], x_n, y_m)] = [y_m k2^{nm}([], x_n, y_m)]$.

remains. We use C_8 again and paramodulate with C'_{14} substituting with

$$\{x_n \mapsto a_n, z \mapsto k1^{nm}([], x_n, y_m)\}$$

which leaves

$C_{15}\quad \neg\psi([y_m k2^{nm}([], a_n, y_m)])$.

In the last step we resolve C_{15} and C_4 with unifier

$$\{y_m \mapsto b_m, x \mapsto k2^{nm}([], x_n, y_m)\}$$

to get the empty clause.

In the functional translation we can prove instances of formulae with concrete values assigned to the n and m in the modal operators. There are examples of formulae for which the proofs with symbolic arithmetic terms instead of concrete values work as well. However, this approach may not always work. The formula (5) below provides an example of a theorem which is true for all n and m (that satisfy the required restriction), but which can be proved in our system only for concrete instances of n and m. The situation may be worse. It may be the case that the proof of a formula for a particular concrete instance n depends on the instance of the formula for $n-1$, and the proof of this instance depends on the formula for $n-2$, and so on, to the formula for 0. Call this 'induction on foot'. We now demonstrate a process of how a schema like (5) can be proved in our system by (ordinary) induction for all values followed by a translation step which yields a lemma we add to our theory.

Suppose there are at least twenty objects in p and at least twenty objects in q and in all thirty objects exist. Then we expect the intersection of p and q to contain at least 10 objects. Our intuition is captured by the following formula

(15) $\Diamond_n\varphi \wedge \Diamond_m\psi \wedge \Box_j\neg(\varphi\wedge\psi) \rightarrow \Diamond_{n+m+1-j}(\varphi\vee\psi)$

with $n + m + 1 - j \geq 0$, if we let $p \equiv \varphi$, $q \equiv \psi$, $n = m = 19$ and $j = 9$. In Example 18 we showed (5) for the case that $j = 0$. Unfortunately there is no resolution-based proof for the general case. In the next theorem we use induction to prove (5).

THEOREM 19 *(5) is a theorem in* $\overline{\mathbf{K}}$.

Proof. The proof is by induction on j. We proved the base case in Example 18. Let $j\rangle 0$. As induction hypothesis assume

$$\Diamond_n\varphi \wedge \Diamond_m\psi \wedge \Box_{j-1}\neg(\varphi \wedge \psi) \rightarrow \Diamond_{n+m+1-(j-1)}(\varphi \vee \psi)$$

holds. Assume further $\Diamond_n\varphi$, $\Diamond_m\psi$, and $\Box_j\neg(\varphi \wedge \psi)$ hold. That $\Box_j\neg(\varphi \wedge \psi)$ holds implies $(\neg\Box_{j-1}\neg(\varphi \wedge \psi) \wedge \Box_j\neg(\varphi \wedge \psi)) \vee \Box_{j-1}\neg(\varphi \wedge \psi)$ holds.

Suppose $\Box_{j-1}\neg(\varphi \wedge \psi)$ holds. Then we apply the induction hypothesis. We get $\Diamond_{n+m+1-(j-1)}(\varphi \vee \psi)$ which implies, by A2, $\Diamond_{n+m+1-j}(\varphi \vee \psi)$ holds.

For the second case assume $\neg\Box_{j-1}\neg(\varphi \wedge \psi) \leftrightarrow \Diamond_{j-1}(\varphi \wedge \psi)$ holds. Let $k = n + m$, $\varphi = \varphi \wedge \neg\psi$ and $\psi = \varphi \wedge \psi$ in A10. Then

$$\Diamond_k\varphi \wedge \Box_k\neg(\varphi \wedge \psi) \rightarrow \Diamond_{k-n}(\varphi \wedge \neg\psi).$$

From $\Diamond_n\varphi$, respectively $\Diamond_m\psi$, and $\Box_j\neg(\varphi \wedge \psi)$ we infer that $\Diamond_{n-j}(\varphi \wedge \neg\psi)$, respectively $\Diamond_{m-j}(\neg\varphi \wedge \psi)$, holds. Hence, by A12,

$$\Diamond_{m+n+1-j-1}((\varphi \wedge \neg\psi) \vee (\neg\varphi \wedge \psi))$$

holds. Using A12 again, this time applied to the formulae

$$\Diamond_{m+n+1-j-1}((\varphi \wedge \neg\psi) \vee (\neg\varphi \wedge \psi)) \quad \text{and} \quad \Diamond_{j-1}(\varphi \wedge \psi),$$

we conclude $\Diamond_{m+n+1-j}(\varphi \vee \psi)$ holds. This proves the theorem. ■

The next result shows we can replace the axiom N8 in $\overline{\mathbf{K}}_E$ by the corresponding $\overline{\mathbf{K}}_E$-formulation of the formula (5).

THEOREM 20 *Axiom* N8 *of* $\overline{\mathbf{K}}_E$ *can be replaced by*

(16) $\langle n\rangle\Box\varphi \wedge \langle m\rangle\Box\psi \wedge [j]\Diamond\neg(\varphi \wedge \psi) \rightarrow \langle n + m + 1 - j\rangle\Box(\varphi \vee \psi)$

for $n + m + 1 - j \geq 0$.

Proof. In Theorem 19 we proved (5), its $\overline{\mathbf{K}}$-formulation, is a theorem in $\overline{\mathbf{K}}$. Thus, by Theorem 6, (20) holds in $\overline{\mathbf{K}}_E$. It remains to show (20) implies N8. This is immediate if we let $j = 0$ and substitute $\varphi \wedge \psi$ for φ and $\varphi \wedge \neg\psi$ for ψ exploiting $[0]\Diamond\top \leftrightarrow \top$ (N4). ■

Although replacing N8 with (20) does not increase the number of provable formulae, we avoid the induction argument necessary for proving (20) which we would have to provide by hand as we don't have an induction theorem prover at our disposal. Also, we avoid proving instances of (20).

The functional translation of (20) into predicate logic is somewhat more complicated than that of N8. It is given by

$$\begin{aligned}\forall\varphi\psi\forall x\ &[[(\neg de_n(x) \wedge \exists\gamma{:}n\ \forall\delta{:}E\ \varphi(\downarrow([\gamma\delta], x))) \wedge\\ &(\neg de_m(x) \wedge \exists\gamma{:}m\ \forall\delta{:}E\ \psi(\downarrow([\gamma\delta], x))) \wedge\\ &(\neg de_j(x) \rightarrow \forall\gamma{:}j\ \exists\delta{:}E\ \neg(\varphi(\downarrow([\gamma\delta], x)) \wedge \psi(\downarrow([\gamma\delta], x))))] \rightarrow\\ &[(\neg de_{n+m+1-j}(x) \wedge\\ &\quad \exists\gamma{:}n{+}m{+}1{-}j\ \forall\delta{:}E\ (\varphi(\downarrow([\gamma\delta], x)) \vee \psi(\downarrow([\gamma\delta], x))))]].\end{aligned}$$

Like $\Pi_f(\text{N7})$ this formula cannot be reduced to a first-order formula. We swap the quantifiers $\exists\gamma{:}n{+}m{+}1{-}j$ and $\forall\delta{:}E$. The quantification elimination algorithm SCAN produces then for this input the following clauses:

P8 $\quad de_{\max(n,m)}(w_*),\ \neg de_{n+m+1-j}(w_*),$
$\quad [w_* f7_j^{nmj}(w_*, x_n, x_m)u] = [w_* x_n f5^{nmj}(w_*, x_n, u)]$

P9 $\quad de_{\max(n,m)}(w_*),\ \neg de_{n+m+1-j}(w_*),$
$\quad [w_* f7_j^{nmj}(w_*, x_n, x_m)u] = [w_* x_m f6^{nmj}(w_*, x_m, u)]$

P10 $\quad de_{\max(n,m)}(w_*),\ \neg de_j(w_*),$
$\quad [w_* f1_{n+m+1-j}^{nmj}(w_*, v, x_n)v] = [w_* x_n f2^{nmj}(w_*, v, x_n)],$
$\quad [w_* f3_{n+m+1-j}^{nmj}(w_*, v, x_m)v] = [w_* x_m f4^{nmj}(w_*, v, x_m)]$

P11 $\quad de_{\max(n,m)}(w_*),$
$\quad [w_* f1_{n+m+1-j}^{nmj}(w_*, v, x_n)v] = [w_* x_n f2^{nmj}(w_*, v, x_n)],$
$\quad [w_* f3_{n+m+1-j}^{nmj}(w_*, v, x_m)v] = [w_* x_m f4^{nmj}(w_*, v, x_m)],$
$\quad [w_* f7_j^{nmj}(w_*, x_n, x_m)u] = [w_* x_n f5^{nmj}(w_*, x_n, u)]$

P12 $\quad de_{\max(n,m)}(w_*),$
$\quad [w_* f1_{n+m+1-j}^{nmj}(w_*, v, x_n)v] = [w_* x_n f2^{nmj}(w_*, v, x_n)],$
$\quad [w_* f3_{n+m+1-j}^{nmj}(w_*, v, x_m)v] = [w_* x_m f4^{nmj}(w_*, v, x_m)],$
$\quad [w_* f7_j^{nmj}(w_*, x_n, x_m)u] = [w_* x_m f6^{nmj}(w_*, x_m, u)]$

together with the clause

(17) $de_n(w),\ de_m(w),\ \neg de_j(w),\ \neg de_{n+m+1-j}(w),$

which is implicit in P1–P7. We can show that, for any positive integers n, m and j,

$$\exists k\ l\ (k, l) \in \{n, m\} \times \{j, n{+}m{+}1{-}j\} \quad \text{such that} \quad k \geq l.$$

For, suppose not. Suppose n, m and j exist such that for any k and l with $(k,l) \in \{n,m\} \times \{j, n{+}m{+}1{-}j\}$ we have $k\langle l$. Then, $n\langle j$, $m\langle j$ and $n\langle n+m+1-j$. Hence, $j\langle m+1$, and thus, $m\langle j\langle m+1$, which cannot be for j a positive integer.

If the values n, m and j are such that we can choose k and l with k strictly larger than l then (5) is subsumed by P3. Otherwise, if the values are such that we can choose identical k and l, then (5) is a tautology. In either case (5) is redundant.

We conclude this section with an example (supplied to us by Werner Nutt) in which we exhibit the computational effect of using the clauses P8–P12.

EXAMPLE 21 *Suppose the universe consists of at most thirty objects. If there are at least twenty objects in p and there are at least twenty objects in q, then there are at least ten objects in $p \wedge q$. A standard tableaux system for the number operators would generate twenty witnesses for p, twenty witnesses for q and then it would need to identify ten of them in order not to exceed the limit of thirty. But there are combinatorically many ways for identifying ten of them.*

In our system we prove the conjecture by showing the following set of $\overline{\mathbf{K}}$-formulae is inconsistent:

$$\{\Diamond_{19}p,\ \Diamond_{19}q,\ \Box_{30}\bot,\ \Box_{9}\neg(p \wedge q)\}.$$

We can choose any other suitable combination of numbers. This will not change the structure of the proof at all. The translation into PL_M is:

$$\{\neg de_{19}([]) \wedge p([a_{19}x]),\ \neg de_{19}([]) \wedge q([b_{19}x]),\ de_{30}([]),\\ de_{9}([]) \vee \neg p([y_9 c]) \vee \neg q([y_9 c])\}.$$

The corresponding set of clauses consists of:

C_1	$\neg de_{19}([])$	C_4	$de_{30}([])$
C_2	$p([a_{19}x])$	C_5	$de_9([]),\ \neg p([y_9c]),\ \neg q([y_9c])$
C_3	$q([b_{19}y])$		

Resolve C_5 with P3 *and C_1 and eliminate the $de_9([])$ literal from C_5 leaving:*

$$C_5' \quad \neg p([y_9c]),\ \neg q([y_9c])$$

We resolve the instance of P9 *with $n = m = 19$, $j = 9$, namely*

$$\text{P9'} \quad de_{19}([]),\ \neg de_{30}([]),\\ [f7_9^{19\,19\,9}([], x_{19}, x_{19}')u] = [x_{19} f6^{19\,19\,9}([], x_{19}', u)],$$

with C_1 and C_4 and obtain

$$C_6 \quad [f7_9^{19\,19\,9}([], x_{19}, x_{19}')u] = [x_{19}' f6^{19\,19\,9}([], x_{19}', u)].$$

Applying the unifier $\{y_9 \mapsto f7_9^{19\,19\,9}([], x_{19}, x_{19}'), u \mapsto c\}$, we can use this in a paramodulation step with C_5' resulting in

$$C_7 \quad \neg p([x'_{19} f6^{19\,19\,9}([], x'_{19}, c)]), \ \neg q([f7_9^{19\,19\,9}([], x_{19}, x'_{19})c])$$

Unify in C_2 and C_7 with $\{x'_{19} \mapsto a_{19}, x \mapsto f6^{19\,19\,9}([], x'_{19}, c)\}$. Resolving C_2 and C_7 yields

$$C_8 \quad \neg q([f7_9^{19\,19\,9}([], x_{19}, a_{19})c]).$$

Now we use the following instance of P8:

$$\text{P8'} \quad de_{19}(w_*), \ \neg de_{30}(w_*),$$
$$[w_* f7_9^{19\,19\,9}(w_*, x_{19}, x'_{19})u] = [w_* x_{19} f5^{19\,19\,9}(w_*, x_{19}, u)]$$

This can be reduced with C_1 and C_4 to the equation

$$C_9 \quad [f7_9^{19\,19\,9}([], x_{19}, x'_{19})u] = [x_{19} f5^{19\,19\,9}([], x_{19}, u)],$$

which we can now use in a paramodulation step with C_8. We get

$$C_{10} \quad \neg q([x_{19} f5^{19\,19\,9}([], x_{19}, c)]).$$

The empty clause is obtained if we resolve C_{10} with C_3 using the appropriate unifier.

6 FROM CONCEPT DESCRIPTION LANGUAGES TO GRADED MODALITIES

A knowledge representation system in the KL-ONE-style [4] usually consists of a so called T-Box and an A-Box [5]. The T-Box axiomatizes the part of the world that is to be modelled in the system whereas the A-Box is more or less a classical database containing information, in general ground facts, about the actual situation.

Most T-Box (or terminological) languages have as syntactic primitives *concept names* and *rule names*. Concept names denote sets of objects and role names denote binary relations between these objects. Using concept forming connectives, like $\neg$, $\sqcap$, $\sqcup$, some and all, compound concept terms can be built which also denote sets of objects.

A prototypical concept description language is the $\mathcal{ALC}$ language (short for 'attributive concept description language with complement'). It has a well-defined model-theoretic semantics and its computational behaviour is completely understood. The terminological language of $\mathcal{ALC}$ uses only the concept-forming operators $\neg$, $\sqcap$, $\sqcup$ with the usual meaning (complement, union, intersection) as well as role quantifications (all R C) and (some R C). (all R C) denotes the set of all objects whose R-successors (R-fillers in the KL-ONE terminology) are all in C. (some R C) denotes the set of all objects with some R-successor in C. Typical examples for concept definitions in $\mathcal{ALC}$ are:

$$\begin{aligned}
\text{man} &= \text{person} \sqcap (\text{some sex male}) \\
\text{parent} &= \text{person} \sqcap (\text{some child } \top) \\
\text{father} &= \text{parent} \sqcap \text{man} \\
\text{grandfather} &= \text{father} \sqcap (\text{some child parent}) \\
\text{woman} &= \text{person} \sqcap \neg\text{man}
\end{aligned}$$

$\top$ denotes the set of all objects.

Given a set T of concept equations, a concept C is *coherent* if there is a model for T in which C denotes a nonempty set. Furthermore, a concept description C *subsumes* a concept description D in T, if C denotes in *every* model of T a superset of D. Deciding coherence and subsumption is the basic reasoning service of the knowledge representation systems based on $\mathcal{ALC}$. According to the above definitions, for example, it is possible to infer that grandfathers are fathers and persons and men as well, i.e. man subsumes father and grandfather.

In [28], Schmidt-Schauß and Smolka show that deciding coherence and subsumption of concept descriptions is P-SPACE-complete and can be decided with linear space. Many variants and extensions of $\mathcal{ALC}$ have now been investigated [19, 20, 21, 27, 9, 8, 35] and are used in implementations of knowledge representation languages [3]. We focus on a language very much like $\mathcal{ALCN}$ which includes numerical quantification operators atleast and atmost. For example, the concept term (atleast 3 has-child blond) represents the set of individuals who have at least three children who are blond. The term (atmost 2)has-parent $\top$ represents the set of individuals who have at most two parents. The language we consider is slightly more expressive than $\mathcal{ALCN}$. Our version, referred to as $\mathcal{ALCN}^+$, allows for arbitrary concepts to be included in other concepts, whereas in $\mathcal{ALCN}$ only atomic concepts can be included in other concepts.

Now, we define the syntax of $\mathcal{ALCN}^+$. The signature of the terminological language of $\mathcal{ALCN}^+$ consists of a set Σ_R of *role names* and a disjoint set Σ_C of *concept names*. From role names $\mathsf{Q} \in \Sigma_R$ and concept names $\mathsf{A} \in \Sigma_C$ compound concept terms C are formed according to the following rules:

$$\begin{aligned}
\mathsf{C}, \mathsf{D} \quad \longrightarrow \quad & \mathsf{A} \mid \neg\mathsf{C} \mid \mathsf{C} \sqcap \mathsf{D} \mid \mathsf{C} \sqcup \mathsf{D} \mid (\text{some R C}) \mid (\text{all R C}) \mid \\
& (\text{atleast } n \text{ R C}) \mid (\text{atmost } n \text{ R C}) \mid \mathsf{C} \sqsubseteq \mathsf{D} \mid \mathsf{C} = \mathsf{D}.
\end{aligned}$$

n is a non-negative integer. Most authors define the symbols $\sqsubseteq$ and $=$ to be sentential symbols. We define them to be connectives just as $\sqcap$ and $\sqcup$ are. Note, we consider terminological sentences of the form $\mathsf{C} \sqsubseteq \mathsf{D}$ and $\mathsf{C} = \mathsf{D}$ to be concept terms. In $\mathcal{ALCN}$ terminological sentences are constrained to be of the form $\mathsf{A} \sqsubseteq \mathsf{C}$ and $\mathsf{A} = \mathsf{C}$, where A are concept names. A *T-Box* is defined as a set of concept terms.

The semantics of $\mathcal{ALCN}^+$ is specified by an interpretation $\mathcal{I} = (U, V)$ with U a non-empty set U (the domain of interpretation) and a signature assignment V. The signature assignment maps role names to binary relations on U and it maps concept names to subsets of U. The interpretation of concept terms C and D specified by:

$$\begin{aligned}
\mathsf{C}^{\mathcal{I}} &= V(\mathsf{C}) \qquad \text{if C is a concept name} \\
(\neg\mathsf{C})^{\mathcal{I}} &= U \setminus \mathsf{C}^{\mathcal{I}} \\
(\mathsf{C} \sqcap \mathsf{D})^{\mathcal{I}} &= \mathsf{C}^{\mathcal{I}} \cap \mathsf{D}^{\mathcal{I}} \\
(\mathsf{C} \sqcup \mathsf{D})^{\mathcal{I}} &= \mathsf{C}^{\mathcal{I}} \cup \mathsf{D}^{\mathcal{I}} \\
(\mathsf{C} \sqsubseteq \mathsf{D})^{\mathcal{I}} &= (U \setminus \mathsf{C}^{\mathcal{I}}) \cup \mathsf{D}^{\mathcal{I}} \\
(\mathsf{C} = \mathsf{D})^{\mathcal{I}} &= (U \setminus (\mathsf{C}^{\mathcal{I}} \cup \mathsf{D}^{\mathcal{I}})) \cup (\mathsf{C}^{\mathcal{I}} \cap \mathsf{D}^{\mathcal{I}})) \\
(\mathsf{some\ R\ C})^{\mathcal{I}} &= \{x \in U \mid \exists y \in U\ \mathsf{R}^{\mathcal{I}}(x,y) \wedge y \in \mathsf{C}^{\mathcal{I}}\} \\
(\mathsf{all\ R\ C})^{\mathcal{I}} &= \{x \in U \mid \forall y \in U\ \mathsf{R}^{\mathcal{I}}(x,y) \rightarrow y \in \mathsf{C}^{\mathcal{I}}\} \\
(\mathsf{atleast}\ n\ \mathsf{R\ C})^{\mathcal{I}} &= \{x \in U \mid |\{y \in \mathsf{C}^{\mathcal{I}} \mid \mathsf{R}^{\mathcal{I}}(x,y)\}| \geq n\} \\
(\mathsf{atmost}\ n\ \mathsf{R\ C})^{\mathcal{I}} &= \{x \in U \mid |\{y \in \mathsf{C}^{\mathcal{I}} \mid \mathsf{R}^{\mathcal{I}}(x,y)\}| \leq n\}
\end{aligned}$$

Atomic concept names in a T-Box T are interpreted as the entire domain and are all equivalent to the *top concept* $\top$. $\top$ is the largest element in the subsumption ordering. The complement of $\top$ is $\bot$ and represents the empty set.

An interpretation $\mathcal{I} = (U, V)$ with $\mathsf{C}^{\mathcal{I}} = U$ for all concept terms C in the T-Box T is a *model* of T. A concept term C is *universal* iff $\mathsf{C}^{\mathcal{I}} = U$ for *all* interpretations $\mathcal{I}$. C is *empty* iff $\mathsf{C}^{\mathcal{I}} = \varnothing$ for *all* interpretations $\mathcal{I}$.

The entailment relation $\models$ between concept terms is defined by:

$$\mathsf{C} \models \mathsf{D} \quad \text{iff} \quad \mathsf{D}^{\mathcal{I}} = U \text{ for every interpretation } \mathcal{I} \text{ of C.}$$

Then $\mathsf{C} \models \mathsf{D}$ iff $\mathsf{C} \sqsubseteq \mathsf{D}$ is universal iff $\mathsf{C} \sqcap \neg\mathsf{D}$ is empty.

We treat sets $\{\mathsf{C}_1, \ldots, \mathsf{C}_n\}$ of concept terms in the same way as the conjunction $\mathsf{C}_1 \sqcap \ldots \sqcap \mathsf{C}_n$. Thus, a given T-Box T will be treated as the conjunction of its elements.

In contrast to other terminological languages the language $\mathcal{ALCN}^+$ includes no role-forming operators. Roles that occur are all atomic. To simplify our presentation, without loss of generality we assume there is one atomic role R.

Now we show that we can embed $\mathcal{ALCN}^+$ in $\overline{\mathbf{K}}$. Define a mapping Π from the language of $\mathcal{ALCN}^+$ to the language of $\overline{\mathbf{K}}$ by:

$$\begin{aligned}
\Pi(\mathsf{C}) &= C \qquad \text{if C is a concept name or } \top \text{ or } \bot \\
\Pi(\neg\mathsf{C}) &= \neg\Pi(\mathsf{C}) \\
\Pi(\mathsf{C} \sqcap \mathsf{D}) &= \Pi(\mathsf{C}) \wedge \Pi(\mathsf{D}) \\
\Pi(\mathsf{C} \sqcup \mathsf{D}) &= \Pi(\mathsf{C}) \vee \Pi(\mathsf{D}) \\
\Pi(\mathsf{C} \sqsubseteq \mathsf{D}) &= \Pi(\mathsf{C}) \rightarrow \Pi(\mathsf{D}) \\
\Pi(\mathsf{C} = \mathsf{D}) &= \Pi(\mathsf{C}) \leftrightarrow \Pi(\mathsf{D}) \\
\Pi(\mathsf{some\ R\ C})) &= \Diamond_0 \Pi(\mathsf{C}) \\
\Pi(\mathsf{all\ R\ C}) &= \Box_0 \Pi(\mathsf{C}) \\
\Pi(\mathsf{atleast}\ n\ \mathsf{R\ C}) &= \Diamond_{n-1} \Pi(\mathsf{C}) \\
\Pi(\mathsf{atmost}\ n\ \mathsf{R\ C}) &= \Box_n \neg\Pi(\mathsf{C})
\end{aligned}$$

It is easy to verify that Π is well-defined. The following is the main statement of this section.

THEOREM 22 (Soundness and Completeness of Π)

A concept term C *is universal* *iff* $\Pi(\mathsf{C})$ *is a tautology.*

Proof. Let $\mathcal{I} = (U, V)$ be any interpretation of a T-Box of $\mathcal{ALCN}^+$. Let $\mathcal{M}$ be the modal model $(U, \mathsf{R}^{\mathcal{I}}, V)$. By induction on the structure of C prove, for every $x \in U$:

$$x \in \mathsf{C}^{\mathcal{I}} \quad \text{iff} \quad \mathcal{M}, x \models \Pi(\mathsf{C}).$$

We omit the details. ■

7 CONCLUSION

In the logic of graded modalities it is possible to express properties of finite sets. The usual inference calculi for this logic generate for all sets used in the proof at least as many constants (witnesses) as the cardinality of each set. Even for moderate values a vast number of witnesses are generated which are processed by case distinctions in the proof.

In this paper we present an alternative method which avoids case distinctions, instead our method uses limited arithmetical reasoning. It arises in a series of transformation steps. First, we translate the logic of graded modalities $\overline{\mathbf{K}}$ into a new normal multi-modal logic, called $\overline{\mathbf{K}}_E$. Unfortunately, $\overline{\mathbf{K}}_E$ does not reduce by the standard relational translation to first-order logic. One of the axioms of $\overline{\mathbf{K}}_E$ is second-order. We solved this irreducibility problem by, instead of using the relational translation, using a functional translation with a particular optimization which exploits the richer structure of the functional models.

Our method can also be applied in the field of knowledge representation. The terminological logic $\mathcal{ALCN}$ is closely related to the graded modal logic $\overline{\mathbf{K}}$. In fact, there is an exact correspondence between terminological operators and modal operators. Our approach provides a viable alternative inference mechanism to the constraint algorithms commonly used, which also suffer from the overhead of evaluating case distinctions.

Our approach must be viewed as a first step toward efficient reasoning with finite sets. There are a number of open problems which need to be addressed.

(i) A general completeness result for $\overline{\mathbf{K}}_E$ would allow us to use the full expressivity of this system. As long as this is not proved, we can guarantee completeness only for the original $\overline{\mathbf{K}}$ formulae. This is what we wanted from the beginning, but a stronger result would be preferable.

(ii) Our first-order theory is represented by a set of axiom schemas which are understood to be conjunctions of all its instances with the numerical variables instantiated with concrete values. The implementation of the calculus will rely on theory resolution. The axiom schemas will be encoded as inference rules. Since the axiom schemas contain equations the realization will not be easy, but it is certainly solvable.

(iii) The original logic of graded modalities is decidable. Accordingly, we expect a resolution strategy for the translated formulae can be developed that is complete and terminates. This has yet to be done.

(iv) Our calculus is still limited in reasoning with arithmetical terms. It remains to be investigated whether and how this capability can be enhanced.

(v) We have applied our methods to KL-ONE-type reasoning but only for reasoning within the T-Box. This corresponds directly to that in modal logic. We haven't accounted for A-Box reasoning about concrete instantiations of concepts/sets and roles/relations. The functional translation applied to A-Box terms generates many equations. It is not immediate how these can be treated efficiently.

(vi) The correspondence properties for the axioms of $\overline{\mathbf{K}}_E$ except one are first-order. This does not rule out that $\overline{\mathbf{K}}_E$ is complete with respect to a first-order model theory. If this were the case we can get a translation into predicate logic that avoids some equational reasoning.

ACKNOWLEDGEMENTS

This work was supported by the ESPRIT-Project MEDLAR (Nr. 6471) and by the LOGO-Project ITS 9102 funded by the BMFT.

Max Planck Instiute, Saarbrücken, Germany.

REFERENCES

[1] H. Andréka, I. Németi, I. Sain. On interpolation, amalgamation, universal algebra and Boolean algebras with operators. Unpublished manuscript, Univ. of Budapest, 1995.

[2] Y. Auffray, P. Enjalbert. Modal theorem proving: An equational viewpoint. *Journal of Logic and Computation*, **2**, 247–297, 1992.

[3] F. Baader, B. Hollunder. A terminological knowledge representation system with complete inference algorithms. In: M. M. Richter, H. Boley, (eds.), *Proc. of the International Workshop on Processing Declarative Knowledge (PDK'91)*, Univ. of Kaiserslautern, 1991.

[4] R. J. Brachman, J. G. Schmolze. An overview of the KL-ONE knowledge representation system. *Cognitive Science*, **9**, 171–216, 1985.

[5] R. J. Brachman, R. E. Fikes, H. J. Levesque. KRYPTON: A functional approach to knowledge representation. *IEEE Computer*, **16**, 67–73, 1983.

[6] C. Cerrato. General canonical models for graded normal logics (Graded modalities IV). *Studia Logica*, **49**, 241–252, 1990.

[7] F. de Caro. Graded modalities II. *Studia Logica*, **47**, 1–10, 1988.

[8] F. M. Donini, M. Lenzerini, D. Nardi, W. Nutt. The complexity of concept languages. In: J Allen, R Fikes, E Sandewall, (eds.), *Proc. of KR'91*, pp. 151–162. Morgan Kaufmann, 1991.
[9] F. M. Donini, M. Lenzerini, D. Nardi, W. Nutt. Tractable concept languages. In *Proc. of IJCAI'91*, pp. 458–463, 1991.
[10] L. Fariñas del Cerro, A. Herzig. Quantified modal logic and unification theory. Rapport LSI 293, Languages et Systèmes Informatique, Univ. Paul Sabatier, Toulouse, 1988.
[11] M. Fattorosi-Barnaba, C. Cerrato. Graded modalities III. *Studia Logica*, **47**, 99–110, 1988.
[12] M. Fattorosi-Barnaba, F. de Caro. Graded modalities I. *Studia Logica*, **44**, 197–221, 1985.
[13] K. Fine. *For Some Propositions and so Many Possible Worlds*. PhD thesis, Univ. of Warwick, 1969.
[14] K. Fine. In so many possible worlds. *Notre Dame Journal of Formal Logic*, **13**, 516–520, 1972.
[15] D. M. Gabbay, H. J. Ohlbach. Quantifier elimination in second-order predicate logic. *South African Computer Journal*, **7**, 35–43, 1992. Also in: *Proc. of KR'92*, 425–436. Also available as Technical Report MPI-I-92-231, Max-Planck-Institut für Informatik, Saarbrücken, July 1992.
[16] D. M. Gabbay. An irreflexivity lemma with applications to axiomatizations of conditions on linear frames. In: U. Mönnich, (ed.), *Aspects of Philosophical Logic*, pp. 67–89. Reidel, 1981.
[17] L. F. Goble. Grades of modality. *Logique et Analyse*, **13**, 323–334, 1970.
[18] A. Herzig. *Raisonnement automatique en logique modale et algorithmes d'unification*. PhD thesis, Univ. Paul-Sabatier, Toulouse, 1989.
[19] H. J. Levesque, R. J. Brachman. Exprèssiveness and tractability in knowledge representation and reasoning. *Computational Intelligence*, **3**, 78–93, 1987.
[20] B. Nebel, G. Smolka. Representation and reasoning with attributive descriptions. In: K.-H. Bläsius, U. Hedtstück, C.-R. Rollinger, (eds.), *Sorts and Types in Artificial Intelligence, Lecture Notes in Artificial Intelligence*, Vol. 418, pp. 112–139. Springer-Verlag, Berlin, 1990.
[21] B. Nebel. *Reasoning and Revision in Hybrid Representation Systems, Lecture Notes in Artificial Intelligence*, Vol. 422, Springer-Verlag, 1990.
[22] H. J. Ohlbach, R. A. Schmidt. Functional translation and second-order frame properties of modal logics. Technical Report MPI-I-95-2-002, Max-Planck-Institut für Informatik, Saarbrücken, January 1995.
[23] H. J. Ohlbach. A resolution calculus for modal logics. In: E. Lusk and R. Overbeek, (eds.), *Proc. of CADE'88, Lecture Notes in Computer Science*, Vol. 310, pp. 500–516. Springer-Verlag, 1988.
[24] H. J. Ohlbach. Semantics based translation methods for modal logics. *Journal of Logic and Computation*, **1**, 691–746, 1991.
[25] A. N. Prior. Egocentric logic. *Nôus*, **2**, 191–207, 1968.
[26] H. Sahlqvist. Completeness and correspondence in the first and second order semantics for modal logics. In: S. Kanger, (ed.), *Proceedings of the 3rd Scandinavian Logic Symposium, 1973*, pp. 110–143. North Holland, Amsterdam, 1975.
[27] K. Schild. A correspondence theory for terminological logics: Preliminary report. In: *Proc. of IJCAI'91*, pp. 466–471, 1991.
[28] M. Schmidt-Schauß, G. Smolka. Attributive concept description with complements. *Artificial Intelligence*, **48**, 1–26, 1991.
[29] J. van Benthem. *Modal Logic and Classical Logic*. Bibliopolis, Naples, 1983.
[30] J. van Benthem. Correspondence theory. In: D. Gabbay, F. Guenther, (eds.), *Handbook of Philosophical Logic*, Vol. II, pp. 167–247. Reidel, Dordrecht, 1984.
[31] W. van der Hoek, M. de Rijke. Generalized quantifiers and modal logic. ITLI Prepublication Series for Logic, Semantics and Philosophy of Language LP-91-01, Institue for Language, Logic and Information, Univ. of Amsterdam, 1991.
[32] W. van der Hoek, M. de Rijke. Counting objects. *Journal of Logic and Computation*, **5**, 325–345, 1995.
[33] W. van der Hoek. *Modalities for Reasoning about Knowledge and Quantities*. PhD thesis, Vrije Univ. Utrecht, 1992.
[34] W. van der Hoek. On the semantics of graded modalities. *Journal of Applied Non-Classical Logics*, **2**, 81–123, 1992.
[35] W. A. Woods, J. G. Schmolze. The KL-ONE family. *Computers and Mathematics with Applications*, **23**, 133–177, 1992.
[36] N. K. Zamov. Modal resolutions. *Izvestiya VUZ. Mathematika*, **33**, 22–29, 1989. Also published in *Soviet Mathematics*, Allerton Press.

OLIVIER GASQUET AND ANDREAS HERZIG

FROM CLASSICAL TO NORMAL MODAL LOGICS

1 INTRODUCTION

Classical modal logics (Segerberg [27], Chellas [2]) are weaker than the well-known normal modal logics: The only rule that is common to all classical modal logics is

$$RE: \quad \frac{F \leftrightarrow G}{\Box F \leftrightarrow \Box G}$$

(We nevertheless note that this principle raises problems in systems containing equality (Hughes and Cresswell [14]).)

Classical modal logics do not necessarily validate

$$RN: \quad \frac{F}{\Box F}$$

$$RM: \quad \frac{F \to G}{\Box F \to \Box G}$$

$$C: \quad (\Box F \wedge \Box G) \to \Box(F \wedge G)$$

$$K: \quad (\Box F \wedge \Box(F \to G)) \to \Box G$$

which are valid in any normal modal logic. [1]

In an epistemic reading, adopting one of the above formulas as an axiom means to close knowledge under a particular principle. Precisely, RE corresponds to the principle of knowledge closure under logical equivalences, RN under logical truth,

[1] Under RE, the principles RN, RM, C, and K are respectively equivalent to

$$N: \quad \Box\top$$

$$M: \quad \Box(F \wedge G) \to (\Box F \wedge \Box G)$$

$$C': \quad (\Box F \wedge \Diamond(F \to G)) \to \Diamond G$$

$$K': \quad (\Box F \wedge \Diamond G) \to \Diamond(F \wedge G)$$

H. Wansing (ed.), Proof Theory of Modal Logic, 293–311.

RM under logical consequence, C under conjunction, and K under material implication.

These principles, in an epistemic reading, as well as in a deontic one, are not always desirable (see e.g. (Fagin and Halpern [3]), (Jones and Pörn [16, 17])). In fact, each closure principle expresses an aspect of what has been called the omniscience problem.

Each of these principles can be adopted independently of the others. Nevertheless, under RE and RM, C is equivalent to K, and under RE and RN, K is equivalent to RM and C.

Classical modal logics have possible worlds semantics in terms of minimal models (cf. Scott-Montague structures in (Montague [20])), involving a neighbourhood function mapping worlds to sets of sets of worlds. Now each of the above closure principles identifies a particular class of minimal models. For several of these principles, specialisations of minimal models have been given such as augmented minimal models, models with queer worlds or models with inaccessible worlds. They are more tailored to these principles and give a better account of the logic associated to the principle.

While standard Kripke semantics can be translated straightforwardly into classical first-order logic, this is *a priori* not the case of the above modal logics. The reason is that the neighbourhood function of minimal models cannot be represented directly by a first-order formula.

In this paper (section 4) we show that nevertheless semantics in terms of minimal models can be expressed in first-order logic. We show this indirectly: What we prove is that minimal models can be translated into standard Kripke models (which on their turn can be translated into first-order logic).

A major advantage of such translations - and in fact our main motivation - is that they allow to reuse the proof systems that have been developed for normal modal logics.

In the rest of this paper we first briefly present several classes of minimal models and their simplified semantics (section 3). Then for each class we present a translation into normal multimodal logics (section 4).

2 GENERAL POINTS

2.1 *Language*

The *language* of modal logic is built on a set of propositional variables, classical connectives and a modal operator $\Box$. F, G, and H denote formulas, and $\top$ and $\bot$ respectively stand for logical truth and falsehood. $\Diamond F$ is an abreviation of $\neg\Box\neg F$. $\mathcal{FOR}$ denotes the set of formulas.

2.2 Semantics

Semantics is stated in terms of frames and models. Generally, a *frame* is composed of a set W (whose elements are called worlds) and some structure $\mathcal{S}$ on W. In the well-known case of K-frames (or Kripke frames), the structure is just some binary relation r on W: $r \subseteq W \times W$.[2] Then a *model* is composed of a frame and a meaning function m mapping propositional variables to sets of worlds.

Given a model, it is the truth conditions which uniquely determine a ternary forcing relation between models, worlds, and formulas. In the case of propositional variables and classical connectives, the truth conditions are the usual ones:

- $M, w \models F$ iff $w \in m(F)$ if F is a propositional variable.
- $M, w \models F \wedge G$ iff $M, w \models F$ and $M, w \models G$.
- $M, w \models \neg F$ iff *not*$(M, w \models F)$.

The structure of W is exploited when it comes to the truth condition for the modal operator. In the case of standard Kripke semantics, the forcing relation must satisfy the following one:

- $M, w \models \Box F$ iff for all $v \in r(w)$, $M, v \models F$.

Structures that are richer have more complex truth conditions. In order to avoid confusions we shall add a superscript to $\models$ designating the type of the non-normal semantics.

Generally, in a given semantics, we say that a formula F is *true in a model* $M = (W, \mathcal{S}, m)$ if $M, w \models F$ for every $w \in W$. A formula F is *true in a frame* $(W, \mathcal{S})$ if for every every meaning function m, F is true in $(W, \mathcal{S}, m)$. A formula F is *valid in a class of frames* $\mathcal{F}$ (noted $\models_{\mathcal{F}} F$) if F is true in every frame of $\mathcal{F}$.

A particular semantics will always be identified by a condition on the structure type together with the truth condition for the modal operator. An example in standard Kripke semantics is condition t : $w \in r(w)$ for every $w \in W$. (As in (Chellas [2]), conditions on structures are denoted by small letters.)

We confuse conditions and the class of frames satisfying them: k being the case of standard Kripke models, the class of standard Kripke frames satisfying condition t is called $k + t$ (or kt for short).

2.3 Axiomatics

When a class of frames can be characterized by some axiom system we denote the latter by the corresonding capital letters. E.g. the basic normal modal logic is called K, and $\vdash_K F$ expresses that F is a theorem of K. The class of frames kt being

[2]We shall often write $w' \in r(w)$ instead of $(w, w') \in r$.

characterized by the T-axiom $\Box F \to F$, the corresponding logic is called $K + T$ (or KT for short).

2.4 Multimodal Logics

Multimodal languages generalize modal languages by allowing indexed modal operators. The target logics of our translations being bi- or trimodal logics, what we need is a language containing three modal operators [1], [2], [3].

The multimodal logics we need are normal ones having a standard Kripke semantics. We index every accessibility relation and condition on it by the number of the corresponding modal operator. E.g. $t[1]$ expresses that the accessibility relation r_1 is reflexive. As before, classes of multimodal frames are refered to by sums of conditions. $k[1] + k[2]$ denotes the class of bimodal frames, and $k[1] + k[2] + k[3]$ the class of trimodal frames. (Sometimes we write $k[1, 2]$ and $k[1, 2, 3]$ for short.)

On the axiomatical side, we index axioms and inference rules by the corresponding number. E.g. $K[1]$ is the axiom $([1]F \wedge [1](F \to G)) \to [1]G$, and $D[2]$ is $[2]F \to \langle 2\rangle F$. As before, multimodal systems are referred to by sums of axiom or inference rule names. Our basic normal bimodal logic is $K[1]+K[2]$, and the basic trimodal one is $K[1] + K[2] + K[3]$ ($K[1, 2]$ and $K[1, 2, 3]$ for short).

3 MINIMAL FRAMES AND THEIR CHILDREN

In this section we present the class of minimal frames as well as three subclasses of it: supplemented minimal frames, quasi-filters, and singleton minimal frames. (The last subclass corresponds to Humberstone's inaccessible worlds logic.)

3.1 Minimal Frames

Minimal frames are the basic semantical tool for classical modal logics.

A *minimal frame* (Chellas [2]) is a couple (W, N), where

- W is a set of worlds, and
- $N : W \to 2^{2^W}$ maps worlds to sets of sets of worlds.

Sets of worlds being usually called propositions, we may also say that N associates a set of propositions to every world. $N(w)$ is sometimes called the neighbourhood of w. We recall that then a *minimal model* is a triple (W, N, m) where (W, N) is a minimal frame, and m is a meaning function.

The forcing relation $\models^{min}$ results from the following truth condition for the modal connective:

- $M, w \models^{min} \Box F$ iff there is $V \in N(w)$ such that ($v \in V$ **iff** $M, v \models^{min} F$)

Hence $\Box F$ is true in a world if its neighbourhood contains the extension of F.

The class of minimal frames is noted e. Subclasses of e can be obtained by adding conditions on the mapping N. An example of a condition is t: $w \in V$ for all $V \in N(w)$ and $w \in W$. The class of minimal frames satisfying condition t is called $e + t$, or et for short.

The class of minimal frames e can be characterized by some axiom system for classical propositional logic plus the inference rule RE (Chellas [2]). This basic classical modal logic is called E.

The class of frames et is characterized by the T-axiom $\Box F \rightarrow F$. As expected, the corresponding logic is called $E + T$, or ET for short.

REMARK 1 *There is an isomorphic form of minimal frames (see e.g. (Fitting [5])), where there is a set of accessibility relations instead of the neighbourhood function N. In order to account for worlds with empty neighbourhoods, just as in regular modal logics the concept of a queer world is used.* [3]

Formally, Fitting's frames are triples (W, Q, R), where W and Q are sets of worlds such that $Q \subseteq W$ and R is a set of relations on W. Q is called the set of queer worlds. Here, the forcing relation $\models^{min}$ must fulfil the following truth condition for the modal operator:

- $M, w \models \Box F$ *iff* $w \notin Q$, *and there is* $r \in R$ *such that* $v \in r(w)$ **iff** $M, v \models F$

Hence in a queer world all formulas or the form $\Box F$ are false.

3.2 Singleton Minimal Frames

Singleton minimal frames correspond to inaccessible world frames that were introduced and axiomatized in (Humberstone ([15]).

A *singleton minimal frame* is a minimal frame, where for every world w, the set $N(w)$ is a singleton. In other words, there is exactly one propostion that is necessary in a given world.

The only principle of our list that supplemented minimal frames validate is that of closure under conjunction:

$C : (\Box F \wedge \Box G) \rightarrow \Box(F \wedge G)$

There are countermodels for RN, RM, and K.

Clearly, singleton minimal frames are isomorphic to standard Kripke frames of the form (W, r). Hence the only difference from the standard Kripke semantics is the modal truth condition:

- $M, w \models^{iw} \Box F$ iff ($v \in r(w)$ **iff** $M, v \models^{iw} F$)

---

[3] Note that although Fitting uses these frames only for monotonic modal logics, they can be used as well to give semantics to classical modal logics.

Note that RE and C are not enough to completely characterize singleton minimal frames, and that Humberstone's ([15]) axiomatization is infinitary. This logic has also been studied formally in (Goranko [11], Goranko and Passy [13]). The concept of inaccessible worlds is fundamental in the logics of knowledge and belief of Levesque ([19]). There, $\Box F$ is read "I only know F".

3.3 Supplemented Minimal Frames

Supplemented minimal frames are at the base of monotonic modal logics.

A *supplemented minimal frame* is a minimal frame where for every world w, the set $N(w)$ is closed under supersets. [4]

Supplemented minimal frames can be characterized by the principle of closure under logical consequences:

$$RM: \frac{F \to G}{\Box F \to \Box G}$$

(from which RE can be derived). There are countermodels for RN, C, and K. Modal logics containing RE and RM are called *monotonic*, and the basic monotonic logic is called EM.

There is a slightly different semantics where the closure under supersets is implicit in the truth condition. It has been used e.g. in Fitting's ([5]) underlying logic U. There, frames are just the minimal ones. It is in the forcing relation (that we note $\models^{sup}$) where the truth condition for the modal operator is different from that of classical modal logics:

- $M, w \models^{sup} \Box F$ iff there is $V \in N(w)$ such that ($v \in V$ **implies** $M, v \models^{sup} F$)

In other words, we can simulate superset closure by replacing "iff" by "implies" in the truth conditions.

For the same semantics, the terms *local reasoning frames* and *logic of local reasoning* have been employed by Fagin and Halpern ([3]). They have employed these logics to model implicit and explicit belief: Explicit implies implicit belief, and the operator of explicit belief has a monotonic modal logic, whereas that of implicit belief has a normal modal logic.

3.4 Quasi-Filters

Quasi-filters are the base of regular modal logics.

A supplemented minimal frame is a *quasi-filter* if $N(w)$ is closed under supersets and under finite intersection, for every world w. Closure under finite intersection can be strengthened to closure under arbitrary intersections (v. (Chellas [2]), p.

[4] An equivalent condition is: If $U \cap V \in N(w)$ then $U \in N(w)$ and $V \in N(w)$.

255). Hence quasi-filters are isomorphic to local reasoning frames where for every world w, the set $N(w)$ is either a singleton or the empty set.

Such a presentation can be specialised to the following one in terms of augmented Kripke frames (Fitting [5], [6]). An *augmented Kripke frame*[5] is a quadruple (W, Q, r, m) where

- W is a set of worlds,
- $Q \subseteq W$ is a set of worlds (called queer worlds), and
- $r \subseteq W^2$ is a relation on W.

Thus queer worlds correspond to worlds in minimal frames having an empty neighbourhood. In this semantics the truth condition for the modal connective is:

- $M, w \models^{aug} \Box F$ iff ($w \notin Q$ and for every $v \in r(w) : M, v \models^{aug} F$)

We are quite close to standard Kripke semantics here, except that there are some worlds that do not satisfy any formula of the form $\Box F$.

Augmented Kripke frames can be characterized (Fitting [5]) by the principles of closure under logical consequences and material implication:

$$RM: \quad \frac{F \to G}{\Box F \to \Box G}$$

$$K: \ (\Box F \wedge \Box(F \to G)) \to \Box G$$

They can also be characterized by RM and C (closure under conjunction). There are countermodels for RN (closure under logical truth).

Modal logics containing RM and C are called *regular*, and the basic monotonic modal logic is called EMC.

4 TRANSLATIONS

In this section, for each class of classical modal logics that we have presented we give a translation into normal multimodal logics.

Monotonic, regular, and inaccessible world modal logics are translated into bimodal logics, whereas classical modal logics are translated into a trimodal modal logic. As the latter translation combines the principles of the previous ones, we give it at the end.

[5]Augmented Kripke frames should not be confused with augmented minimal frames (Chellas [2]) which are minimal frames that are isomorphic to standard Kripke frames.

4.1 Monotonic Modal Logics

The Translation

The idea of the translation is to give a world status to those propositions that are associated to worlds via the neighbourhood function N: Every element of $N(w)$ is viewed as a world accessible from w. Thus "there is a $V \in N(w)$" can be expressed via some existential modal operator $\langle 1 \rangle$. The truth of a formula in every element of V can then be expressed via a second universal modal operator $[2]$. In this way we can transform a minimal model into a standard Kripke model, and the logic of $\langle 1 \rangle$ and $[2]$ is a normal one.

Formally we define a translation τ from monotonic modal logics into normal multi-modal logics as follows:

- $\tau(F) = F$ if F is a propositional variable
- $\tau(\Box F) = \langle 1 \rangle [2] \tau(F)$

and homomorphic for the cases of the classical connectives.

The same translation has been given independently by F. Wolter ([28]) in order to prove the completeness of a large class of monotonic modal logics.

THEOREM 2 $\models_{em} F$ *iff* $\models_{k[1,2]} \tau(F)$.

Hence by soundness and completeness of EM and $K[1,2]$ we also have that a formula F is a theorem of basic monotonic modal logic EM if and only if $\tau(F)$ is a theorem of normal bimodal logic $K[1,2]$.

REMARK 3 *We can also translate the operator $\mathcal{I}$ of implicit belief of (Fagin and Halpern [3]) by adding the supplementary case $\tau(\mathcal{I}F) = [1][2]\tau(F)$.*

REMARK 4 *At least for the basic monotonic modal logic EM we are able to strengthen our translation to $\tau(\Box F) = \langle 1 \rangle [1] \tau(F)$ and thus to translate into mono-modal logic.*

In the next theorem we give a more general result for those monotonic modal logics that are characterized by axioms among the following standard ones:

$D : \Box F \to \Diamond F$

$T : \Box F \to F$

$4 : \Box F \to \Box\Box F$

The systems $EMD, EMT, EM4, EMD4, EMT4$ are complete (Chellas [2]). We denote by $emd, emt, em4, emd4, emt4$ the corresponding classes of frames. Completeness of these logics is used in the following theorem:

THEOREM 5 *Let κ be any sum of conditions on minimal frames among $d, t, 4$. Let $\mathcal{A}$ be the corresponding combination of axioms $D, T, 4$. Then $\models_{em+\kappa} F$ iff $\vdash_{K[1,2]+\tau(\mathcal{A})} \tau(F)$.*

Hence using the completeness of $EMD, EMT, EM4, EMD4, EMT4$ we can prove now theorems of these monotonic modal logics via our translation into particular multi-modal logics.

Note that the translations $\tau(D), \tau(T), \tau(4)$ of the standard modal axioms D, T, and 4 become multi-modal axioms in the style of Sahlqvist [26] for which completeness results (Catach [1], Kracht [18]) and automated deduction methods (Ohlbach [23], [24], Gasquet [7], [8], Nonnengart [21], Fariñas and Herzig [4]) are known. Note also that the translations of the modal axioms 5 and B would become axioms which have not been studied yet in the literature.

Examples

EXAMPLE 6 The formula $(\Box p \wedge \Box q) \rightarrow \Box(p \wedge q)$ is translated into
$(\langle 1\rangle[2]p \wedge \langle 1\rangle[2]q) \rightarrow \langle 1\rangle[2](p \wedge q)$,
which can be proved neither in EM nor in $EMD, EMT, EM4, EMD4$, $EMT4$.

Proofs

We do not give the proofs, because the technique is a particular case of that for classical modal logics (cf. subsection 4.4).

4.2 Regular Modal Logics

The Translation

Having at our disposal the preceding translation for monotonic modal logics, an immediate way to translate regular modal logics is to prove that $\models_{emc} F$ iff $\vdash_{K[1,2]+\tau(C)} \tau(F)$, where C is the axiom of closure under conjunction, and τ is the same as in subsection 4.1. Nevertheless,
$\tau(C) = (\langle 1\rangle[2]\tau(F) \wedge \langle 1\rangle[2]\tau(G)) \rightarrow \langle 1\rangle[2](\tau(F) \wedge \tau(G))$is a rather complex axiom. Here we give an optimization by associating a special axiom just to the first modal operator [1]. This will also permit to state a more general exactness theorem.

What we shall prove is that the translation is exact when the modal logics for [1] and [2] are normal, and moreover that for [1] satisfies an axiom that we call T_c (the converse of the standard T-axiom):

$T_c:\ F \rightarrow [1]F$

Semantically, this corresponds to the condition t_c that the accessibility relation r_1 is a subset of the diagonal of W: $r_1 \subseteq \delta_W$. (In other words, for every world w,

$r_1(w)$ is either a singleton w or the empty set). The idea behind this axiom T_c is that, in our proof, queer worlds will not be in r_1, while $\Box A$ will be translated into $\langle 1\rangle[2]A$; queer worlds will correspond to worlds without r_1-successors and hence will invalidate $\langle 1\rangle[2]A$.

Clearly, $\tau(C)$ is a theorem of $K[1,2]+T_c[1]$. In the following theorem we give a general result for systems of regular modal logics:

THEOREM 7 *Let κ be any semantical condition on accessibility relations r of augmented Kripke frames. Let $\kappa(1,2)$ be the translation of κ on accessibility relations r_1 and r_2 of standard Kripke frames ($u \notin Q$ being translated into $(u,v) \in r_1$ and $(u,v) \in r$ into $(u,v) \in r_2$.*
Then $\models_{emc+\kappa} F$ iff $\models_{k[1,2]+t_c[1]+\kappa(2)} \tau(F)$.

Hence if κ can be characterized by an axiom $\mathcal{A}$ and $\kappa(1,2)$ by a bimodal axiom $\mathcal{A}^{\tau}$, then we can prove theorems of regular modal logic $EMC + \mathcal{A}$ by proving theorems of the normal bimodal logic $K[1,2] + T_c[1] + \mathcal{A}^{\tau}$.

THEOREM 8 *Let κ be any semantical condition on accessibility relations r of augmented Kripke frames that is characterized by some axiom $\mathcal{A}$. Let $\kappa(1,2)$ be as in the previous theorem. Then $\models_{emc+\kappa} F$ iff $\vdash_{K[1,2]+T_c[1]+\tau(\mathcal{A})} \tau(F)$.*

Examples

EXAMPLE 9 The formula $\tau(C) = (\Box p \wedge \Box q) \rightarrow \Box(p \wedge q)$ is translated into $\tau(C) = ((\langle 1\rangle[2]p \wedge \langle 1\rangle[2]q) \rightarrow \langle 1\rangle[2](p \wedge q)$.
As $((\langle 1\rangle F \wedge \langle 1\rangle G) \rightarrow \langle 1\rangle(F \wedge G)$ can be proved in $K[1,2] + T_c[1]$, the formula $\tau(C)$ is a theorem.

Proofs

LEMMA 10 *For augmented Kripke models of the form $M_{aug} = (W, Q, r, m)$ let f be a mapping such that $f(M_{aug}) = (W, R_1, R_2, m)$ where*

- $R_1 = \delta_Q = \{(w,w) : w \notin Q\}$,
- $R_2 = R$.

Then f is an isomorphism between augmented Kripke models and standard Kripke $K[1,2] + T_c[1]$-models.

Proof. It is easy to see that for every augmented Kripke model M_{aug}, $f(M_{aug})$ is a $K[1,2]+T_c[1]$-model, and $M_{aug}, w \models^{aug} F$ iff $M, w \models \tau(F)$, for every $w \in W$.

In the other sense, for every $K[1,2]+T_c[1]$-model M, we have that $f^{-1}(M)$ is defined and is an augmented Kripke model, and $M, w \models \tau(F)$ iff M_{aug}, $w \models^{aug} F$, for every $w \in W$. ■

Thus we also have an isomorphism between $EM + \mathcal{A}$-models and $K[1,2] + T_c[1] + \tau(\mathcal{A})$-models. Theorem 7 follows immediately from that.

4.3 *Humberstone's Inaccessible Worlds Logic*

The Translation

The idea of the translation has been given in (Goranko [12]). It is the following: The truth condition is equivalent to

- $M_{iw}, w \models^{iw} \Box F$ iff for all $v \in W$
  - if $v \in R(w)$ then $M_{iw}, v \models^{iw} F$, **and**
  - if $v \in \bar{R}(w)$ then $M_{iw}, v \models^{iw} \neg F$.

where $\bar{R}$ denote the complement of R. What we do is to introduce two modal connectives, [1] to access R-successors, and [2] to access $\bar{R}$-successors. This leads to the following translation τ from logics with inaccessible worlds into normal bimodal logics:

- $\tau(F) = F$ if F is a propositional variable
- $\tau(\Box F) = [1]\tau(F) \wedge [2]\neg\tau(F)$

and homomorphic for the cases of the classical connectives.

In order to get an exact translation, we must choose $KT5[1 \cup 2]$ [6] as our target logic:

THEOREM 11 $\models^{iw} F$ *iff* $\vdash_{KT5[1\cup 2]} \tau(F)$.

Hence F is valid in Humberstone's inaccessible worlds logic iff $\tau(F)$ is valid in normal bimodal logic $KT5[1 \cup 2]$.

REMARK 12 *Contrarily to what could be expected, our target logic is not required to have accessibility relations for* [1] *and* [2] *with an empty intersection. In fact, it is sufficient to prove that* $KT5[1 \cup 2]$ *is characterized by the frames where:* $R_1 \cup R_2$ *is an equivalence relation* **and** $R_1 \cap R_2 = \varnothing$. *(This key lemma is given in the next section.)*

---

[6]By $KT5[1 \cup 2]$ we mean a normal multi-modal logic where both modal connectives [1] and [2] have the axioms of modal logic K, plus the axioms T and 5, stated for the modal operator $[1 \cup 2]$. ($[1 \cup 2]F$ is an abreviation for $[1]F \wedge [2]F$.) In other words, $KT5[1 \cup 2]$ is axiomatized by some axiomatization of classical logic, necessitation rules for [1] and [2], plus the axioms $[1 \cup 2]F \to F$ and $\neg[1 \cup 2]F \to [1 \cup 2]\neg[1 \cup 2]F$. It is well-known that $KT5[1 \cup 2]$ is characterized by the class of Kripke frames (W, R_1, R_2) where $R_1 \cup R_2$ is an equivalence relation over W (see e.g. (Catach [1])).

REMARK 13 *The axiomatization given in (Humberstone [15]) is infinitary. On the contrary, the above translation provides an indirect (because the modal connectives are not the original ones) but still modal axiomatization of this logic. Most of all, this axiomatization is finitary.*

Examples

EXAMPLE 14 The formula $\Box\top$ is translated into $[1]\top \wedge [2]\neg\top$, which is not valid in $KT5[1 \cup 2]$. Hence $\Box\top$ is not valid in inaccessible worlds logic.

EXAMPLE 15 The formula $\Box(p \wedge q) \rightarrow \Box p$ is translated into

$$[1](p \wedge q) \wedge [2]\neg(p \wedge q) \rightarrow [1]p \wedge [2]\neg p.$$

In $KT5[1\cup2]$, this formula is equivalent to the conjunction of $(([1](p\wedge q)\wedge[2]\neg(p\wedge q)) \rightarrow [1]p)$ and $(([1](p \wedge q) \wedge [2]\neg(p \wedge q)) \rightarrow [2]\neg p)$. Now the first conjunct is a theorem of $KT5[1 \cup 2]$ (because of $[1](p \wedge q) \rightarrow [1]p$), but the second is not.

EXAMPLE 16 The formula $(\Box p \wedge \Box q) \rightarrow (p \leftrightarrow q)$ is translated into

$$([1]p \wedge [2]\neg p \wedge [1]q \wedge [2]\neg q) \rightarrow (p \leftrightarrow q).$$

In $KT5[1 \cup 2]$, the antecedent is equivalent to $([1](p \leftrightarrow q) \wedge [2](p \leftrightarrow q)$, and now

$$([1](p \leftrightarrow q) \wedge [2](p \leftrightarrow q) \rightarrow (p \leftrightarrow q)$$

is an instance of the $T[1 \cup 2]$-axiom.

Proofs

The proof has been first given in (Goranko [12]).

LEMMA 17 *Let $M_{iw} = (W, R, m)$ be an inaccessible worlds model wherein a formula F is satisfied at w_0. Let $R_1 = R$ and $R_2 = W^2 \setminus R$. Then (W, R_1, R_2, m) is a $KT5[1 \cup 2]$-model satisfying $\tau(F)$ at w_0.*

Proof. First note that the $KT5[1 \cup 2]$ axioms are true in (W, R_1, R_2) because $R_1 \cup R_2$ is an equivalence relation. Let $M = (W, R_1, R_2, m)$. We prove that for every formula G and $w \in W, M_{iw}, w \models^{iw} G$ iff $M, w \models \tau(G)$ by induction on the structure of G. The only non-trivial case is

- $M_{iw}, w \models^{iw} \Box H$
- iff for all $v \in W$, ($v \in R(w)$ iff $M_{iw}, v \models^{iw} H$)
- iff for all $v \in W$, ($v \in R(w)$ iff $M, v \models \tau(H)$), by induction hypothesis
- iff for all v in W,
  - if $v \in R_1(w)$ then $M, v \models \tau(H)$ by construction of R_1, and
  - if $v \in R_2(w)$ then $M, v \models \neg\tau(H)$, by construction of R_2

- iff $M, w \models [1]\tau(H) \wedge [2]\neg\tau(H)$
- iff $M, w \models \tau(\Box H)$

■

LEMMA 18 *$KT5[1 \cup 2]$ is characterized by the class of frames (W, R_1, R_2) where $R_1 \cup R_2$ is universal (i.e. $R_1 \cup R_2 = W^2$) and $R_1 \cap R_2 = \varnothing$.*

Proof. As the converse is trivial, we only have to prove that if F is $KT5[1 \cup 2]$-satisfiable then it is also satisfiable in a frame (W, R_1, R_2) where $R_1 \cup R_2$ is an equivalence relation and $R_1 \cap R_2 = \varnothing$. Suppose given a standard Kripke model $M = (W, R_1, R_2, m)$ such that $M, w_0 \models F$ for some $w_0 \in W$. First, let W' be the connected part of W that contains w_0, i.e. the set of worlds that can be reached from w_0 via $R_1 \cup R_2$. Let R'_1, R'_2, and m' be the respective restrictions of R_1, R_2, and m to W'. It is well-known that $M' = (W', R'_1, R'_2, m')$ is still a model of $KT5[1 \cup 2]$, and $M', w_0 \models F$.
Now, we build a new model which will simulate (W', R'_1, R'_2, m'), but where the two relations will have an empty intersection. Let V_1 and V_2 be two sets such that V_1, V_2 and W' are pairwise disjoint, V_1 is isomorphic to W' via an isomorphism f_1, and V_2 is isomorphic to W' via an isomorphism f_2.(For example, V_i could be $W \times \{i\}$.)Let V denote $V_1 \cup V_2$, and let R''_1 and R''_2 be relations on V as follows:

1. $R''_1 = \{(f_i(v), f_i(w)) : w \in W,\ i = 1, 2 \text{ and } (v, w) \in R_1\} \cup$
   $\{(f_i(v), f_{3-i}(w)) : w \in W,\ i = 1, 2,\ (v, w) \in R_1 \text{ and } (v, w) \notin R_2\}$
2. $R''_2 = \{(f_i(v), f_i(w)) : w \in W,\ i = 1, 2 \text{ and } (v, w) \in R_2\} \cup$
   $\{(f_i(v), f_{3-i}(w)) : w \in W,\ i = 1, 2,\ (v, w) \in R_2 \text{ and } (v, w) \notin R_1\}$

Graphically,

Will give:

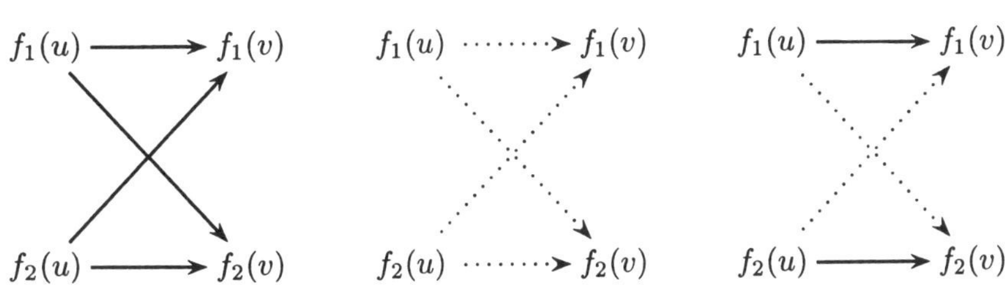

Where dotted lines denote R2 and full lines R1.

Finally, let m'' be defined by: $f_i(w) \in m''(F)$ iff $w \in m(F)$, for F a propositional variable.

Then let $\varphi = f_1^{-1} \cup f_2^{-1}$. φ is a pseudo-epimorphism from (V, R_1'', R_2'', m'') onto (W', R_1', R_2', m) (Hughes and Cresswell [14]). This ensures that F is satisfiable in (V, R_1'', R_2'', m''). Moreover, it can easily be shown that $R_1'' \cup R_2''$ is an equivalence relation over V, and even that $R_1'' \cup R_2'' = V^2$, and that $R_1'' \cap R_2'' = \varnothing$. Hence the lemma is proved. ■

LEMMA 19 *Let (W, R_1, R_2) be a frame for $KT5[1 \cup 2]$ wherein $\tau(F)$ is satisfiable. Then there is an inaccessible worlds frame wherein F is satisfiable.*

Proof. The proof is similar to that in (Humberstone [15]). ■

Suppose $M = (W, R_1, R_2, m)$, and $M, w_0 \models \tau(F)$ for some $w_0 \in W$. By the above Lemma 18, we can suppose that $R_1 \cap R_2 = \varnothing$, and that $R_1 \cup R_2 = W^2$, i. e. R_2 is the complement of R_1.

Let $M_{iw} = (W, R_1, m)$. Clearly M_{iw} is an inaccessible worlds model. We prove by induction that for every $w \in W$ and every formula F, we have $M, w \models \tau(F)$ iff $M_{iw}, w \models^{iw} F$. The only non-trivial case is: $F = \Box G$, i. e. $\tau(F) = [1]\tau(G) \wedge [2]\neg\tau(G)$.

- From the left to right, suppose $M, w \models [1]\tau(G) \wedge [2]\neg\tau(G)$. Let v be any world from W. If $v \in R_1(w)$, as $M, w \models [1]\tau(G)$, we have that $M, v \models \tau(G)$, and by induction hypothesis, $M_{iw}, v \models^{iw} G$. If $v \notin R_1(w)$ then $v \in R_2(w)$ (because $R_1 \cup R_2 = W^2$). As $M, w \models [2]\neg\tau(G)$, we have that $M, v \models \neg\tau(G)$, and by induction hypothesis, $M_{iw}, v \models^{iw} \neg G$. Putting both together we get that $M_{iw}, w \models^{iw} \Box G$.

- From the right to the left, suppose $M_{iw}, w \models^{iw} \Box G$. Then v is in $R_1(w)$ iff $M, v \models^{iw} G$. By induction hypothesis, $v \in R_1(w)$ iff $M, v \models \tau(G)$. Hence $M, w \models [1]\tau(G)$. As $R_1 \cap R_2 = \varnothing$, we have that $M, w \models [2]\neg\tau(G)$. Putting both together we get that $M, w \models [1]\tau(G) \wedge [2]\neg\tau(G)$.

REMARK 20 *It is clear that this lemma cannot handle extensions of the basic inaccessible world logic e.g. with axiom D. The reason is that we cannot ensure that our construction of the inaccessible world model preserves the accessibility relation properties.*

Now the theorem follows immediately from Lemma 17 and 19.

4.4 Classical Modal Logics

The Translation

The translation τ from classical modal logics into multi-modal logics combines the above translations of monotonic modal logics and Humberstone's inaccessible worlds logic. It goes as follows:

- $\tau(F) = F$ if F is a propositional variable
- $\tau(\Box F) = \langle 1\rangle([2]\tau(F) \wedge [3]\neg\tau(F))$

and homomorphic for the cases of the classical connectives.
Then we have

THEOREM 21 $\models_e F$ *iff* $\vdash_{K[1,2,3]} \tau(F)$.

Hence a formula F is valid in classical modal logic E iff $\tau(F)$ is valid in normal multi-modal logic $K[1,2,3]$.

REMARK 22 *Note that at least for the basic classical modal logic* E *we are able to strengthen our translation to* $\tau(\Box F) = \langle 1\rangle([2]\tau(F) \wedge [1]\neg\tau(F))$.

In the next theorem we extend our result to ET. We translate into $K[1,2,3] + B_{1,2}$, where $B_{1,2}$ is the axiom $\langle 1\rangle[2]G \rightarrow G$. $B_{1,2}$ axiomatizes a condition on frames that we call $b_{1,2} : R_1 \subseteq R_2^{-1}$.

THEOREM 23 $\models_{et} F$ *iff* $\vdash_{K[1,2,3]+B_{1,2}} \tau(F)$.

Note that axiom $B_{1,2}$ is in fact the monotonic modal logic translation of the T-axiom.
The generalization towards extensions of E by other axioms such as $D, 4, B, T$ seems to be much more difficult. This is due to the fact that the monotonic translations of these axioms do not completely axiomatize the translated models (see Lemma 25 in the proofs). On the other hand, the classical translation of e.g. axiom T yielding $\langle 1\rangle([2]G \wedge [3]\neg G) \rightarrow G$ would also be a candidate for the target logic axioms[7], but it is difficult to devise a semantics for such complex multimodal axioms. (Note nevertheless that we would only need a soundness result for the multimodal axiom).

Examples

EXAMPLE 24 The formula $\Box\top$ is translated into
$\langle 1\rangle([2]\top \wedge [3]\neg\top)$,
which in $K[1,2,3]$ is equivalent to $\langle 1\rangle[3]\bot$. This is clearly not a theorem of $K[1,2,3]$. Hence $\Box\top$ is not valid in classical modal logic.

[7]and not (i): $\langle 1\rangle([2]\tau(G)\wedge[3]\neg\tau(G)) \rightarrow \tau(G)$ as a particular instance of T is $\Box G \rightarrow G$, for G propositional variable, whose translation is $\langle 1\rangle([2]G \wedge [3]\neg G \rightarrow G$ which give (by substitution) $\langle 1\rangle([2]G \wedge [3]\neg G) \rightarrow G$ for any formula G.

Proofs

LEMMA 25 *Let $M_{min} = (W, N, m)$ be a minimal model.*
Let $M = (V, R_1, R_2, R_3, m)$ be a model such that

- $V = W \cup \bigcup_{w \in W} N(w)$,
- $R_1(w) = N(w)$, *for every* $w \in W$,
- $R_2(U) = U$, *for every* $w \in W$ *and* $U \in N(w)$,
- $R_3(U) = W - U$, *for every* $w \in W$ *and* $U \in N(w)$,

Then for every formula F and world $w \in W$, $M_{min}, w \models^{min} F$ iff $M, w \models \tau(F)$.

Proof. First, M is a model of $K[1,2,3]$. Then the proof is straightforward by induction on F. The only non-trivial case is that of $F = \Box G$. We have: $M_{min}, w \models^{min} \Box G$

iff there is $U \in N(w)$ such that for every $v \in W$, ($v \in U$ iff M_{min}, $v \models^{min} G$)

iff (by induction hypothesis) there is $U \in N(w)$ s. th. for every $v \in W$, ($v \in U$ iff $M, v \models \tau(G)$)

iff there is a $U \in N(w)$ s. th. for every $v \in W$

- if $v \in R_2(U)$ then $M, v \models \tau(G)$ (by the definition of R_2), and
- if $v \in R_3(U)$ then $M, v \models \neg\tau(G)$) (by the definition of R_3)

iff there is $U \in R_1(w)$ such that $M, U \models [2]\tau(G) \wedge [3]\neg\tau(G)$

iff $M, w \models \tau(F)$.

■

LEMMA 26 *(Chellas [2], p. 261) If $\models_{et} F$ then $\vdash_{ET} F$.*

The next lemma is proven syntactically, because we can state it thus in a general way for extensions of E by any axiom $\mathcal{A}$.

LEMMA 27 *Let $\mathcal{A}$ be any axiom schema. If $\vdash_{E+\mathcal{A}} F$ then $\vdash_{K[1,2,3]+\tau(\mathcal{A})} \tau(F)$.*

Proof. We use induction on the proof of F: Whenever the axiom $\mathcal{A}$ is used, we can replace it by $\tau(\mathcal{A})$. The other axioms are classical. Whenever the non-normal inference rule RE:

$$\frac{F \leftrightarrow G}{\Box F \leftrightarrow \Box G}$$

is used, we can replace it by

$$\frac{\tau(F) \leftrightarrow \tau(G)}{\langle 1\rangle([2]\tau(F) \wedge [3]\neg\tau(F)) \leftrightarrow \langle 1\rangle([2]\tau(G) \wedge [3]\neg\tau(G))}$$

which is a derived inference rule of $K[1,2,3]$ (by substitution of equivalences). ■

LEMMA 28 *(Catach [1])* $\vdash_{K[1,2,3]+B_{1,2}} F$ *implies* $\models_{k[1,2,3]+b_{1,2}} F$.

THEOREM 29 $\models_e F$ *iff* $\vdash_{K[1,2,3]} \tau(F)$.

Proof. The proof follows from Lemmas 25 and 27, using the completeness of E (Lemma 26). ■

THEOREM 30 $\models_{et} F$ *iff* $\vdash_{K[1,2,3]+B_{1,2}} \tau(F)$.

Proof. From the left to the right, the proof follows from Lemmas 27 and 26.

From the right to the left, the proof is semantical: Given a minimal model, we must warrant that the normal model that we construct in Lemma 25 is a model for

$\tau(T) = (\langle 1\rangle([2]\tau(F) \wedge [3]\neg\tau(F))) \to F$.

Now the translated model of Lemma 25 already satisfies the required property.[8] Then we take advantage of the soundness of the normal modal logic $K[1,2,3] + B_{1,2}$.

■

5 CONCLUSION

We have given a translation from classical modal logics into normal modal logics. For particular classes of non-normal modal logics we have given specialised translations. Precisely, we have proved the exactness of the translation for the following logics:

- the basic classical modal logic E,
- the basic monotonic modal logic EM,
- the basic regular modal logic EMC,
- Humberstone's inaccessible worlds logic.

We have also given exactness proofs for the extension of E with axiom T and for extensions of EM with axioms

$D : \Box F \to \Diamond F$

[8] Note that this is not always the case (a simple example being the axiom $\Diamond\top$). This makes it difficult to prove the exactness of other extension of E e.g. by axioms 5 or B.

$T : \Box F \rightarrow F$

$4 : \Box F \rightarrow \Box\Box F$

Moreover we have proved exacteness of the translation for any extension of the basic regular modal logic EMC.

ACKNOWLEDGEMENTS

This work has been partially supported by the Esprit projects DRUMS II and MEDLAR II. It has been inspired by Andrew Jones's presentations in the frame of MEDLAR on work done by Stig Kanger, Ingmar Pörn and himself in deontic logic. Thanks to him, and to Philippe Balbiani, Luis Fariñas del Cerro, Stephan Merz, and Hans Jürgen Ohlbach for their comments. Previous versions of parts of this paper have been presented in (Gasquet and Herzig 1993a, 1993b). When we wrote these versions, we were not completely aware of the work of Valentin Goranko, Solomon Passy, Tinko Tinchev and Dimiter Vakarelov who had published beautiful results on Humberstone's inacessible worlds logic which were more general than ours. Thanks to Tinko Tinchev and Dimiter Vakarelov for kindly discussing these points with us.

Université Paul Sabatier, France.

REFERENCES

[1] L. Catach. Normal multimodal logics. In: *Proc. Nat. Conf. on AI (AAAI 88)*, pp. 491–495, 1988.
[2] B. F. Chellas. *Modal Logic: An Introduction*. Cambridge University Press, Cambridge, 1980.
[3] R. Fagin, J. Y. Halpern. Belief, Awareness, and Limited Reasoning. *Journal of Artificial Intelligence*, **34**, 39–76, 1988.
[4] L. Fariñas del Cerro, A. Herzig. Automated Deduction for Epistemic and Temporal Logics. In: *Handbook of Logic in AI and Logic Programming, Vol. 4: Epistemic and Temporal Reasoning*, D. Gabbay, C.J. Hogger, J. A. Robinson, (eds.), pp. 499–594. Oxford University Press, Oxford, 1995.
[5] M. Fitting. *Proof Methods for Modal and Intuitionistic Logics*. Reidel, Dordrecht, 1983.
[6] M. Fitting. Modal Logics. In: *Handbook of Logic in Artificial Intelligence and Logic Programming, Vol. 1: Logical Foundations*, D. Gabbay, C. J. Hogger, J. A. Robinson, (eds.), 365–448. Oxford University Press, Oxford, 1994.
[7] O. Gasquet. Deduction for Multimodal Logics. In: *International Conference on Applied Logic (Logic at Work'92), Amsterdam*, 1992.
[8] O. Gasquet. Optimised Deduction for Multimodal Logics. In: *Applied Logic: How, What and Why?*, M. Masuch, L. Polos, (eds.), Kluwer, Dordrecht, to appear.
[9] O. Gasquet, A. Herzig. Translating Inaccessible Worlds Logic into Bimodal Logic. In: *Proc. Eur. Conf. on Symbolic and Quantitative Approaches to Reasoning and Uncertainty (ECSQARU-93)*, Granada. LNCS, Springer-Verlag, Berlin, 1993.
[10] O. Gasquet, A. Herzig. Translating Non-Normal Modal Logics into Normal Modal Logics. In: A. I. J. Jones, M. Sergot, (eds.), *Proc. Int. Workshop on Deontic Logic*. 1993.
[11] V. Goranko. Modal Definability in Enriched Languages. *Notre Dame Journal of Formal Logic*, **31**, 81–105, 1990.
[12] V. Goranko. Completeness and Incompleteness in the Bimodal Base L(R,R-). In: P. Petkov, (ed.), *Mathematical Logic*. Plenum Press, New York, 1990.

[13] V. Goranko, S. Passy. Using the universal modality: gains and questions. *J. Logic Computat.*, **2**, 5–30, 1992.
[14] G. Hughes, M. J. Cresswell. *An introduction to modal logic*. Methuen, London, 1968.
[15] L. I. Humberstone. Inaccessible worlds. *Notre Dame Journal of Formal Logic*, **24**, 346–352, 1983.
[16] A. J. I. Jones, I. Pörn. Ideality, Sub-Ideality and Deontic Logic. *Synthese*, **65**, 1985.
[17] A. J. I. Jones, I. Pörn. Ought and Must. *Synthese*, **66**, 1986.
[18] M. Kracht. How completeness and correspondence theory got married. In: M. de Rijke, (ed.), *Diamonds and Defaults*. Kluwer, Dordrecht, 1993.
[19] H. Levesque. All I know: A study in autoepistemic logic. *Artificial Intelligence*, **42**, 263–309, 1990.
[20] R. Montague. Logical necessity, physical necessity, ethics, and quantifiers. *Inquiry*, **4**, 259–269, 1960.
[21] A. Nonengart. First-Order Modal Logic Theorem Proving and Functional Simulation. In: *International Joint Conference on Artificial Intelligence*, Chambéry, 1993.
[22] H. J. Ohlbach. A Resolution Calculus for Modal Logics. In: E. Lusk, R. Overbeek (eds.), *Proceedings of CADE '88*, Lecture Notes in Computer Science, Vol. 310. Springer Verlag, Berlin, 1988.
[23] H. J. Ohlbach. Semantics-Based Translation Method for Modal Logics. *Journal of Logic and Computation*, **1**, 691–746, 1991.
[24] H. J. Ohlbach. Optimized Translation of Multi-Modal Logic into Predicate Logic. In: A. Voronkov, (ed.), *Proc. of Int. Conf. on Logic Programming and Automated Reasoning (LPAR)*, Lecture Notes in Artificial Intelligence, pp. 253–264. Springer Verlag, Berlin, 1993.
[25] H. J. Ohlbach, A. Herzig. Parameter Structures for Parametrized Modal Operators. *International Joint Conference on Artificial Intelligence*, 1991.
[26] H. Sahlqvist. Completeness and correspondence in the first and second order semantics for modal logics. In: S. Kanger (ed.), *Proc. 3rd Scand. Logic Symp.*, pp. 110–143. North-Holland, Amsterdam, 1973.
[27] K. Segerberg. An essay in classical modal logic. *Filosofiska Studier*, 13, Uppsala, 1971.
[28] F. Wolter. Interpretations of monotonic modal logics as normal modal logics. Personal Communication, Dept. of Mathematics, FU Berlin, Jan. 1994.

Index

$GS4$, 18
$GS4*$, 28
$RS4$, 24
S categories, 204
$S\Box$ categories, 205
$S4_F$, 20
β-reduction, 188, 214, 231
η-reduction, 188
2-LL, 37
A scheme, 82, 84
A′ scheme, 90
GA scheme, 89
Kt, 93, 97, 101, 105, 124, 137
LL, 34
P scheme, 82, 84
RRA, 57
S4, iii, 3, 17, 33, 45, 46, 168, 176, 256
b arrow, 175
c arrow, 175
k arrow, 175
w arrow, 175
RI-semantics, 62
R-**LL**, 36
Υ, 272
$\lambda S4$, 237
$\lambda^\ell K$, 218, 239
$\lambda^\ell K4$, 219, 239
$\lambda^\ell KT$, 218, 239
$\lambda^\ell S4$, 237, 239
λ^ℓ, 217
$\mathcal{L}_n^=$, 244
$\mathcal{L}_n$, 243
${}^r\mathcal{L}_n^=$, 245

$\overline{\mathbf{K}}$, 257, 289
$\overline{\mathbf{K}}_E$, 260–262, 289
$[n]\Diamond\varphi$, 261
$\langle n\rangle\Box\varphi$, 261
$\Box_n\varphi$, 254, 257
$\Diamond_n\varphi$, 253, 259
$\Diamond_n$, 257
$\Diamond!_0\varphi$, 257
$\Diamond!_n\varphi$, 257
$N_m\varphi$, 256
$[n]$, 255, 260
$\langle n\rangle$, 255, 260
$\mathsf{Alg}\mathcal{L}$, 247
CA_n, 247
QPEA_n, 247
RCA_n, 247
2-sequent, 215, 225

adjunction, 173
A-Box, 286
admissible formula, 274
$\mathcal{ALC}$, 253, 287
$\mathcal{ALCN}$, 253, 265, 287, 290
$\mathcal{ALCN}^+$, 287
Allwein, G., 89, 91
Anderson, A. R., 79
Andréka, H., 243, 248, 251
'apply' function, 270
arrow, 169
augmented frame, 299
automated deduction, 20, 254

Balbiani, P., 310
Belnap, N. D., 79, 93, 103, 124, 136

Benevides, M. R. F., 239
bicartesian categories, 174, 203
bicartesian closed categories, 167, 168, 189, 191, 192, 205
Bierman, G., 240
bifunctor, 173
Blackburn, P., iv
Boolean algebra, 46, 55
bottom sort, 270
Brachman, R. J., 253
Burris, S., 247

canonical formula, 109
cardinality function, 254
cartesian categories, 167, 168, 189, 192, 193, 206
cartesian closed categories, 167, 168, 183, 189, 191, 192, 209
Catach, L., 301, 303, 309
category, 170
Cerrato, C., 141, 256
Chellas, B. F., 142, 293, 295–300, 308
clausal sequent, 5
combinatorial completeness, 167
comonad, 178
compositionality, 56
compressed proofs, 116, 118
compressed structures, 117
concept, 253
concept names, 286
confluence, 33, 42, 214, 231, 232, 237, 239
correspondence theory, iii
cotriple, 178
Cresswell, M. J., 293, 306
Čubrić, D., 203
Curry, H. B, 79
Curry–Howard isomorphism, 167, 240
cut-elimination, 33, 34, 37, 42, 80, 93, 94, 102, 105, 107, 114, 115, 128, 129, 135, 139, 141, 143, 148, 149, 154, 158, 164, 203, 213

de Caro, F., 253, 256
de Paiva, V., 240
de Quieroz, R. J. G. B., 240
de Rijke, M., 256
decidability, 115
decisive sets, 109, 110
decisive valuations, 110
decomposition rules, 64
Deduction Theorem, 167
deductive system, 170
deep formula, 11
dischargeable, 27
discriminator variety, 247
display equivalence, 83, 125
display factor, 120
display logic, iii, 79, 93, 95, 124, 127
display problem, 79, 86
display property, 79
Display Theorem, 80, 95, 97, 127
displayable, 135, 136
distant formula, 14
Došen, K., 141, 225
Došen principle, 225
duality map, 98
duality rules, 145
dualization, 96
Dunn, J. M., 83, 89, 91

Easy scheme, 86
elementary subframe logics, 111
essentially modal formula, 223
exponentials, 33

Fagin, R., 294, 298, 300
Fariñas del Cerro, L., 301, 310
Fattorosi-Barnaba, M., 165, 253, 256
Fine, K., 253, 256
first-order definable, 255, 270
first-order logic with n variables, 244
Fitting, M., 297–299
focusing strategy, 28
Font, J. M., 210
Fuhrmann, A., 51

functional completeness, 123, 124, 133, 134, 136, 167–169, 191, 194, 204, 209
functional language, 256
fundamental sequences, 66
functional translation, 255, 270, 271, 274, 275, 282

Gabbay, D., iv, 240, 243, 244, 263, 265
Gabbay-style proof systems, 244, 246
Galois connection, 89
Gasquet, O., 301, 310
generalized quantifiers, 256
generalized restricted quantifiers, 113
generated submodels, 113
Gentzen duals, 93
Gentzen, G., 79–83, 210, 225
Gentzenization, 93
Girard, J.-Y., 141
Goble, L. F., 253, 256
Goranko, V., iv, 298, 303, 304, 310
Goré, R., iv, 94, 141, 149
graded modalities, 253, 256, 289
Grefe, C., 115
Grishin–Ono translation, 36

Habel, C., iv
Halpern, J. Y., 294, 298, 300
Hartonas, C, 89, 91
Henkin, L., 244, 247
hereditarily universal, 110
Herzig, A., 301, 310
hexagonal equation, 180, 187, 198, 199, 202, 204
higher-arity proof systems, iii
higher-dimensional proof systems, iii
higher-level proof systems, iii
Hilbert-style proof systems, 142, 220, 243, 246, 254
Hughes, G. E., 293, 306
Humberstone, L. I., 296–298, 303, 304, 306, 309
hyperresolution, 26

ideal relation, 58
inaccessible worlds, 294, 298, 303, 304, 306, 309
inference system, 246
initial modal clause, 23
internal language, 167, 209
inverse method, 20
inverse method strategy, 27
inversion strategy, 22, 27
invertible, 4, 22
irreflexivity-rule, iii, 265

Jaspars, J., iv
Jones, A. J. I., 294, 310

Kanger, S., 310
KL-ONE, 253, 256, 286, 287, 290
König's lemma, 156
Kanazawa, M., iv
Kleene, S. C., 148
Kracht, M., iv, 128, 136, 137, 140, 301
Kripke, S., 158, 214

labelled deductive systems, iii
labels, 20
Lambek, J., iv, 168, 171
level index, 217
level substitution, 229
Levesque, H., 298
Lindenbaum–Tarski algebra, 250
linear decorations, 33
LM, 14
LM_0, 4
LM_1, 7
LM_2, 8
LM_3, 9
LM_4, 11
LM_5, 13
Łukasiewicz, J., 45

Mac Lane, S., 168, 169, 173, 178–180, 210
Maddux, R., 58

Maibaum, T. S. E., 239
many-sorted predicate logic, 270
McKinsey's axiom, 270
McKinsey, J. C. C., 270
Masini, A., iv
Maslov transformation, 25, 27–29
Maslov, S., 17, 20, 27
Matsumoto, K., 123
membership operators, 260
Meré, C., 240
Merz, S., 310
minimal frame, 296
Minsky, M., 253
Mints, G., iv, 83, 136
modal clause, 5, 23
Modal Deduction Theorem, 194
modal functional completeness, 167, 201, 204, 205, 209
Modal Functional Completeness Theorem, 183, 189, 193, 194, 199, 202
modal literal, 5, 23
modal skeleton, 35
modal translation, 33, 168
modal tree-sequents, 165
Monk, D., 58, 247
monoid, 55, 181
monoidal categories, 168, 173, 179, 180, 210
Montague, R., 294
Muchnik, A., 115

NL category, 171
NL□ categories, 174
Németi, I., 243, 251
negation normal form, 259, 272
Nonnengart, A., 301
normal form, 116, 117
normalisation, 167, 214, 222, 223, 224, 231, 239
number restriction, 253
numerical modalities, 256
numerical operators, 260
Nutt, W., 285

octagonal equation, 169, 181, 187, 190–193, 200, 204
Ohlbach, H.-J., 30, 263, 270–273, 301, 310
Ohnishi, M., 123
Orevkov, V., 26, 28
Orlowska, E., 60

p-inversion strategy, 22
p-inverted, 19
p-invertible, 22
p-morphism, 111, 114, 265, 306
Passy, S., 298, 310
pentagonal equation, 179, 186, 187, 198, 202, 204
Pereira, L. C., iv
polynomial categories, 168, 182, 205
Pörn, I., 294, 310
Post operations, 60
Prawitz, D., 214, 222–224, 240
Preller, A., 210
prenex normal form, 272
primitive formula, 107, 109, 110, 114
proof-theoretic semantics, 123, 129
properly displayable, 105, 109, 111, 128, 135
propositional clauses, 23
pruned derivation, 18
pseudotriangular equation, 201

quantifier exchange rule, 272
quasi-filter, 298
queer worlds, 294, 297, 302

Rasiowa–Sikorski style proof systems, 64
rectangular, 248
rectangularly dense, 248
reduced structure, 116, 117
refutation calculi, 45
relation algebras, 57

relation-algebraic semantics, 55, 58, 62
relational logic, 61
relational proof systems, iii, 56, 76
replacement theorem, 133
residuation, 58, 59, 75, 89
resolution, 17, 23
Restall, G., 94
Restricted first-order logic, 245
restricted formula, 113
right ideal relation, 58, 60, 62, 73
role, 253
role filler, 253
rule skeleton, 103

Sahlqvist, H., 301
Sahlqvist formula, 109, 110
Sahlqvist Theorem, 264, 271
Sain, I., 243, 244, 247, 251
Sankappanavar, H. P., 247
SCAN, 263, 284
Schmidt, R., 270–272
Schmidt-Schauß, M., 287
Schütte translation, 146, 153, 162
Scott, D., 45
Segerberg, K., iv, 213, 193, 293
semantic modal sequent, 150, 165
semantics-based proof system, iii
set variable, 254
shallow formula, 11
Shehtman, V., iv, 115
signed formula, 143
Simon, A., 243
skeleton, 33
Skolem function, 273, 275
Smolka, G., 287
special rule, 109
Spurr, J., iv
standard translation, 263
strong normalization, 33, 37, 42, 214, 224, 231, 237, 238
subformula property, 24, 102, 107, 114, 115, 128, 141, 222, 224
substitution, 247
Substructural Functional Completeness Theorem, 208, 209
substructural logics, 80, 124, 125, 167, 168
subsumption, 22
subsumption strategy, 23

T-Box, 286
Tait, W., 232
Takano, M., 141, 149
Takeuti, G., 141, 158
Tammet, T., 27
Tarski, A., 58, 244, 247
temporal completeness, 124
theory resolution, 290
Thompson, R. J., 247
Thue-process, 115
Tinchev, T., 310
tree sequents, 150, 159
triangular equation, 179, 188, 189, 198, 200, 201
Tschernig, A., iv
two-dimensional proof systems, 214

uniform substitution rule, 257
unique occurrence strategy, 27
Urchs, M., iv

Vakarelov, D., 310
van Benthem, J. F. A. K., 263
van der Hoek, W., iv, 253, 256, 257
variety, 58
Venema, Y., 243, 244, 247, 251
Voronkov, A., 27

Wansing, H., 79, 82, 83, 93, 94, 137, 138, 210, 225
weakly serial, 264
Wolter, F., 94
world path notation, 273
worlds, 258

Zach, R., iv